COMPANION IN ELECTRONICS

Ashton Jairam

Prentice Hall
Upper Saddle River, New Jersey   Columbus, Ohio

Editor: Linda Ludewig
Production Editor: Rex Davidson
Design Coordinator: Karrie M. Converse
Cover Designer: Rod Harris
Production Manager: Patricia A. Tonneman
Marketing Manager: Ben Leonard

This book was printed and bound by Courier/Kendallville, Inc. The cover was printed by Phoenix Color Corp.

Simon & Schuster/A Viacom Company
Upper Saddle River, New Jersey 07458

Printed in the United States of America

10 9 8 7 6 5 4 3 2 1

ISBN: 0-13-080401-0

Prentice-Hall International (UK) Limited, *London*
Prentice-Hall of Australia Pty. Limited, *Sydney*
Prentice-Hall Canada Inc., *Toronto*
Prentice-Hall Hispanoamericana, S. A., *Mexico*
Prentice-Hall of India Private Limited, *New Delhi*
Prentice-Hall of Japan, Inc., *Tokyo*
Simon & Schuster Asia Pte. Ltd., *Singapore*
Editora Prentice-Hall do Brasil, Ltda., *Rio de Janeiro*

PREFACE

In the teaching of electronics I have found that most students find difficulty not in the basic theoretical principles but mainly in the application of the theoretical principles to the solution of practical problems.

This book attempts to bridge the gap between theoretical principles and practical application by presenting carefully detailed solutions to selected problems and in some cases give alternative methods.

Brief theoretical notes have been introduced in selected chapters on particular topics, focusing on specific aspects of the subject matter only; no attempt has been made to produce a conventional type of textbook but rather a companion sequence providing a guide to the methods and procedures in the solution to examination problems and for comparison with one's own personal efforts.

The range of problems provided covers most of the syllabuses in electronic programs conducted by Technical Institutes, Technical Colleges, Colleges of Applied Arts and Technology and Polytechnics at the introductory, advanced undergraduate and graduate levels. This apart, this book may be considered a suitable companion in continuing education and supplemental studies.

Some of the problems have been selected in part from past papers of examining bodies, and in this connection I wish to express my gratitude to the authorities for using extracts from their past examination papers.

Moreover, I wish to express my appreciation to the final year students of the electrical engineering technician program of the Gulf Polytechnic, Bahrain, Arabian Gulf, the final year electrical technician engineering students of the Government Technical Institute, Georgetown, Guyana, and the graduate technician engineers in training at the G.E.C. Training Department, Georgetown, Guyana, for their tireless efforts in researching sources and in finding solutions to many of the past examination papers.

It is hoped that this book will provide direction to students in Technical Institutions and those working on their own to broaden their horizons in electronic engineering .

A.J.

ABOUT THE AUTHOR

Born in Guyana, educated and trained in the U.K., and now living with his wife Barbara in Toronto, Canada, Ashton Jairam is a Chartered Electrical Engineer and administrator with an international background of diversified experience in electronics and the electric utility industry, with special emphasis on Human Resources development. Ashton has been involved in professional training and development in both electronics and electric utility services for more than 20 years. His experience includes Training Manager G.E.C. Training Department; Senior Lecturer, Gulf Polytechnic and Guyana Government Technical Institute; Product Supervision Engineer, Smiths Industries; and Technical Adviser and Chief Design Engineer Crossfield Electronics. He is currently involved in system analysis programs; personal computers, and Human Resources development and uses both MS-DOS and Microsoft Windows to write a wide range of simple modular training programs supported by CorelDRAW for graphic production.

D.J.L.

ABOUT THE BOOK

This book is the first in a major series of three books covering the fundamental aspects of electrical technology. The author's carefully chosen examples demonstrate the application of theory to practical problems. The Companion Series should prove to be a valuable source of reference for students and will be retained by them through their technician and engineering careers.

D.J. Longman O.B.E.
B.Sc.Eng. (London), F.I.C.E.
M.I.Struct.E, C.Eng.

CONTENTS

I often say that when you can measure what you are speaking about, and express it in numbers, you know something about it; but when you cannot express it in numbers, your knowledge is of a meager and unsatisfactory kind; it may be the beginning of knowledge, but you have scarcely, in your thoughts, advanced to the stage of science, whatever the matter may be.

LORD KELVIN

CHAPTER 1

CIRCUIT THEOREMS AND NETWORK ANALYSIS

Introduction

Kirchhoff's Laws

First Law: The sum of the current moving toward a junction in an electrical circuit is equal to the algebraic sum of the currents moving away from the junction; i.e., the current in the junction at any instant of time is zero.
Refer to Fig. 1.0-0

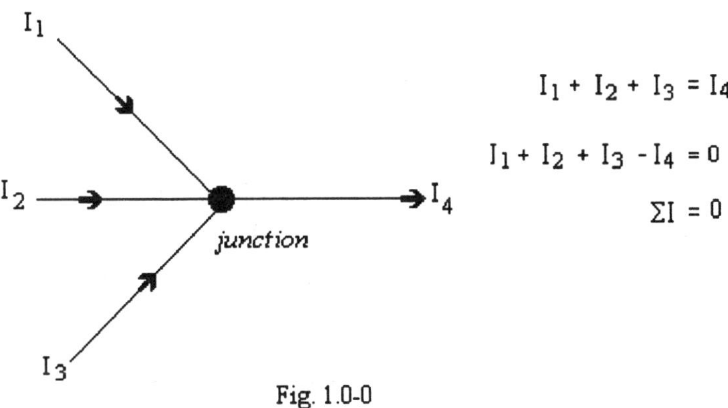

Fig. 1.0-0

Second Law: The algebraic sum of the IR drops (current x resistance) around any closed loop of an electrical circuit is equal to the resultant e.m.f in the network.
Refer to Fig. 1.0-1

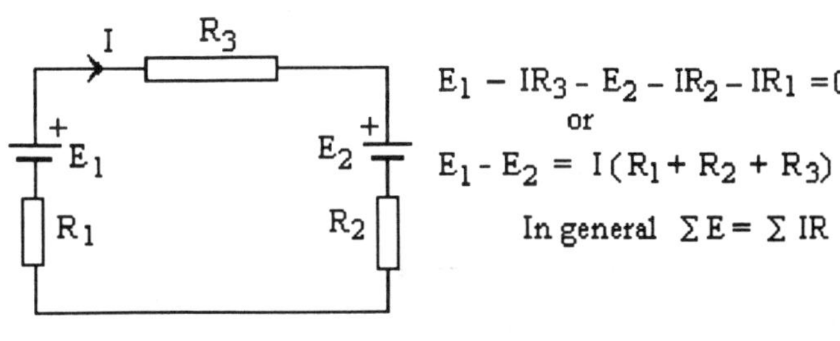

Fig. 1.0-1

Superposition Theorem

The resultant current in any branch of an electrical network containing more than one source of e.m.f is the algebraic sum of the currents that would be produced by each source acting individually while all other sources are short-circuited and replaced by their internal resistances. Refer to Figs. 1.0-2 and 1.0-3

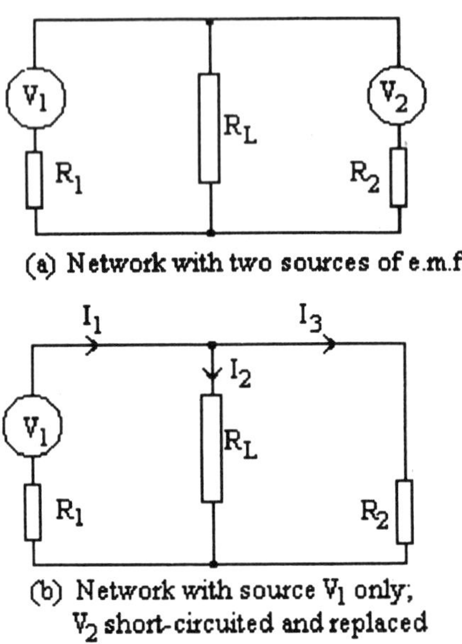

(a) Network with two sources of e.m.f

(b) Network with source V_1 only; V_2 short-circuited and replaced by its internal resistance, R_2

Fig. 1.0-2

Refer to Fig. 1.0-2(b)

$$I_1 = \frac{V_1}{R_1 + \left(\frac{R_2 R_L}{R_2 + R_L}\right)}$$

$$I_2 = I_1 \times \frac{R_2}{R_2 + R_L}$$

Note: Current divides between parallel resistances in inverse proportion to the ratio of resistances.

$$I_3 = I_1 - I_2$$

(contd)

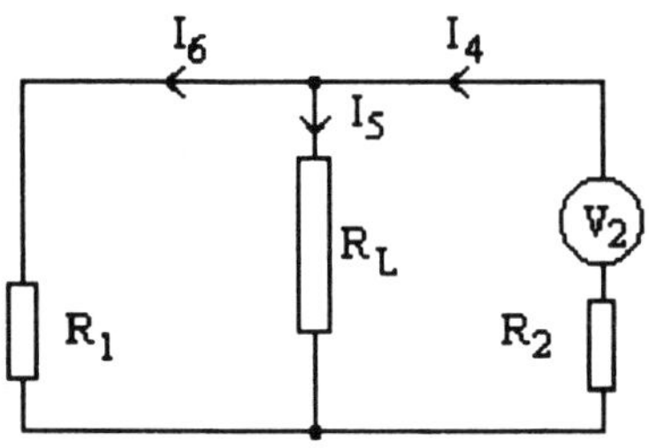

(c) Network with source V_2 only

Fig. 1.0-3

Refer to Fig. 1.0-3

$$I_4 = \frac{V_2}{R_2 + \left(\frac{R_1 R_L}{R_1 + R_L}\right)}$$

$$I_5 = I_4 \times \frac{R_1}{R_1 + R_L}$$

$$I_6 = I_4 - I_5$$

Resultant current through V_1 = $I_1 - I_6$

Resultant current through V_2 = $I_4 - I_3$

Resultant current in R_L = $I_2 + I_5$

A negative resultant current indicates that the direction of current flow is reversed to the direction indicated in the network.

Thévenin's Theorem

An active network having a pair of terminals may be replaced by an equivalent circuit having a constant-voltage generator of e.m.f **E** and an internal resistance **r.** The value of **E** is the open-circuit voltage at the terminals and **r** equals the resistance of the network measured at the terminals looking back into the network with the load disconnected and all sources removed and replaced by their internal resistances. Refer to Fig. 1.0-4

(contd)

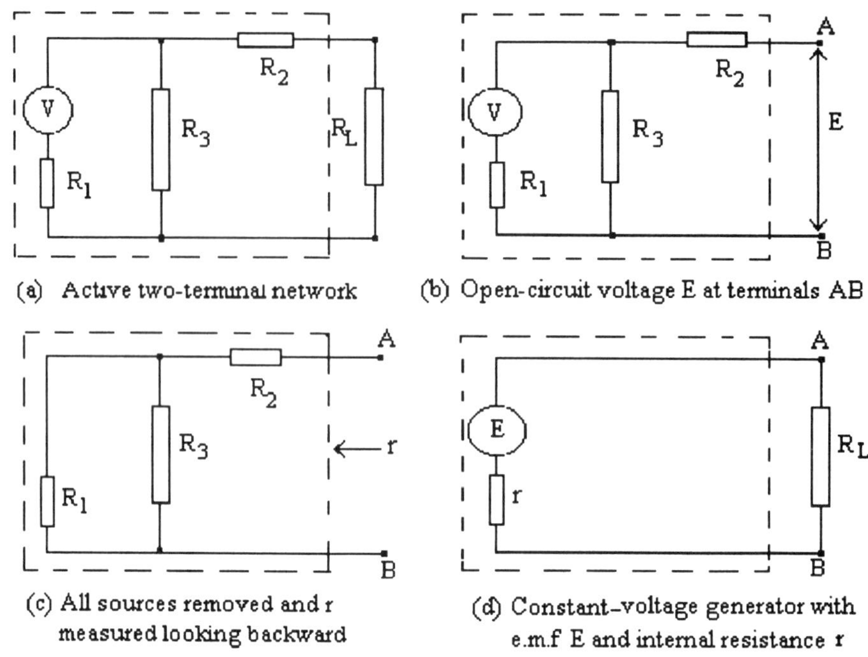

Fig. 1.0-4

<u>Refer to Fig. 1.0-4(b)</u>

$$\text{Voltage across terminals AB} = V \times \frac{R_3}{R_1 + R_3}$$

<u>Refer to Fig. 1.0-4(c)</u>

$$\text{Resistance } r = R_2 + \frac{R_1 R_3}{R_1 + R_3}$$

<u>Refer to Fig. 1.0-4(d)</u>

$$\text{Constant voltage generator e.m.f } E = \frac{V R_3}{R_1 + R_3}$$

and

$$\text{Internal resistance } r = R_2 + \frac{R_1 R_3}{R_1 + R_3}$$

Note: (d) is the same as (b) and (c).

Norton Theorem

An active network having a pair of terminals may be replaced by an equivalent circuit having a constant-current generator equivalent to the short-circuit current I_{sc} at the terminals in parallel with the internal resistance r equal to the resistance seen at the terminals looking backward into the network with all sources of e.m.f removed and replaced by their internal resistances.

Refer to Fig. 1.0-5(a-d)

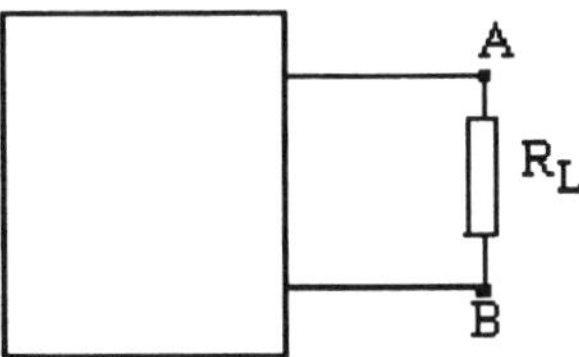

(a) Active two-terminal network

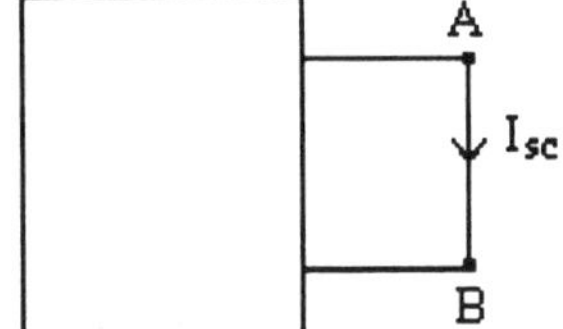

(b) Active two-terminal network with short-circuit current I_{sc}

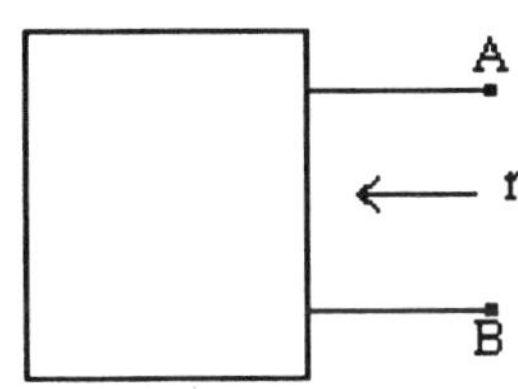

(c) Internal resistance r looking backward into the network; all sources of e.m.f removed and replaced by their internal resistances

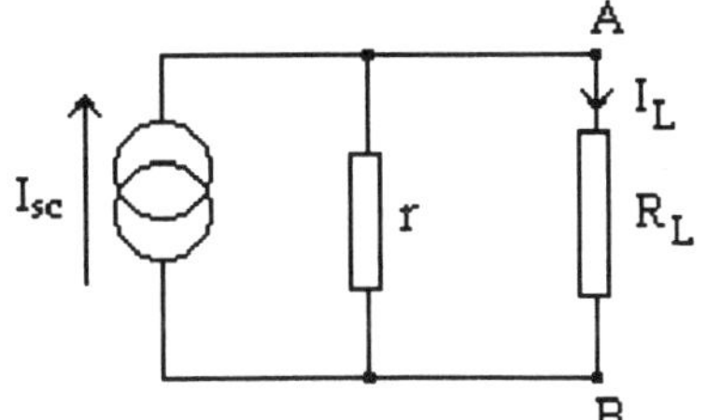

(d) Equivalent constant-current generator I_{sc} with internal resistance r

Fig. 1.0-5

$$\text{Current, } I_L = \frac{I_{sc}\, r}{R_L + r}$$

Example 1.1

Three cells, each having an e.m.f of 2 V and negligible resistance, are connected in series. A resistor of 100 Ω and one of 25 Ω are connected in series across the battery, the 100-Ω resistor being connected to the positive terminal. Calculate the current flowing in a 10-Ω resistor connected between a tapping on the battery 4 V from the positive end and the junction of the two resistors.

Solution

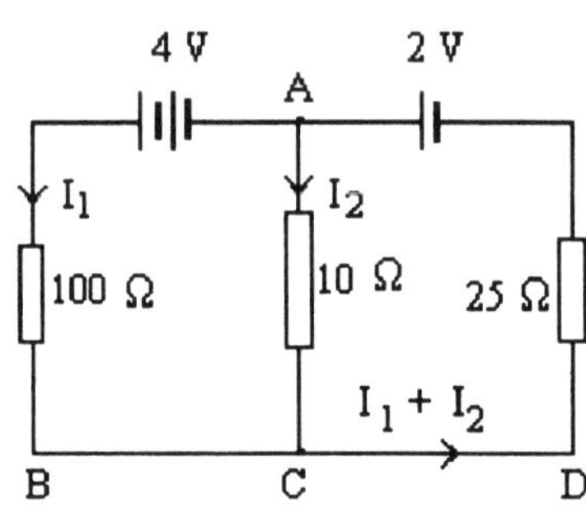

Fig. 1.1-0

Refer to Fig. 1.1-0

Apply Kirchhoff's Law to

Closed loop ABCA : $100I_1 - 10I_2 = 4$ ...(Eq. 1.1-0)

Closed loop ACDA : $10I_2 + 25(I_1 + I_2) = 2$

$25I_1 + 35I_2 = 2$ ...(Eq. 1.1-1)

Subtract Eq.1.1-1 from Eq. 1.1-0

(Eq.1.1-0) x 1 $\quad 100I_1 - 10I_2 = 4$

(Eq.1.1-1) x 4 $\quad \underline{100I_1 + 140I_2 = 8}$

$\underline{-150I_2 = -4}$

$\therefore \quad I_2 = 4/150$

$= \mathbf{\underline{0.02667\ A}}$

$= \mathbf{\underline{26.67\ mA}}$

Current through 10-Ω resistor is 26.67 mA

Example 1.2

A Wheatstone bridge is supplied by a 2 V battery of negligible resistance. The resistances of the arms of the bridge are shown in Fig. 1.2-0. Determine the value and direction of the current in the galvanometer circuit using

(a) Kirchhoff's Laws

(b) Thévenin's Theorem

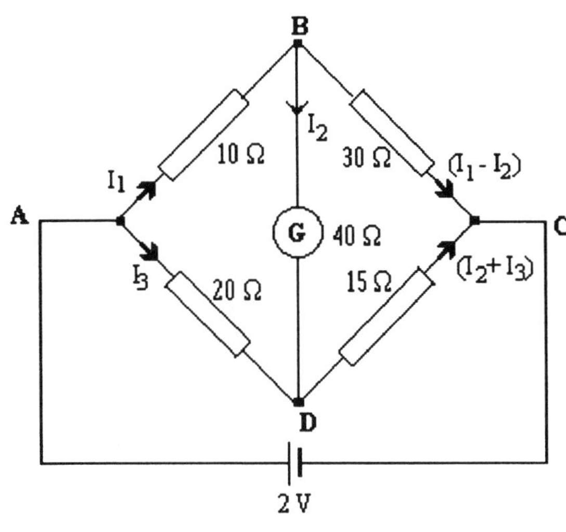

Fig. 1.2-0

Solution

(a) ***By Kirchhoff's Laws*** : (Refer to Fig.1.2-0)

Closed loop **ABDA**: $10I_1 + 40I_2 - 20I_3 = 0$

Divide by 10 $I_1 + 4I_2 - 2I_3 = 0$ .. (Eq.1.2-0)

Closed loop **BCDB**: $30(I_1 - I_2) - 15(I_2 + I_3) - 40I_2 = 0$

$30I_1 - 30I_2 - 15I_2 - 15I_3 - 40I_2 = 0$

$30I_1 - 85I_2 - 15I_3 = 0$

Divide by 5 $6I_1 - 17I_2 - 3I_3 = 0$ ... (Eq. 1.2-1)

Closed loop **ADCA**: $20I_3 + 15(I_2 + I_3) = 2$

$20I_3 + 15I_2 + 15I_3 = 2$

$15I_2 + 35I_3 = 2$ ...(Eq. 1.2-2)

(contd)

Subtract Eq. 1.2-1 from Eq. 1.2-0

(Eq. 1.2-0) x 6	$6I_1 + 24I_2 - 12I_3$	$= 0$
(Eq. 1.2-1) x 1	$6I_1 - 17I_2 - 3I_3$	$= 0$
	$41I_2 - 9I_3$	$= 0$ ....(Eq. 1.2-3)

Add Eq. 1.2-2 and Eq. 1.2-3

(Eq. 1.2-2) x 9	$135I_2 + 315I_3$	$= 18$
(Eq. 1.2-3) x 35	$1435I_2 - 315I_3$	$= 0$
	$1570I_2$	$= 18$

$$\therefore \quad I_2 = {}^{18}/_{1570}$$

$$= \mathbf{0.01146\ A} = \mathbf{11.46\ mA}$$

Current in galvanometer circuit = 11.46 mA

Note: Since the value of I_2 is positive, the direction of the current in the galvanometer circuit is that assumed in Fig. 1.2-0, that is, from **B** to **D**.

(b)

By Thévenin's Theorem

Step 1: Remove the galvanometer from the circuit and determine the voltage at terminals B and D.

Refer to Fig. 1.2-1

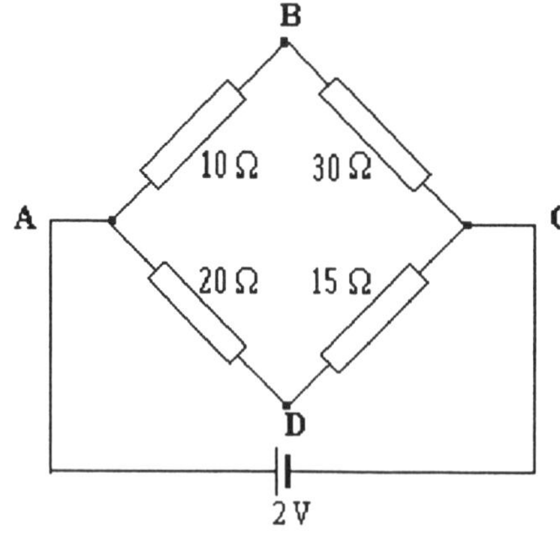

Fig. 1.2-1

$$\text{Voltage drop across } \mathbf{A\,B} = 2 \times \frac{10}{10 + 30} = \underline{0.5\ V}$$

$$\therefore \quad V_B = 2 - 0.5 = \underline{1.5\ V}$$

(contd)

$$\text{Voltage drop across } \mathbf{AD} = 2 \times \frac{20}{20 + 15} = \underline{1.143\ \text{V}}$$

$$\therefore \quad V_D = 2 - 1.143 = \underline{0.857\ \text{V}}$$

V_B is more positive than V_D, hence current flows from B → D

$$\text{Voltage between } \mathbf{BD} = 1.143 - 0.5 = \underline{0.643}$$

Step 2: Remove source of supply and determine **r** between **B** and **D**

Refer to Fig.1.2-2

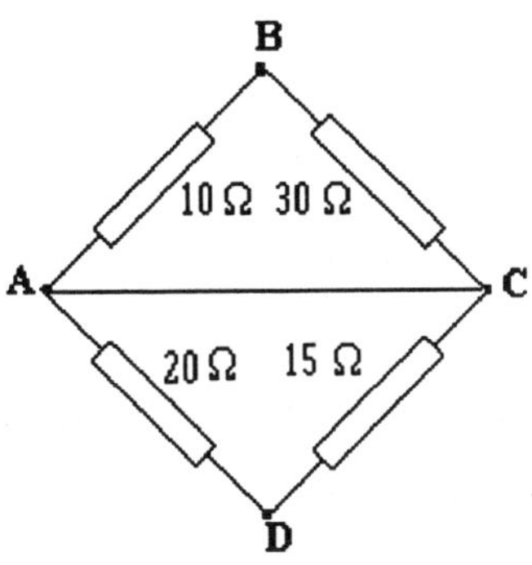

Fig. 1.2-2

The source is removed and replaced by its internal resistance. However, since its internal resistance is negligible, it is replaced by a short circuit.

r = equivalent resistance between BA and BC + equivalent resistance between DA and DC

$$= \frac{10 \times 30}{10 + 30} + \frac{20 \times 15}{20 + 15} = \underline{16.07\Omega}$$

Step 3: Calculate the current in the galvanometer circuit from the equivalent circuit in Fig.1.2-3

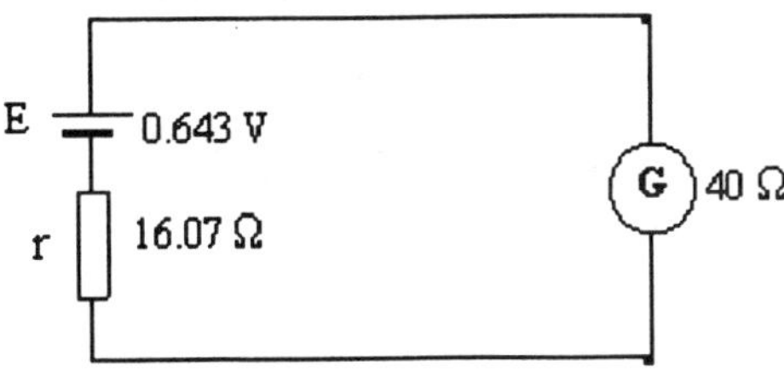

Fig. 1.2-3

(contd)

Current in galvanometer circuit $= \dfrac{E}{40 + r}$

$= 0.643/(40 + 16.07)$

$=$ **0.01147 A**

$=$ **11.47 mA**

Note: This value is not exact as the previous, the small difference is due to rounding of decimals to the nearest three places.

Example 1.3

The network shown in Fig. 1.3-0 has two sources of supply, 6 V and 4 V with internal resistances of 2 Ω and 3 Ω, respectively. The two sources are connected in parallel across a load resistor of 6 Ω. Calculate the current in the 6-Ω resistor using the Superposition Theorem.

Refer to Fig. 1.3-0

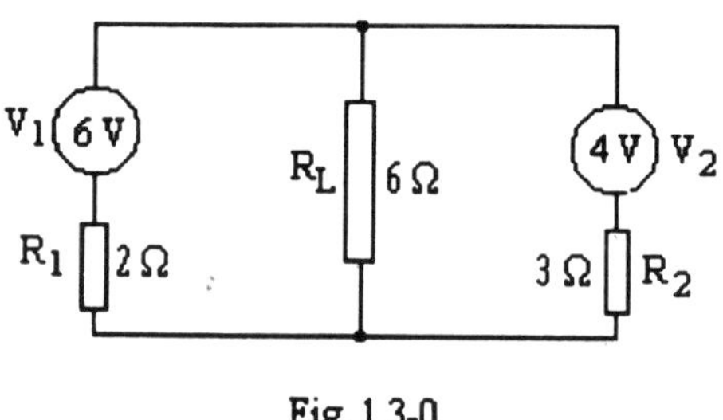

Fig. 1.3-0

Solution

By the Superposition method

This method is best carried out in procedural steps.

Step 1: Determine the current in the load by each source acting individually, the other source removed and replaced by its internal resistance.

Refer to Fig. 1.3-1

(contd)

Network with 6 V source only (Fig. 1.3-1)

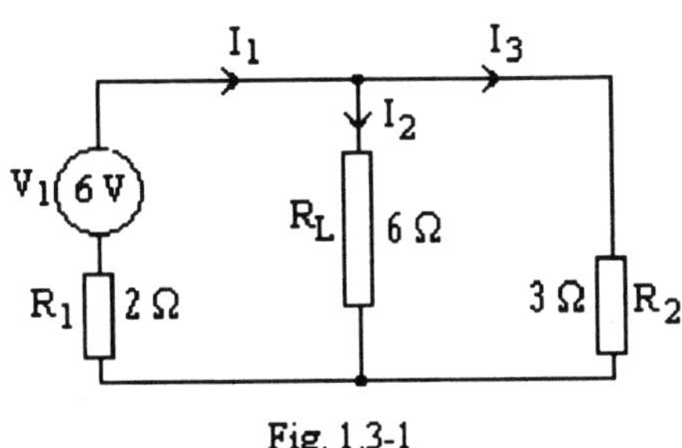

Fig. 1.3-1

$$I_1 = \frac{V_1}{R_1 + \left(\frac{R_2 R_L}{R_2 + R_L}\right)} = \frac{6}{2 + \left(\frac{3 \times 6}{3 + 6}\right)} = \underline{1.5\ A}$$

$$I_2 = I_1 \times \frac{R_2}{R_2 + R_L} = 1.5 \times \frac{3}{3+6} = \underline{0.5\ A}$$

$$I_3 = I_1 - I_2 = 1.5 - 0.5 = \underline{1.0\ A}$$

Network with 4 V source only (Fig. 1.3-2)

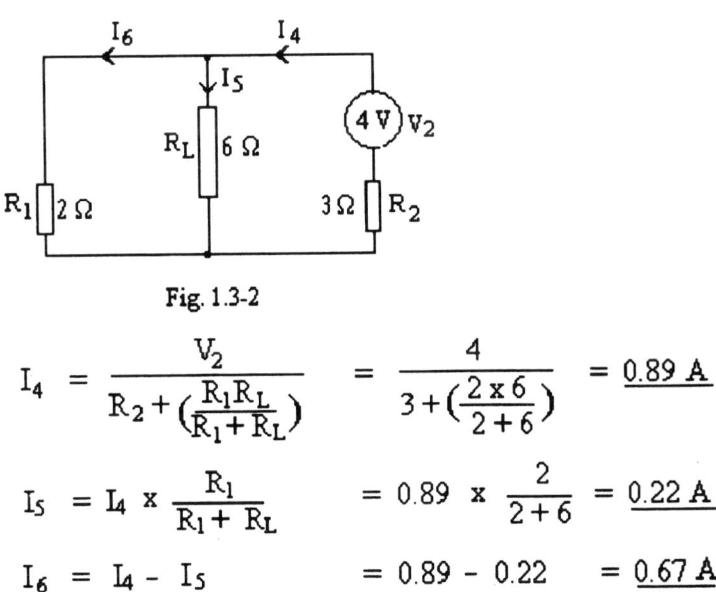

Fig. 1.3-2

$$I_4 = \frac{V_2}{R_2 + \left(\frac{R_1 R_L}{R_1 + R_L}\right)} = \frac{4}{3 + \left(\frac{2 \times 6}{2 + 6}\right)} = \underline{0.89\ A}$$

$$I_5 = I_4 \times \frac{R_1}{R_1 + R_L} = 0.89 \times \frac{2}{2+6} = \underline{0.22\ A}$$

$$I_6 = I_4 - I_5 = 0.89 - 0.22 = \underline{0.67\ A}$$

(contd)

Step 2: Superimpose the currents in the network of Figs. 1.3-1 and 1.3-2. The networks are redrawn below.

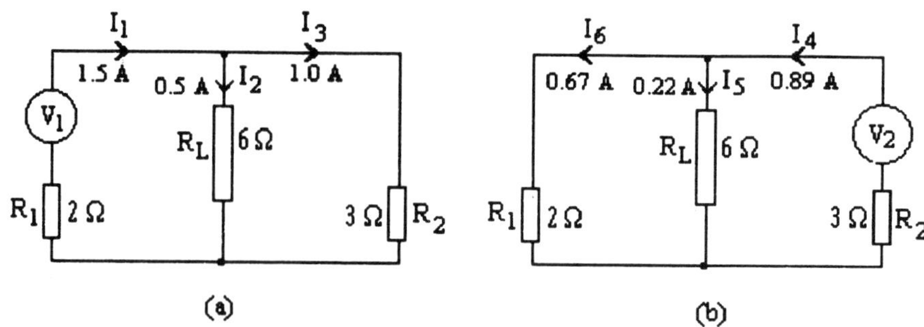

Fig. 1.3-3

Refer to Fig. 1.3-3

Resultant current through $V_1 = I_1 - I_6 = 1.5 - 0.67 = \underline{0.83\text{ A}}$

Resultant current through $V_2 = I_4 - I_3 = 0.89 - 1.0 = -0.11\text{ A}$

The negative sign indicates that V_1 is charging V_2 at 0.11 A.

Resultant current in $R_L = I_2 + I_5 = 0.5 + 0.22 = \mathbf{\underline{0.72\text{ A}}}$

Example 1.4

Calculate the current in each branch of the network shown in Fig. 1.4-0 using the Superposition Theorem.

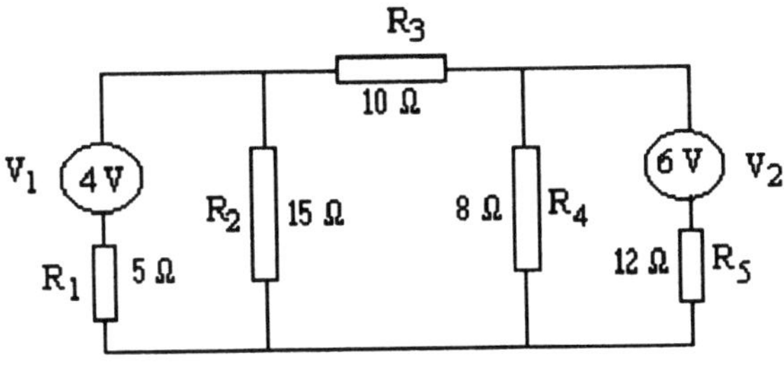

Fig. 1.4-0

Solution

(A) 4-V supply acting alone (Refer to Fig. 1.4-1)

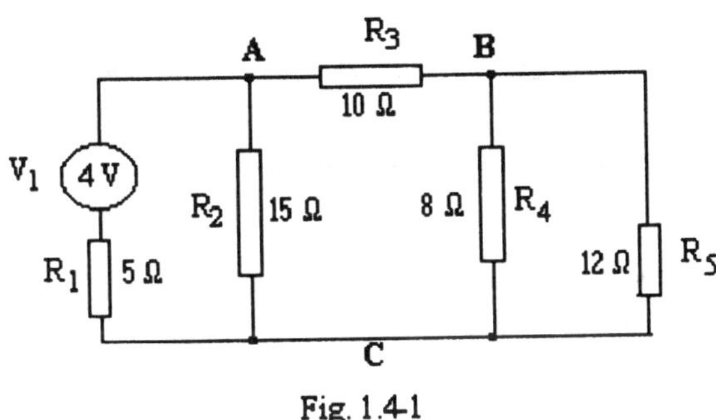

Fig. 1.4-1

Resistance across **BC** $= R_{eq.1}$

$$= \frac{R_4 \times R_5}{R_4 + R_5}$$

$$\therefore \quad R_{eq.1} = \frac{8 \times 12}{8 + 12} = \underline{4.8\ \Omega}$$

(contd)

The circuit is now reduced to the network shown in Fig. 1.4-2.

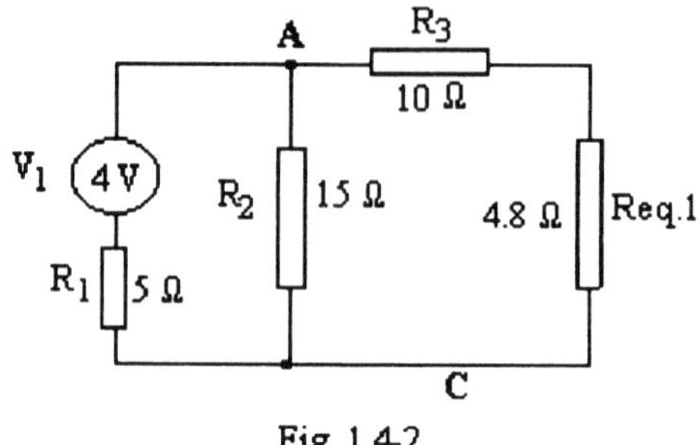

Fig. 1.4-2

Refer to Fig. 1.4-2

Resistance across AC $= R_{eq.2}$

$$= \frac{R_2 \times (R_3 + R_{eq.1})}{R_2 + (R_3 + R_{eq.1})}$$

$$\therefore \quad R_{eq.2} = \frac{15 \times (10 + 4.8)}{15 + (10 + 4.8)} = \underline{7.45\ \Omega}$$

The circuit is now reduced to the network shown in Fig. 1.4-3.

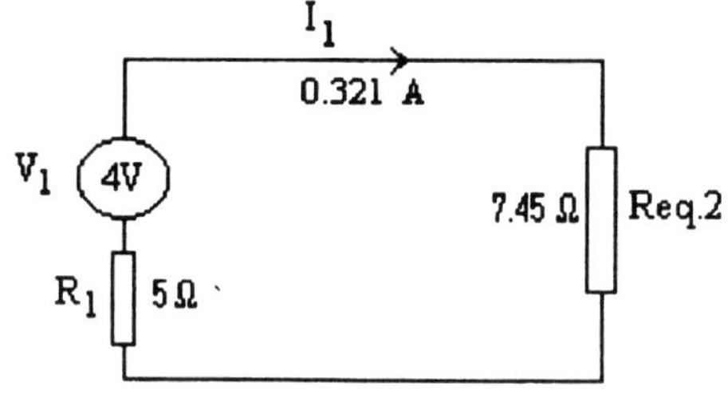

Fig. 1.4-3

Refer to Fig. 1.4-3

$R_T = R_1 + R_{eq.2}$

$= 5 + 7.45 = 12.45\ \Omega$

Current, $I_1 = V_1/R_T$

$= 4/12.45 = \underline{0.321\ A}$

Figure 1.4-4 shows the current distribution with $R_{eq.1}$ in the network.

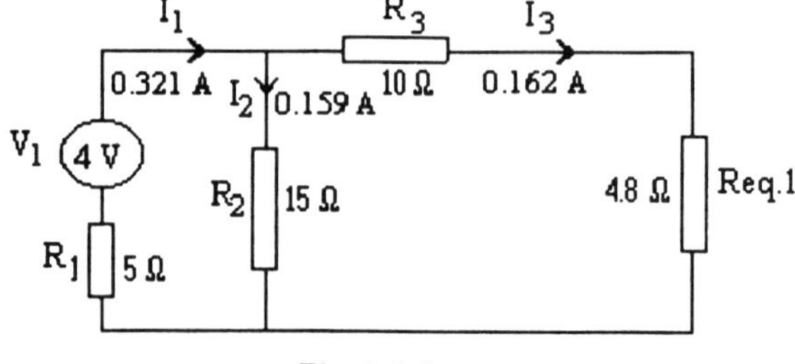

Fig. 1.4-4

(contd)

Refer to Fig. 1.4-4

$$I_2 = I_1 \times \frac{(R_3 + R_{eq.1})}{R_2 + (R_3 + R_{eq.1})}$$

$$= 0.321 \times \frac{(10 + 4.8)}{15 + (10 + 4.8)} = \underline{0.159\ A}$$

$$I_3 = I_1 - I_2 = 0.321 - 0.159 = \underline{0.162\ A}$$

Alternatively

I_3 could have been calculated first and then I_2 obtained.

$$I_3 = I_1 \times {}^{R_2}/(R_2 + R_3 + R_{eq.1})$$

$$= 0.321 \times {}^{15}/(15 + 10 + 4.8)$$

$$= 0.321 \times {}^{15}/29.8 = \underline{0.162\ A}$$

$$I_2 = I_1 - I_3$$

$$= 0.321 - 0.162 = \underline{0.159\ A}$$

*With a **4-V supply acting alone**, the current distribution in each branch of the network is as shown in Fig. 1.4-5.*

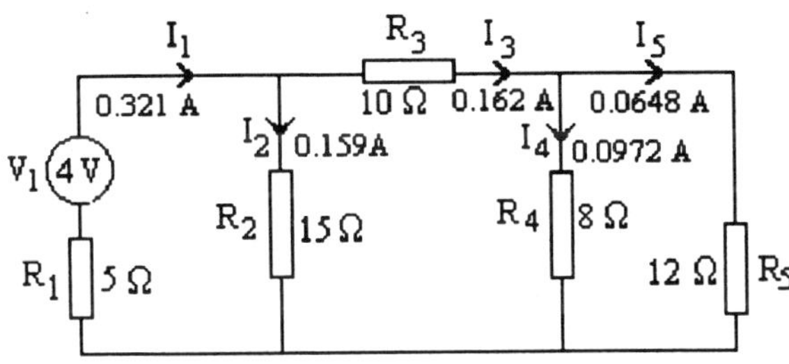

Fig. 1.4-5

$$I_4 = I_3 \times {}^{R_5}/(R_4 + R_5)$$

$$= 0.162 \times {}^{12}/(8 + 12) = \underline{0.0972\ A}$$

$$I_5 = I_3 - I_4 = 0.162 - 0.0972 = \underline{0.0648\ A}$$

(contd)

(B) 6-V supply acting alone (Refer to Fig.1.4-6)

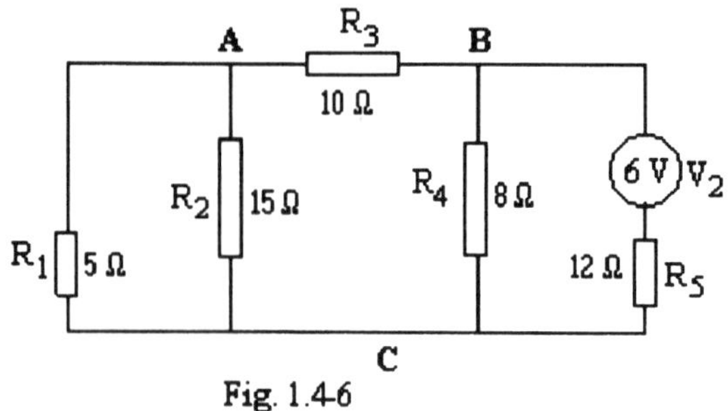

Fig. 1.4-6

Resistance across AC = $R_{eq.3}$

$$= \frac{R_1 \times R_2}{R_1 + R_2}$$

$$\therefore \quad R_{eq.3} = \frac{5 \times 15}{5 + 15} = \underline{3.75\ \Omega}$$

The circuit is now reduced to the network shown in Fig. 1.4-7.

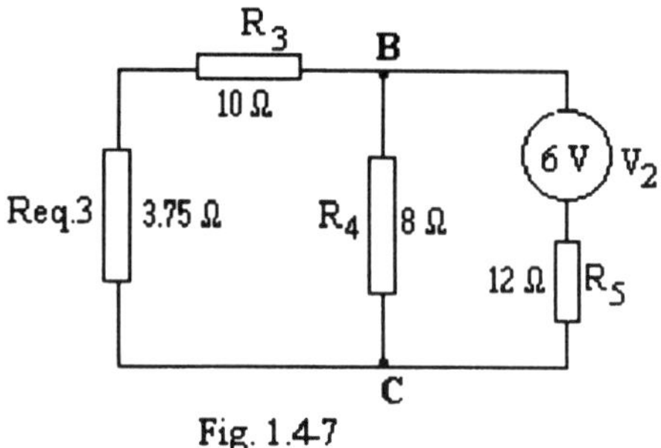

Fig. 1.4-7

Resistance across BC = $R_{eq.4}$

$$= \frac{R_4 \times (R_3 + R_{eq.3})}{R_4 + (R_3 + R_{eq.3})}$$

$$= \frac{8 \times (10 + 3.75)}{8 + (10 + 3.75)} = \underline{5.06\ \Omega}$$

(contd)

The circuit is now reduced to the network shown in Fig. 1.4-8

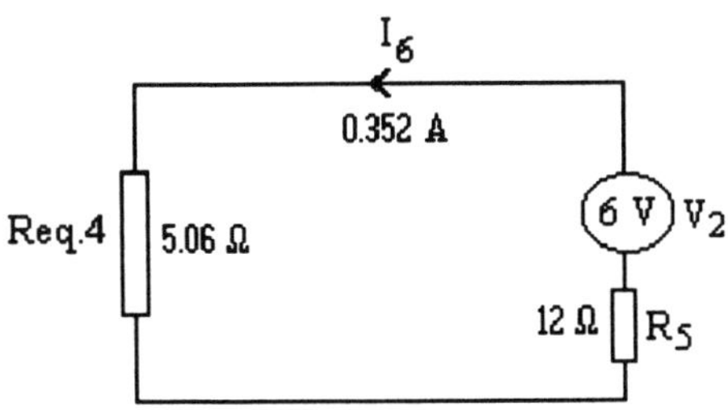

Fig. 1.4-8

Current I_6 in V_2 $= V_2/(R_5 + R_{eq.4})$

$= 6/(12 + 5.06) \quad = \underline{0.352\text{ A}}$

The current distribution with $R_{eq.3}$ in the network is shown in Fig. 1.4-9.

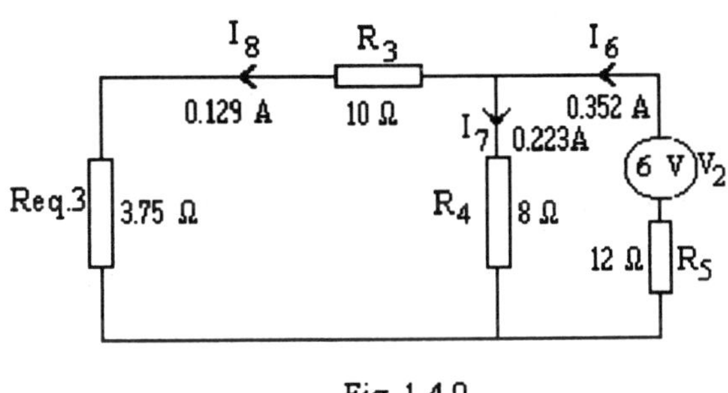

Fig. 1.4-9

$$I_7 = I_6 \times \frac{(R_3 + R_{eq.3})}{(R_3 + R_{eq.3}) + R_4}$$

$$= 0.352 \times \frac{(10 + 3.75)}{(10 + 3.75) + 8} = \underline{0.223\text{ A}}$$

$$I_8 = I_6 - I_7 = 0.352 - 0.223 = \underline{0.129\text{ A}}$$

(contd)

With a ***6-V supply acting alone****, the current in each branch of the network is as shown in Fig. 1.4-10.*

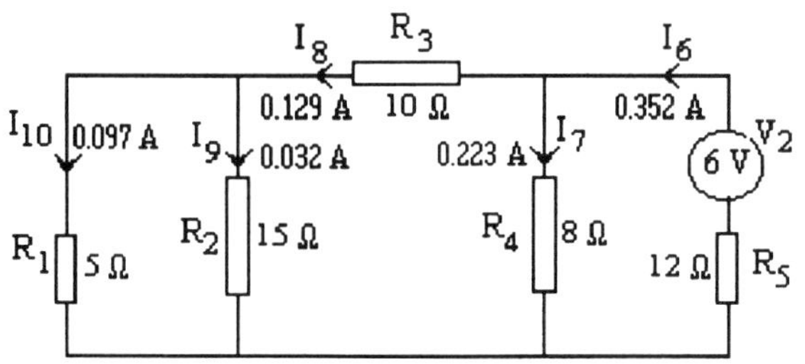

Fig. 1.4-10

$I_9 = I_8 \times R_1/(R_1 + R_2)$

$= 0.129 \times 5/(5 + 15) = \underline{0.032\ A}$

$I_{10} = I_8 - I_9 = 0.129 - 0.032 = \underline{0.097\ A}$

Finally, superimpose the networks of Figs. 1.4-5 and 1.4-10 and determine the current in each branch of the network with both sources of supply.

Superimposing the networks, we get

Current in V_1 (R_1) $= I_1 - I_{10}$

$= 0.321 - 0.097 = \underline{\mathbf{0.224\ A}}$

Current in R_2 (15Ω) $= I_2 + I_9$

$= 0.159 + 0.032 = \underline{\mathbf{0.191\ A}}$

Current in R_3 (10Ω) $= I_3 - I_8$

$= 0.162 - 0.129 = \underline{\mathbf{0.033\ A}}$

Current in R_4 (8Ω) $= I_4 + I_7$

$= 0.0972 + 0.223 = \underline{\mathbf{0.320\ A}}$

Current in V_2 (R_5) $= I_6 - I_5$

$= 0.352 - 0.0648 = \underline{\mathbf{0.287\ A}}$

Example 1.5

A circuit for charging accumulators has two lamps in parallel with it, the resistance of one lamp being 2.2 Ω and the other 1.3 Ω.
Estimate the value of the resistance R (Fig. 1.5-0) that must be connected in series with the accumulators to limit the charging current to 5 A. The circuit is supplied by a source having an E.M.F of 12 V and a resistance of 0.12 Ω, and the accumulators that are being charged have a resistance of 0.4 Ω and an E.M.F of 6 V.

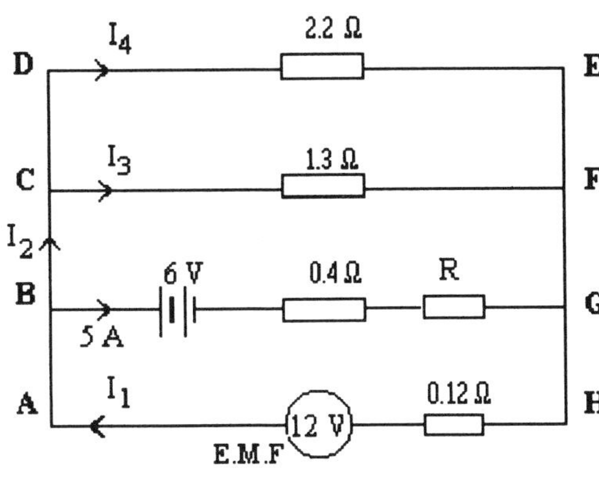

Fig. 1.5-0

Solution

Apply Kirchhoff's second Law to

Closed loop ABGHA: $5(0.4 + R) + 0.12I_1 = 12 - 6$

$$2.0 + 5R + 0.12I_1 = 6$$

$$5R + 0.12I_1 = 4 \quad \text{(Eq. 1.5-0)}$$

Closed loop ACFHA $\quad 1.3I_3 + 0.12I_1 = 12 \quad$ (Eq. 1.5-1)

Subtract Eq. 1.5-1 from 1.5-0

$$5R + 0.12I_1 = 4$$

$$\underline{1.3I_3 \quad + 0.12I_1 = 12}$$

$$\underline{-1.3I_3 + 5R \quad = -8}$$

$$1.3I_3 - 5R = 8 \quad \text{(Eq. 1.5-2)}$$

(contd)

Closed loop ADEHA: $2.2\ I_4 + 0.12\ I_1 = 12$

[Note: $(I_4 = I_2 - I_3)$ $\quad 2.2(I_2 - I_3) + 0.12\ I_1 = 12$

and $(I_2 = I_1 - 5)$] $\quad 2.2[(I_1 - 5) - I_3] + 0.12\ I_1 = 12$

$$2.2\ I_1 - 11 - 2.2\ I_3 + 0.12\ I_1 = 12$$

$$2.32 I_1 - 2.2 I_3 = 23 \quad \text{(Eq. 1.5-3)}$$

Subtract Eq. 1.5-0 from Eq. 1.5-3

(Eq. 1.5-3) x 0.12 $\quad 0.2784 I_1 - 0.264 I_3 = 2.76$

(Eq. 1.5-0) x 2.32 $\quad \underline{0.2784 I_1 + 11.6R = 9.28}$

$$\underline{-0.264 I_3 - 11.6R = -6.52}$$

$$0.264 I_3 + 11.6R = 6.52 \quad \text{(Eq. 1.5-4)}$$

Subtract Eq. 1.5-4 from Eq. 1.5-2

(Eq. 1.5-2) x 0.264 $\quad 0.3432 I_3 - 1.32R = 2.112$

(Eq. 1.5-4) x 1.3 $\quad \underline{0.3432 I_3 + 15.08R = 8.476}$

$$\underline{-16.4R = -6.364}$$

$$\therefore \quad R = \frac{6.364}{16.4}$$

$$= \mathbf{0.388\ \Omega}$$

Example 1.6

A load of resistance 6 Ω is supplied by a parallel combination of two sources, E_1 and E_2. Source E_1 has an open-circuit e.m.f of 42 V and an internal resistance of 12 Ω. Source E_2 has an open-circuit e.m.f of 35 V and internal resistance of 3 Ω. Using the Superposition Theorem, calculate the current supplied by the 35 V source and the power dissipated internally. Refer to Fig. 1.6-0

[Hint: It would be incorrect to calculate the power dissipated in the 3-Ω resistor by superimposing the power dissipated by the current from E_1 and that by the current from E_2. Superposition cannot be applied in power calculations, the reason being that power is not linearly related to voltage or current. Power $P = V^2/R$ or I^2R.]

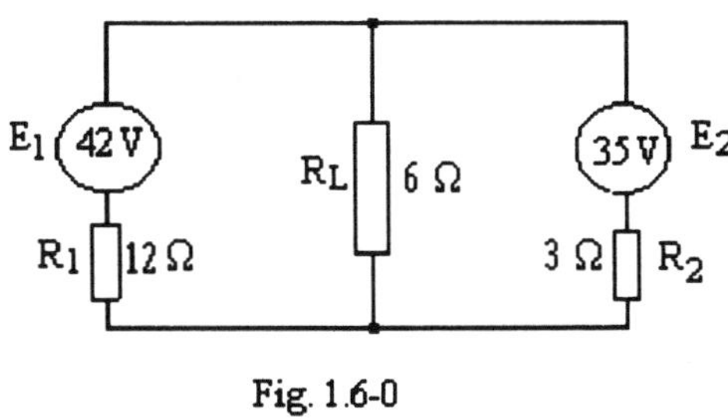

Fig. 1.6-0

Solution

Source E_1 acting alone

Refer to Fig. 1.6-1

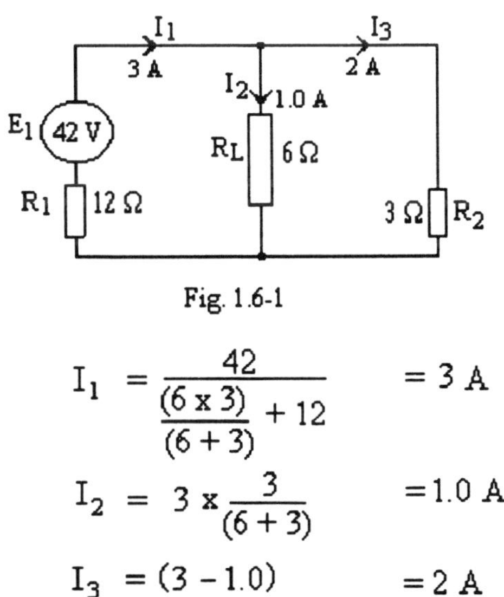

Fig. 1.6-1

$$I_1 = \frac{42}{\frac{(6 \times 3)}{(6+3)} + 12} = 3 \text{ A}$$

$$I_2 = 3 \times \frac{3}{(6+3)} = 1.0 \text{ A}$$

$$I_3 = (3 - 1.0) = 2 \text{ A}$$

(contd)

Source E_2 acting alone

Refer to Fig. 1.6-2

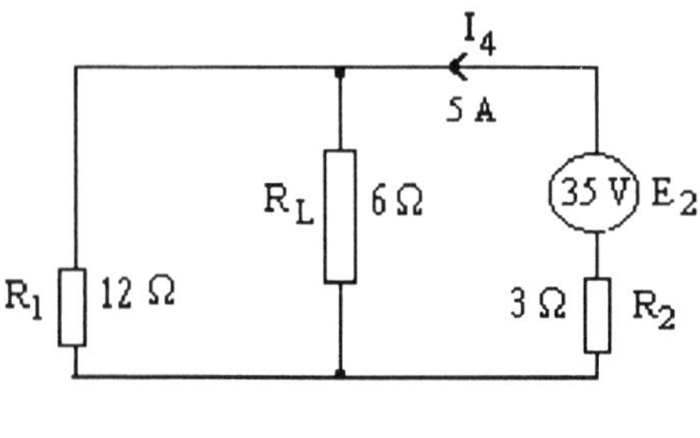

Fig. 1.6-2

Since it is required to determine the current and power supplied and dissipated by E_2, it is only necessary to calculate the current I_4.

$$I_4 = \frac{E_2}{R_2 + \frac{(R_1 R_L)}{(R_1 + R_L)}}$$

$$= \frac{35}{3 + \frac{(12 \times 6)}{(12 + 6)}} = 5 \text{ A}$$

Superimpose Fig. 1.6-1 on Fig. 1.6-2 and calculate the current in E_2.

Current through E_2 $= I_4 - I_3$

$= (5 - 2)$ $=$ **3 A**

Power dissipated in E_2 $= I^2 R_2$

$= 3^2 \times 3$ $=$ **27 W**

CHAPTER 2

SEMICONDUCTOR DIODES AND RECTIFIER CIRCUITS

Introduction

Matter is made up of atoms.

Every atom contains a nucleus of positively charged protons surrounded by electrons moving in orbit around it.

Consider a simple atom which consists of a nucleus with one proton and a single electron in orbit. In this state the atom is complete and has zero electrical charge.

Refer to Fig. 2.0-0

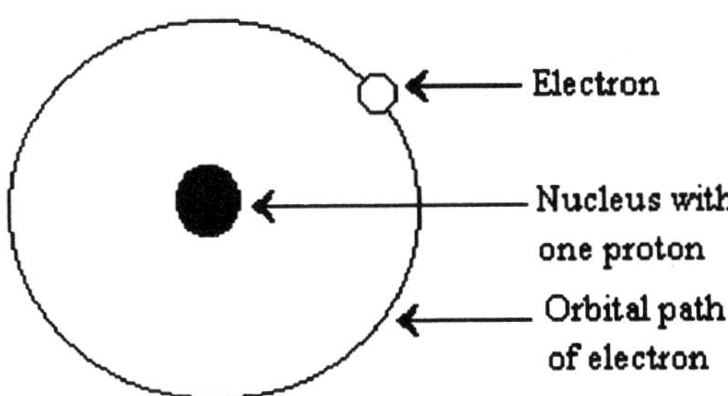

Fig. 2.0-0. A simple atom

Consider the case in which the electron in Fig. 2.0-0 is made to dislodge from its orbital path. The atom may now be considered as being incomplete, having lost its associated electron.

Refer to Fig. 2.0-1

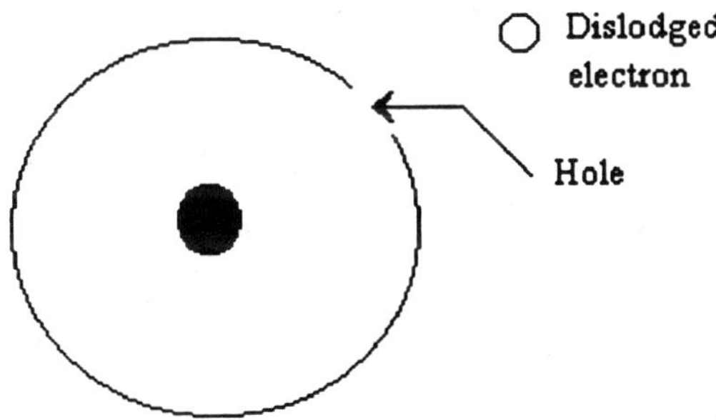

Fig. 2.0-1. Electron dislodged from orbit

(contd)

The vacancy created by the dislodged electron from its orbital path is called a *hole* (incomplete electron) which now acquires a *positive charge*. The *free electron becomes negatively charged* by virtue of breaking away from its orbital path.

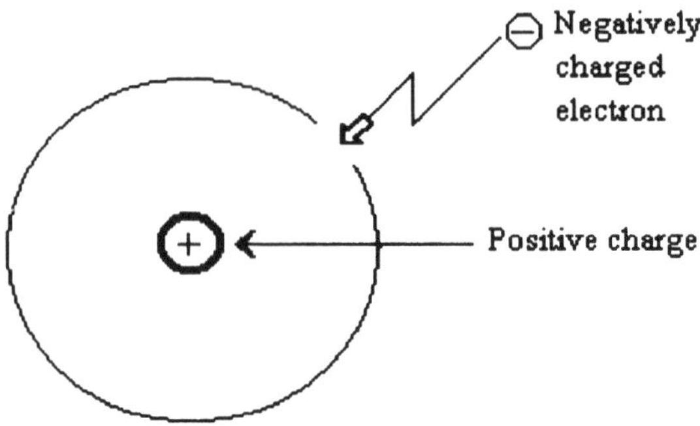

Fig. 2.0-2. Polarity of incomplete atom

The rushing of the electron to fill the hole (that is, to complete the incomplete atom) constitutes the flow of an electric current.
This is shown in Fig. 2.0-2.

Example 2.1

Describe the construction and operation of a ***p-n*** *junction diode. Include in your answer a sketch of a typical forward-biased and reverse-biased characteristic.*

Solution

The *p-n* junction diode

To improve the conduction in an intrinsic semiconductor such as germanium or silicon, impurities (dopant) are added to the material.
A semiconductor material in which additional ***holes*** (positive charges) have been created by a dopant is called a ***p-type material.***
In similar manner, if the semiconductor material is doped with an excess of ***electrons*** (negative charges), it is called an ***n-type material.***

(contd)

To construct a semiconductor junction diode, the *p-type material* is fused to the *n-type material.*

Refer to Fig. 2.1-0

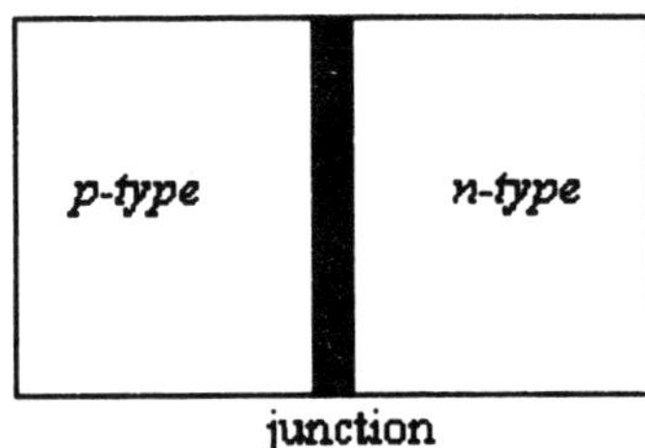

Fig. 2.1-0. A junction diode

The operation of the *p-n junction* diode may best be explained by the movement of *holes and electrons* across the junction when under the influence of an electric field.

Refer to Fig. 2.1-1

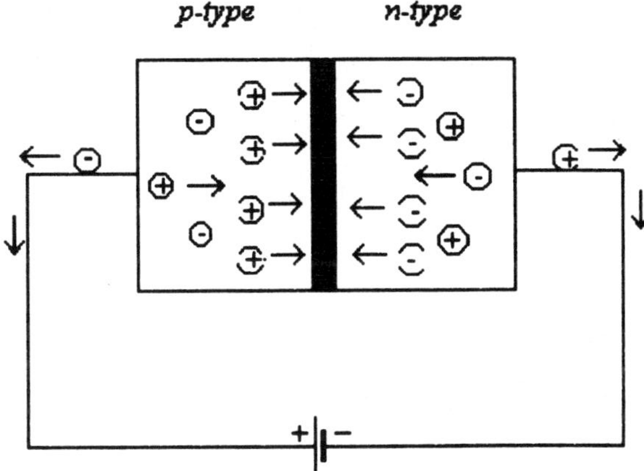

Fig. 2.1-1. Forward-biased junction diode

Circuit Symbol

When a battery is connected in the direction shown in Fig. 2.1-1 called **forward biasing**, the *predominant holes* (positive charges) *move from the p-side to the n-side* and the *excess electrons move from the n-side to the p-side*, causing an easy flow of electric current.

This easy flow of electric current is substantiated by the forward-biased characteristic shown in Fig. 2.1-2; a small voltage produces a large current flow.

(contd)

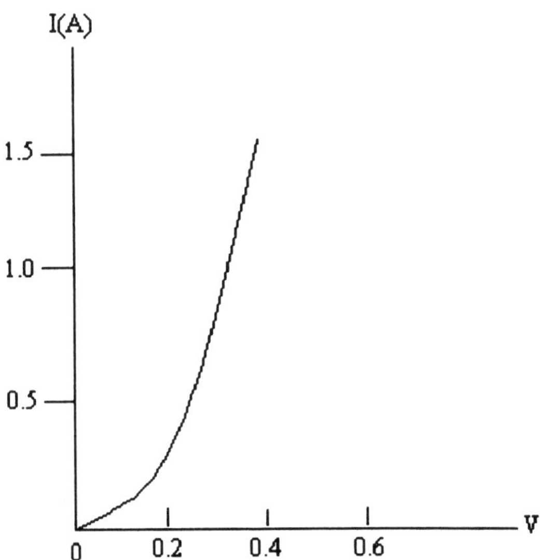

Fig. 2.1-2. Typical forward-biased characteristic

When the battery is connected in the opposite direction (Fig. 2.1-3), called **reverse biasing**, only a small movement of minority holes takes place and only a few electrons are able to cross the junction, causing a reverse current flow of a few microamperes.

A typical reverse-biased characteristic is shown in Fig. 2.1-4.

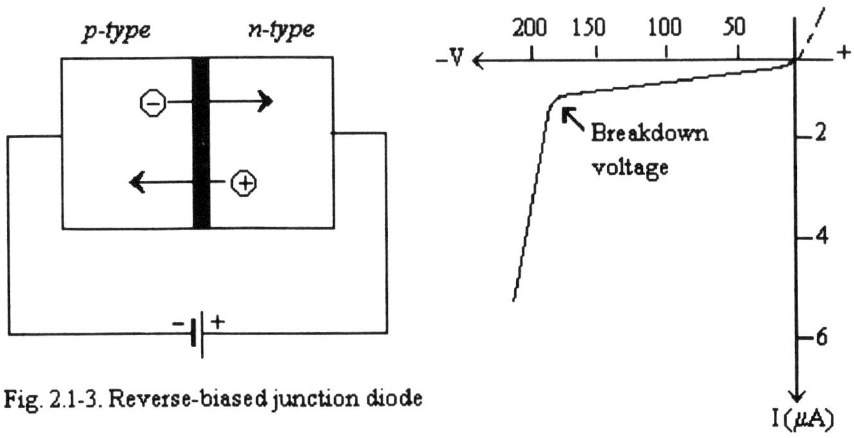

Fig. 2.1-3. Reverse-biased junction diode

Fig. 2.1-4. Typical reverse-biased characteristic

Example 2.2

The current voltage characteristic of a semiconductor diode in the forward-biased direction is given by the data of Table 2.2-0 .

Table 2.2-0

Voltage(V)	0.05	0.10	0.15	0.20	0.25
Current(mA)	0.3	1.0	5.0	25.0	150

Plot the characteristic and use it to determine the a.c. resistance of the diode at the point $V = 0.10$ V.

Solution

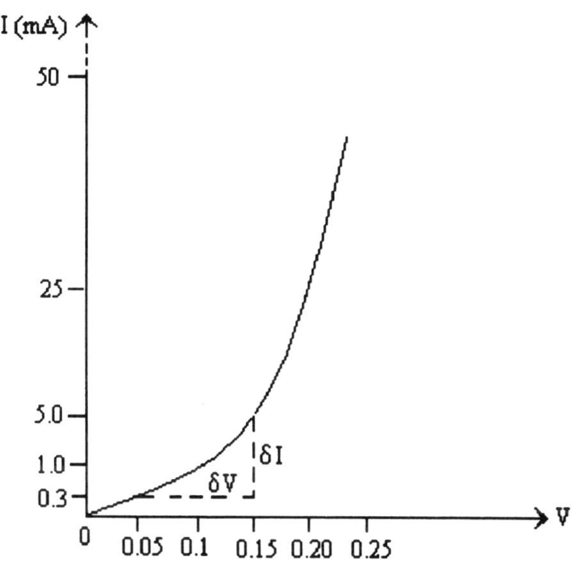

Fig. 2.2-0. Current voltage characteristic

Refer to Fig. 2.2-0

To find the a.c. resistance at the point V = + 0.1 V, it is necessary to select two equidistant points on either side of + 0.1 V, that is, 0.05 V and 0.15V. The corresponding current values are 0.3 mA and 5.0 mA.

$$\text{a.c. resistance, } r_{a.c.} = \frac{\text{change in voltage}}{\text{change in current}}$$

$$= \frac{\delta V}{\delta I}$$

$$= \frac{0.15 - 0.05}{(5.0 - 0.3) \times 10^{-3}} = \underline{\mathbf{21.28\ \Omega}}$$

Example 2.3

(a) *Explain the operation of the single-phase half-wave rectifier circuit shown in Fig. 2.3-0.*
Sketch the associated waveforms for one complete cycle and give an expression for the voltage across the load.

(b) *If the transformer secondary output voltage is 20 V,50 Hz, calculate the voltage across the load. Assume that the source resistance and the diode slope resistance are negligible.*

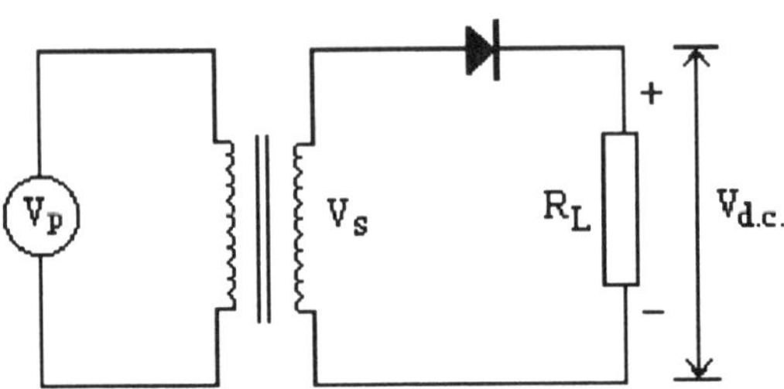

Fig. 2.3-0. Half-wave rectifier circuit

Solution

(a) Consider the circuit shown in Fig. 2.3-0 when V_s is positive; the diode is forward biased and conducts current in this half-cycle.
In the negative half-cycle the diode is reverse biased and only the leakage current flows in the load. In this half-cycle (negative half) the diode is blocking and has to sustain the *peak inverse voltage* (*PIV*). The associated waveforms are shown in Fig. 2.3-1.

The average load voltage, which is the d.c. load voltage, is the peak or maximum value of the line voltage divided by π ($V_{d.c.} = V_m/\pi$). The peak or maximum value and the effective or r.m.s value are related by √2 for sinusoidal waveforms ($V_{r.m.s} = V_{max}/\sqrt{2}$).

(contd)

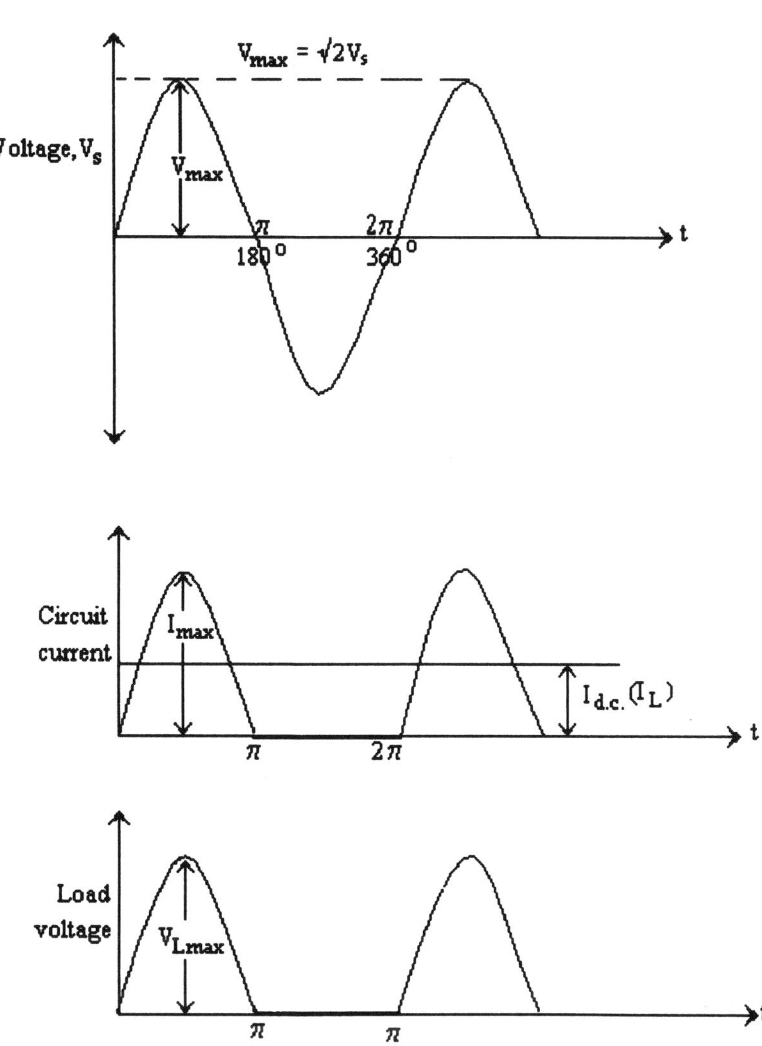

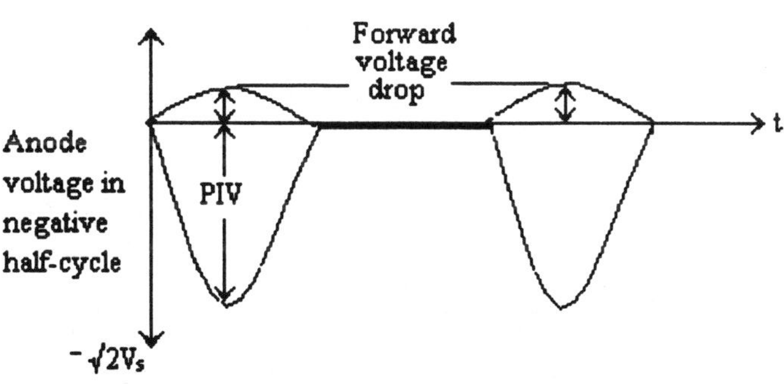

Fig. 2.3-1. Associated waveforms for half-wave rectifier

(contd)

$$V_s = V_m \sin \omega t$$

$$\omega = 2\pi f \text{ rad/sec}$$

The average value or d.c. value of the load voltage is calculated as follows:

$$V_{av} = \frac{1}{2\pi}\int_0^{\pi} V_m \sin \omega t \, dt$$

$$= \frac{V_m}{2\pi}\int_0^{\pi} \sin \omega t \quad dt$$

$$= \frac{V_m}{2\pi}[-\cos \omega t\,]_0^{\pi}$$

$$= \frac{V_m}{2\pi}[(-\cos \pi) - (-\cos 0)]$$

$$= \frac{V_m}{2\pi}[(1) - (-1)]$$

$$= \frac{2V_m}{2\pi}$$

$$= \frac{V_m}{\pi}$$

$$= \mathbf{0.318\, V_m}$$

Similarly $I_L = 0.318\, I_m$

Voltage across load $= I_L \times R_L$

that is $V_L = 0.318\, I_m \times R_L$

$$= 0.318\, V_m$$

$$= 0.318\, \sqrt{2}V_s$$

$$\approx \mathbf{0.45V_s}$$

This is interpreted to mean that the voltage across the load, V_L is approximately 45% of the supply voltage, V_S.

(contd)

Alternatively, the current in the load is

$$I_L = \frac{V_m \sin \omega t}{R_s + r_{a.c.} + R_L}$$

R_s = Source resistance

$r_{a.c}$ = Diode a.c resistance or slope resistance

Since R_s and $r_{a.c}$ are negligible compared with R_L

$$I_L = \frac{V_m \sin \omega t}{R_L}$$

$$= I_m \sin \omega t$$

$$= \frac{1}{2\pi} \int_0^{\pi} I_m \sin \omega t \ dt$$

$$= \frac{I_m}{\pi}$$

$$= \mathbf{0.318\ I_m}$$

The average value of the load voltage

$$V_L = I_L \times R_L$$

$$= 0.318\ I_m \times R_L$$

$$= \mathbf{0.318 V_m}$$

Assume V_s to be sinusoidal. Then

$$V_L = 0.318 \times \sqrt{2} V_s$$

$$\approx \mathbf{0.45 V_s}$$

(b) Given that the transformer secondary voltage is 20 V, i.e.

$$V_s = 20 \text{ V}$$

$$V_L = 0.45\ V_s$$

$$= 0.45 \times 20 \qquad = \underline{\mathbf{9.0\ V}}$$

Example 2.4

Give the meaning of the term 'peak inverse voltage' (PIV) as applied to the diode in a rectifier circuit.

With the aid of waveforms and circuit diagrams, explain the difference between a full-wave rectifier which uses

(a) *two diodes*

(b) *four diodes*

In particular, mention the transformer required and the PIV across the diodes if the d.c. output voltage is the same in each case.

Solution

(a) The PIV of a diode is the maximum reverse voltage that can be sustained by the diode without breaking down. The circuit diagram employing two diodes and associated wave-forms are shown in Fig. 2.4-0(a) and 2.4-0(b).

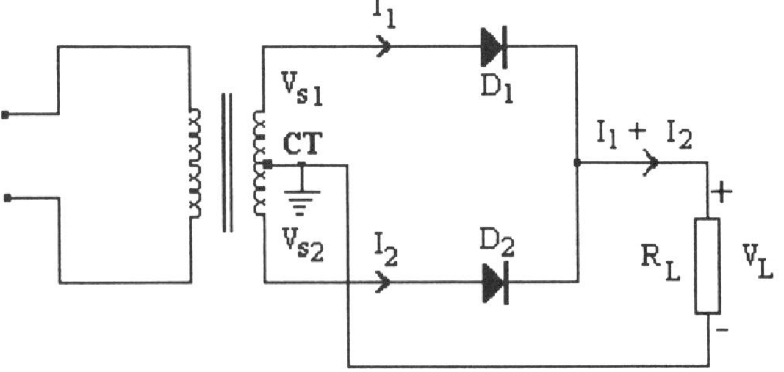

(a) Full-wave rectifier with two diodes

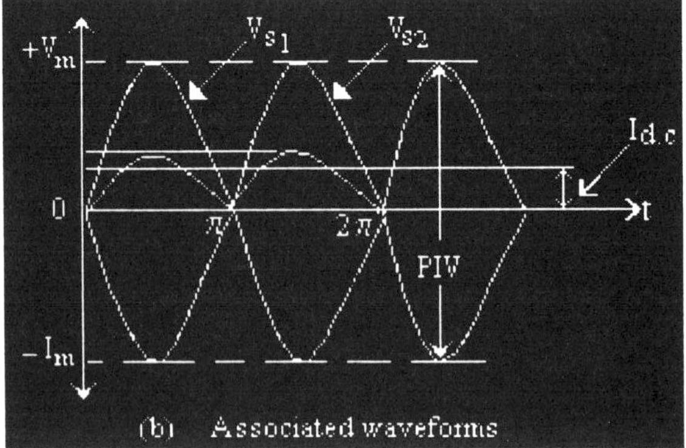

(b) Associated waveforms

Fig. 2.4-0

(contd)

The full-wave, center-tap (CT) or bi-phase (sometimes referred to as a phase inverter) circuit is shown in Fig. 2.4-0(a) employing two diodes.

The output winding of the transformer provides the voltages V_{s1} and V_{s2} which are 180^0 out of phase; such a CT transformer serves as a phase inverter.
On the first half-cycle when V_{s1} is positive, diode D_1 conducts and I_1 is supplied to the load. Meanwhile D_2 is reverse biased and is blocking (cut-off); the PIV is twice the maximum voltage, that is, $2V_m$.

On the second-half cycle V_{s2} is positive and current I_2 is supplied to the load through D_2. Meanwhile D_1 is reverse biased and blocking (cut-off). The potential at the anode of D_1 is $-V_m$ while its cathode is $+V_m$. The PIV sustained when the diode is reverse biased is $2V_m$.
The current through the load in both half-cycles is in the same direction $(I_1 + I_2)$. Since the current flows in both half-cycles, the average load voltage and load current are twice the values of the half wave circuit. That is

$$V_L = 2 \times {}^{V_m}/\pi$$

$$= 0.636\ V_m$$

$$= 0.636 \times \sqrt{2}V_s \qquad = \mathbf{0.90V_s}$$

$$I_L = 2 \times I_m/\pi$$

$$= 0.636\ I_m$$

$$= 0.636\ V_m/R_L$$

$$= 0.636 \times \sqrt{2}{}^{V_s}/R_L \quad = \mathbf{0.90{}^{V_s}/R_L}$$

$$PIV = 2V_m$$

$$= 2\ \sqrt{2}V_s \qquad = \mathbf{2.828V_s}$$

This is the maximum reverse repetitive voltage (VRRM) applied to both diodes.
In practice, a general rule of thumb is that the PIV rating for a diode should be at least 4 times the r.m.s value of the supply voltage.

(contd)

(b) The full-wave bridge rectifier circuit employs four diodes. The arrangement is shown in Fig. 2.4-1; diagonally opposite pairs of diodes conduct simultaneously.

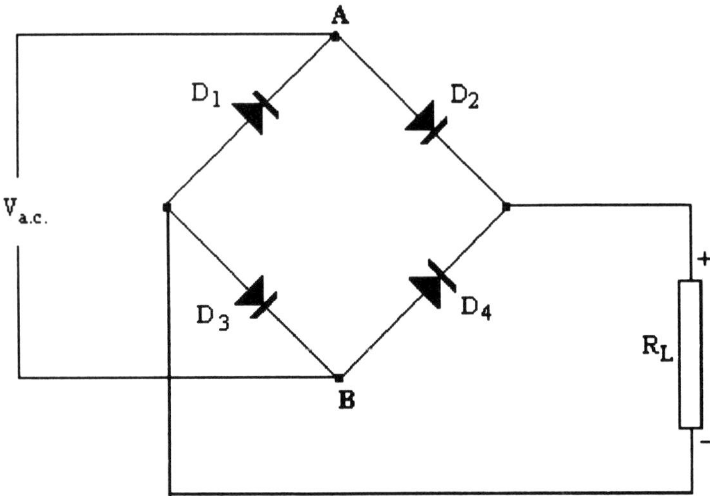

Fig. 2.4-1. Full-wave bridge rectifier

Refer to Fig. 2.4-1

On the first half-cycle of the input voltage, when point **A** is positive with respect to point **B**, current flows in D_2 through the load R_L and out through D_3. Meanwhile D_1 and D_4 are reverse biased.

$$\text{PIV of } D_1 \text{ and } D_4 = \sqrt{2}\, V_s$$

On the second half-cycle D_4 is forward biased and conducts current through load R_L and out through D_1. Meanwhile D_2 and D_3 are reverse biased.

$$\text{PIV of } D_2 \text{ and } D_3 = \sqrt{2}\, V_s.$$

The current in the load is in the same direction with polarity as shown.

(contd)

The main advantages of the bridge circuit over the biphase (full-wave with two diodes) are

(1) it does not require a center-tap transformer
(2) the PIV across each diode is only half of that existing in the biphase rectifier circuit

The main disadvantages are

(1) four diodes are required
(2) in the conducting mode, the slope resistances and forward voltage drops of the two diodes are always in series with the load

When semiconductor diodes are employed in full-wave rectification, the bridge rectifier is more commonly used, as the cost of the two additional diodes is far below the cost of the more expensive center-tap transformer required to produce the same result as the two diodes.

Example 2.5

A 50-μF capacitor is used to smooth the output of a half-wave rectifier. The transformer secondary output is 120 V,50 Hz and the average rectified current is 15 mA. Calculate

(a) the voltage across the load

(b) the percentage ripple

Solution

(a)

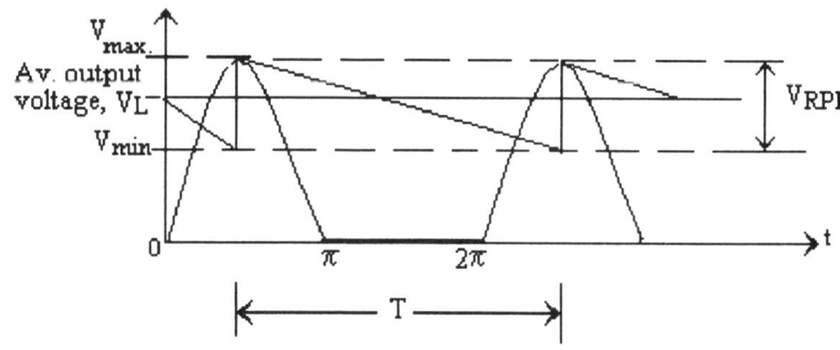

Fig. 2.5-0. Waveform of filtered output

$V_m = \sqrt{2}V_s = 1.414 \times 120 = \underline{169.68\ V}$

Duration of discharge, $T = 1/f = 1/50 = \underline{0.02\ sec}$

Quantity discharge, $Q = I\,T = 15 \times 10^{-3} \times 0.02 = \underline{3 \times 10^{-4}\ C}$

Fall in capacitor voltage, $V_c = Q/C = \dfrac{3 \times 10^{-4}}{50 \times 10^{-6}} = \underline{6\ V}$

Peak-to-peak ripple, $V_{RPP} = V_c = 6V$

Average output voltage $= V_m - V_{RPP}/2$

$= 169.68 - 6/2 = 169.68 - 3 = \mathbf{\underline{166.68\ V}}$

(b) R.M.S value of ripple = R.M.S value of triangular wave

$= 1/\sqrt{3} \times V_{RPP}/2$

$= 1/\sqrt{3} \times 6/2 = 1.732\ V$

% Ripple $= \dfrac{\text{ripple voltage}}{\text{average voltage}}$

$= \dfrac{1.732}{166.68} = 0.01039 = \mathbf{\underline{1.039\%}}$

Example 2.6

Calculate the value of the transformer secondary voltage V_s for the circuit shown in Fig. 2.6-0.

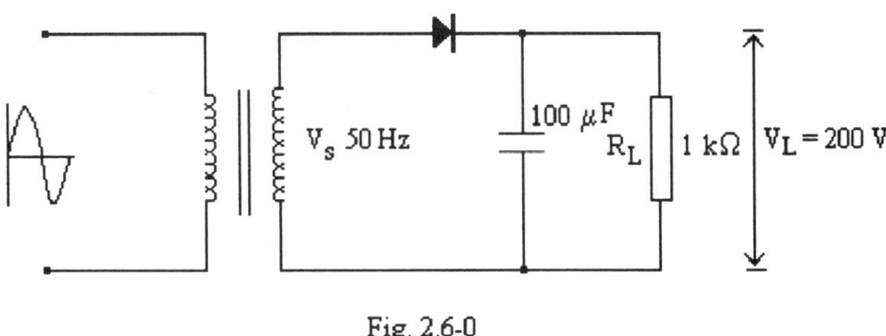

Fig. 2.6-0

Solution

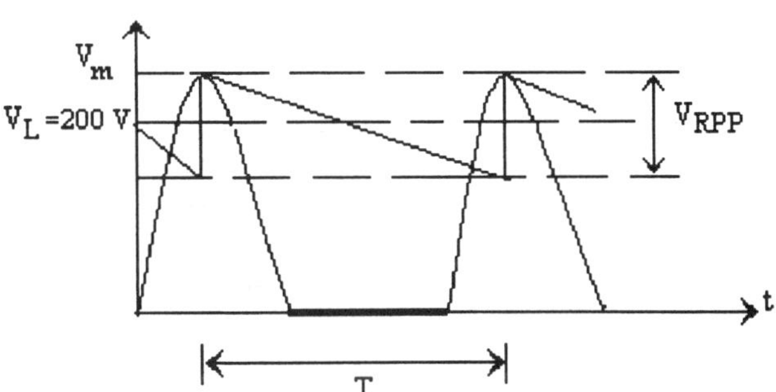

Fig. 2.6-1. Typical output waveform

Refer to Fig. 2.6-1

Let the peak-to-peak ripple be denoted by V_{RPP}.

Then the peak ripple will be $= V_{RPP}/2$

Average voltage, $V_L = V_m - V_{RPP}/2$

Quantity discharged by capacitor, Q = IT coulombs

Fall in capacitor voltage, $V_c = Q/C$ (where $V_c = V_{RPP}$)

$= IT/C$

$= \dfrac{V_m T}{R_L C}$ $\quad (I = V_m/R_L)$

(contd)

$$\therefore \quad V_L = V_m - \frac{V_m T}{2R_L C}$$

$$T = 1/f$$

$$= 1/50 \quad = 0.02 \text{ sec}$$

then $$V_L = V_m - (1 - \frac{T}{2R_L C})$$

and $$V_m = \frac{V_L}{1 - \frac{T}{2R_L C}}$$

$$= \frac{200}{1 - \frac{0.02}{2 \times 1000 \times 100 \times 10^{-6}}}$$

$$= \frac{200}{1 - \frac{0.02}{200 \times 10^{-3}}}$$

$$= \frac{200}{1 - \frac{0.02}{0.2}}$$

$$= \frac{200}{1 - 0.1}$$

$$= \frac{200}{0.9} \qquad = \underline{\mathbf{222.2\ V}}$$

Now $$V_m = \sqrt{2} V_s$$

$$V_s = V_m/\sqrt{2}$$

$$= 222.2/\sqrt{2} \qquad = \underline{\mathbf{157\ V}}$$

$\therefore$ Transformer secondary voltage, V_s $= \underline{\mathbf{157\ V}}$

Note:

In practice, when the transformer is selected, the secondary voltage should be greater than 157 V by about 15% to allow for the voltage drops in the diode and the transformer windings.

Example 2.7

If in Example 2.6 the load is supplied by a full-wave rectifier, shown in Fig. 2.7-0, instead of a half-wave rectifier, determine the transformer secondary voltage.

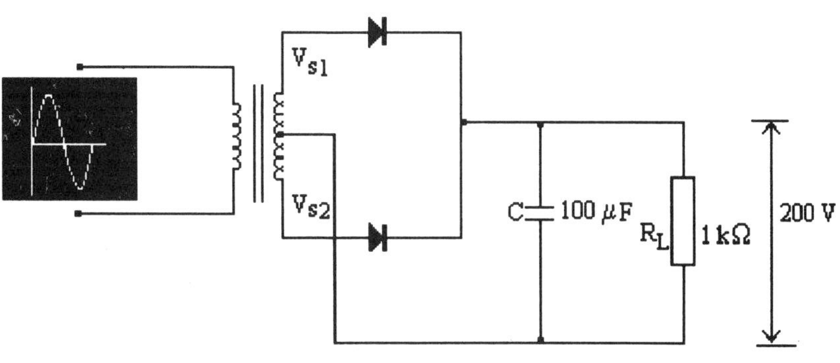

Fig. 2.7-0

Solution

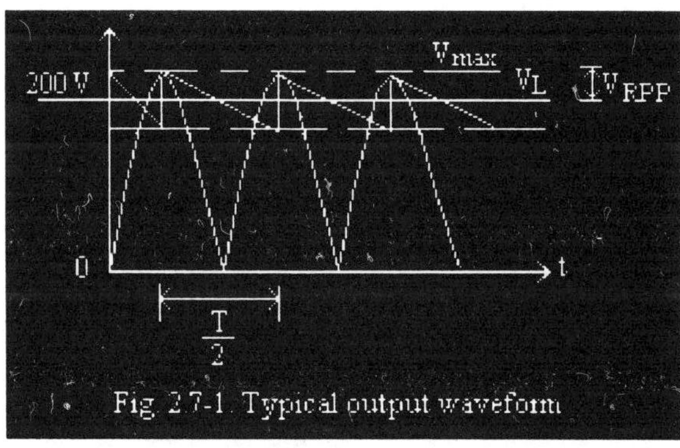

Fig. 2.7-1. Typical output waveform

(contd)

<u>Refer to Fig. 2.7-1</u>

$$V_L = V_m - \frac{V_{RPP}}{2}$$

$$= V_m - \frac{V_m \frac{T}{2}}{2 R_L C}$$

$$= V_m - \frac{V_m T}{4 R_L C}$$

$$\therefore V_m = \frac{V_L}{1 - \frac{T}{4 R_L C}}$$

$$= \frac{200}{1 - \frac{0.02}{4 \times 10^3 \times 100 \times 10^{-6}}}$$

$$= \frac{200}{1 - 0.05} \qquad = \mathbf{210\ V}$$

$$V_s = V_m/\sqrt{2}$$

$$= 210/\sqrt{2} \qquad = \mathbf{148.5\ V}$$

The transformer secondary voltage should be

150 V – 0 – 150 V

Example 2.8

A 117 V, 60 Hz source is the input to a half-wave rectifier that supplies 100 W to a resistive load at 20 V. Determine

(a) the transformer turns ratio

(b) the input power

(c) the transformer rating in voltamperes

Solution

(1)

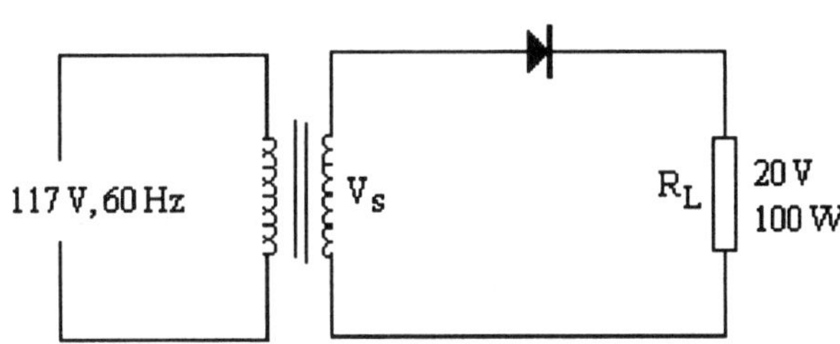

Fig. 2.8-0. Half-wave rectifier circuit

$$V_L = 0.45V_s$$

$$V_s = V_L/0.45 = 20/0.45 = \underline{44.4\ V}$$

Transformation ratio, $n = V_p/V_s = N_p/N_s = I_s/I_p$

$\therefore$ Transformer turns ratio $= V_p/V_s = 117/44.4 = \underline{\mathbf{2.64:1}}$

(2) In the load, total power P_T is the product of $V_{r.m.s} \times I_{r.m.s.}$

Now $V_{r.m.s} = 0.707\ V_s = \sqrt{2}V_s/2 = V_m/2$

and $I_{r.m.s} = V_{r.m.s}/R_L = V_m/2R_L$

$\therefore$ $P_T = V_{r.m.s} \times I_{r.m.s} = V_m/2 \times V_m/2R_L = \mathbf{V_m^2/4R_L}$

(contd)

The d.c. load power $P_{d.c.}$ is the product of $V_{d.c.} \times I_{d.c.}$

$$V_{d.c.} = V_m/\pi$$

$$I_{d.c.} = V_{d.c.}/R_L = V_m/\pi R_L$$

$$P_{d.c.} = V_{d.c.} \times I_{d.c.}$$

$$= V_m/\pi \times V_m/\pi R_L \qquad = \mathbf{V_m^2/\pi^2 R_L}$$

$$\text{Ratio of rectification} = P_{d.c.}/P_T$$

$$= (V_m^2/\pi^2 R_L)/(V_m^2/4R_L)$$

$$= 4/\pi^2 \qquad = \underline{\mathbf{0.406}}$$

$$= \underline{\mathbf{40.6\%}}$$

Output power = 40.6% of input power

$$\therefore \quad \text{Input power} = \frac{1}{0.406} \times 100 \qquad = \underline{\mathbf{246.31\ W}}$$

Alternatively

The ratio of rectification is the operating efficiency of the half-wave rectifier.

$$\therefore \quad \text{Efficiency, } \eta = P_{out}/P_{in}$$

$$P_{in} = P_{out}/0.406 = 100/0.406 = \underline{\mathbf{246.31\ W}}$$

Transformer rating = transformation utilization factor

$$= \frac{P_{d.c.}}{P_{a.c.}}$$

$$P_{d.c.} = \frac{V_m^2}{\pi^2 R_L}$$

$$P_{a.c.} = V_s \times I_{r.m.s}$$

$$= \frac{V_m}{\sqrt{2}} \times \frac{V_m}{2R_L} = \frac{V_m^2}{2\sqrt{2} R_L}$$

$$\text{Transformer rating, } \frac{P_{d.c.}}{P_{a.c.}} = \frac{\dfrac{V_m^2}{\pi^2 R_L}}{\dfrac{V_m^2}{2\sqrt{2} R_L}} = \frac{2\sqrt{2}}{\pi^2}$$

$$\text{and} \quad P_{a.c.} = \frac{P_{d.c.} \times \pi^2}{2\sqrt{2}} = \frac{100 \times 3.14^2}{2\sqrt{2}}$$

$$= \underline{\mathbf{348.6\ kVA}}$$

CHAPTER 3

ZENER DIODE REGULATOR AND VOLTAGE MULTIPLIERS

Introduction

The Zener diode was developed from the property of an avalanche breakdown of a rectifier diode bias in the reverse direction. Consequently, Zener diodes are designed to operate in the reverse breakdown mode. For an overview of the basic operation of the Zener diode, we will briefly consider the following:

(a) the characteristic

(b) the Zener diode as a regulator using typical values (56 V, 8 W Zener whose voltage remains constant down to a Zener current of 1.0 mA)

(c) how stabilization is achieved

(d) maximum and minimum supply voltage between which stabilization is satisfactory for a given load

(e) the minimum value of load resistance at a given supply voltage.

Overview of Basic Theory

(a) Consider the typical Zener diode characteristic shown in Fig. 3.0-0. In the forward direction the characteristic is similar to any silicon diode with a forward voltage drop of about 0.5 V.

In the reverse direction the diode breaks down at the *knee* of the curve, (known as the *Zener breakdown voltage*), an avalanche takes place, and the current rises rapidly with very small changes in voltage.

To limit this rapid rise of current to the diode maximum permissible value I_{zmax}, an external resistance is connected in series with the source supply.

(contd)

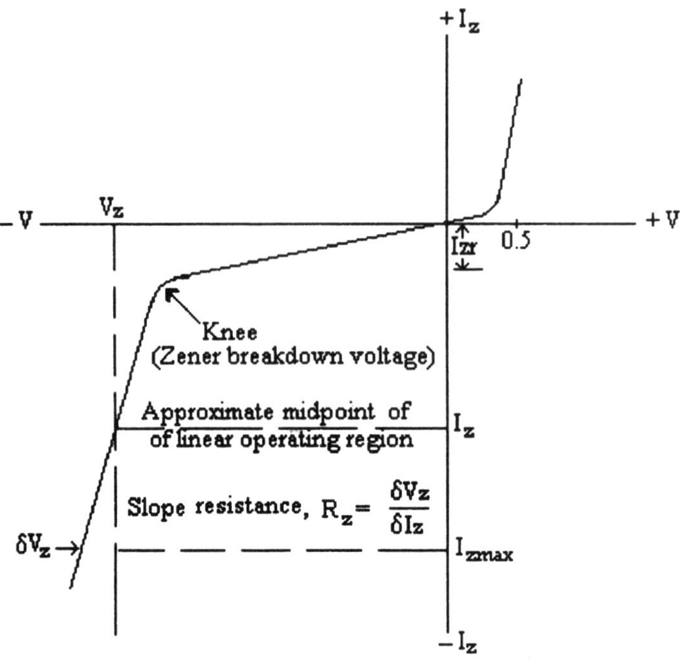

Fig. 3.0-0. Typical Zener diode characteristic

The Zener voltage V_z is the voltage across the diode for a current I_z; this is the approximate midpoint of the linear operating region. The reverse leakage current is denoted by I_{zr}.

Slope resistance, $R_z \quad = \quad {}^{\delta V_z}/_{\delta I_z}$

(b)

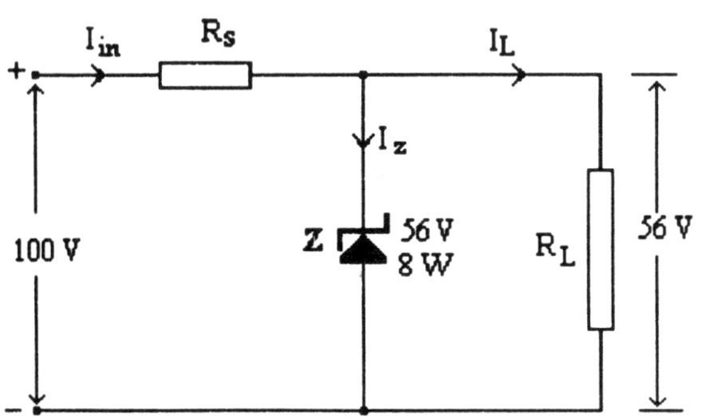

Fig. 3.0-1. A simple Zener diode regulator

On open-circuit load the Zener diode will take its maximum current given by

$$I_{zmax} \quad = \quad {}^{\text{power}}/_{\text{voltage}} \quad = \quad {}^{8}/_{56} \quad = \underline{0.1428\ \text{A}}$$

(contd)

To stabilize 56 V at the output from the supply of 100 V, the voltage difference of 44 V(i.e., 100 V – 56 V) must be dropped across the series resistor R_s, limiting I_{zmax} to the Zener diode permissible value.

$$R_s = (V_{in} - V_{out})/I_{zmax}$$

$$= \frac{100 - 56}{1428 \times 10^{-3}} \qquad = \mathbf{308\ \Omega}$$

(c) Refer to Fig. 3.0-0

(i) Assume the Zener diode is a current *donor* or *acceptor*, operating at the approximate midpoint of the linear region.
When the load current increases, the current in the Zener diode decreases; that is, the Zener diode is a *donor*, donating to the load the increased current required by the load.
Similarly, when the load current decreases, the Zener diode current increases; that is, the Zener diode is an *acceptor*, accepting the amount of current decreased in the load.

(ii) Consider the case for a change in voltage. When the supply voltage increases, the Zener diode will take the extra current while the increase in voltage will drop across R_s.
Likewise, when the supply voltage decreases, the current in the load will fall; however, the Zener diode will shed its current to the load by an amount equal to the voltage drop across R_s, thereby maintaining a stabilized current in the load.

Accordingly, the Zener diode will maintain a constant load voltage for a range of variation in either load current or supply voltage.

(contd)

(d) Refer to Fig. 3.0-2

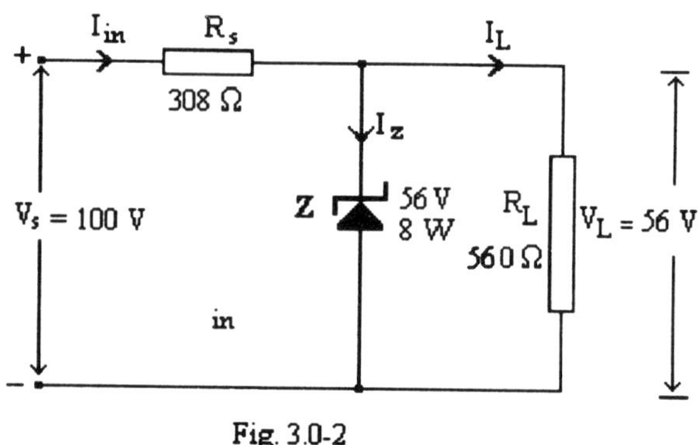

Fig. 3.0-2

To determine V_{smin}

For minimum supply voltage, the Zener diode must still pass 1 mA.

$$I_L = V_L/R_L$$
$$= 56/560 \quad = \underline{0.1\ A} = 1\underline{00\ mA}$$

$$I_{in} = I_L + I_z$$
$$= (100 + 1.0) \quad = 101\ mA$$

Minimum supply voltage V_{smin}

$$= (I_{in} \times R_s) + V_L$$
$$= (101 \times 10^{-3} \times 308) + 56$$
$$= 31.11 + 56 \quad = \mathbf{\underline{87.11\ V}}$$

To determine V_{smax}

For maximum supply voltage, the Zener diode will pass its maximum current, i.e., 143 mA. Hence

$$I_{in} = I_L + I_{zmax}$$
$$= 100 + 143 \quad = \underline{243\ mA}$$

Maximum supply voltage V_{smax}

$$= (I_{in} \times R_s) + V_L$$
$$= (243 \times 10^{-3} \times 308) + 56$$
$$= 74.84 + 56 \quad = \mathbf{\underline{130.84\ V}}$$

(contd)

(e) *To determine R_{Lmin} when supply voltage V_s is 100 V*

R_{Lmin} occurs when I_z is 1.0 mA.

$\therefore \quad I_{in} = (V_s - V_L)/R_s$

$= \frac{100 - 56}{308} = \underline{0.143\ A}$

$= \underline{143\ mA}$

$I_{in} = I_L + I_z$

$\therefore \quad I_L = I_{in} - I_z$

$= 143 - 1.0 = \underline{142\ mA}$

Hence $R_{Lmin} = V_L/I_L$

$= 56/(142 \times 10^{-3})$

$= 56/0.142 = \underline{\mathbf{394.4\Omega}}$

Example 3.1

A 20 V, 0.5 W Zener diode has a maximum current rating of 25 mA. The diode has a slope resistance R_z, of 32 Ω at a Zener current of 5 mA. Calculate the percentage voltage change over the whole range of Zener current variation.

Solution

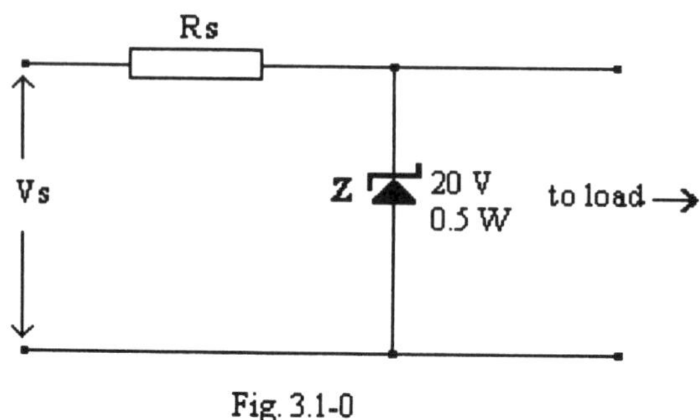

Fig. 3.1-0

Consider Fig. 3.1-0.

When I_z increases by 5 mA, the voltage across the diode will increase by $I_z \times R_z$; that is

$$I_z \times R_z = 5 \times 10^{-3} \times 32 \qquad = \underline{0.16\ V}$$

When the Zener current is 25 mA, the voltage across the Zener diode is

$$= V_z + R_z(I_{zmax} - I_z)$$

$$= 20 + 32(25 - 5) \times 10^{-3}$$

$$= 20 + 0.64 \qquad = \underline{20.64\ V}$$

When the Zener current is 5 mA, the voltage across the diode is

$$= V_z - (I_z \times R_z)$$

$$= 20 - 0.16 \qquad = \underline{19.84\ V}$$

The Zener current variation is

$$= 25 - 5 \qquad = \underline{\mathbf{20\ mA}}$$

The corresponding voltage variation is

$$= 20 - 19.84 \qquad = \underline{\mathbf{0.16\ V}}$$

Expressed as a percentage, the voltage change is 0.8% (0.16/20) over the whole range of current change.

Example 3.2

(a) *A 50 V,50 W Zener diode has a leakage current of 5 mA. The Zener diode is operated at its approximate midpoint of the linear region with an inverse current of 500 mA. Determine the value of the series resistance R_s when the load current is 2.0 A and the supply voltage is 60 V. The diode slope resistance is 1.2 Ω.*

(b) *Calculate the range of input voltage fluctuation for the corresponding load current variations.*

Solution

(a)

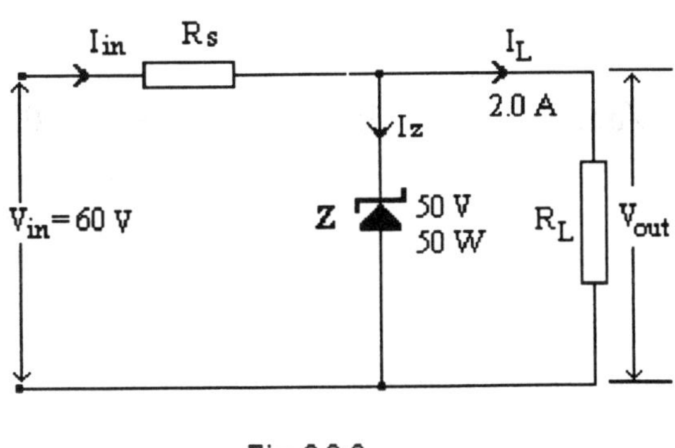

Fig. 3.2-0

Consider the circuit shown in Fig. 3.2-0.

When the load current increases, the current in the Zener diode decreases; when the load current decreases, the current in the diode increases. The Zener diode current can either increase 500 mA or decrease 495 mA and still be within its operating range. Accordingly, the load current can vary from 1.5 A to 2.495 A and the Zener will maintain the load voltage at 50 V.

The supply current, $I_{in} = I_z + I_L$

$= 0.5 + 2.0 = \underline{2.5\ A}$

(contd)

$$R_s = (V_{in} - V_z)/I_{in}$$

The Zener current I_Z varies from 500 mA to 495 mA.
The corresponding load current variation is from (2.0 – 0.5) A to (2.0 + 0.495) A.

That is, the variation of I_L is from 1.5 A to 2.495 A.

When I_L = 1.5 A

$$I_z = I_{in} - I_L$$
$$= 2.5 - 1.5\text{ A} \qquad = \mathbf{\underline{1.0A}}$$
$$\text{Voltage across } R_L = V_z + R_z(1.0 - 0.5)$$
$$= 50 + 1.2(0.5) \qquad = \mathbf{\underline{50.60\ V}}$$

When I_L = 2.495 A

$$I_z = I_{in} - I_L$$
$$= 2.5 - 2.495 \qquad = \mathbf{\underline{0.\ 005\ A}}$$
$$\text{Voltage across } R_L = V_z + R_z(0.005 - 0.5)$$
$$= 50 + 1.2(-0.495)$$
$$= 50 - 0.594 \qquad = \mathbf{\underline{49.41V}}$$

The Zener diode is operated at

$$I_z = 0.5\text{ A}$$
$$\text{and load current } I_L = 2.0\text{ A}$$
$$\therefore \quad R_s = (V_{in} - V_z)/I_{in}$$
$$= (V_{in} - V_z)/(I_z + I_L)$$
$$= \frac{(60 - 50)}{(0.5 + 2.0)} \qquad = \mathbf{\underline{4.0\ \Omega}}$$

(b) When the supply voltage increases, the Zener diode will accept the increase in current while maintaining the voltage across the load constant.

At maximum value of I_Z the supply voltage is

$$
\begin{aligned}
V_{in} &= I_{in} R_s + V_{out} \\
&= (I_Z + I_L) R_s + V_{out} \\
&= (1.0 + 2.0)\,4 + 50.60 \\
&= 12 + 50.60 \qquad = \underline{\mathbf{62.60\ V}}
\end{aligned}
$$

When the supply voltage decreases, the diode will shed current to the load up to a limitation of I_Z = 5 mA and still stabilize the load voltage.
Consequently

$$
\begin{aligned}
V_{in} &= I_{in} R_s + V_{out} \\
&= (I_Z + I_L) R_s + V_{out} \\
&= (0.005 + 2.0)\,4 + 49.41 \\
&= 8.02 + 49.41 \qquad = \underline{\mathbf{57.43\ V}}
\end{aligned}
$$

Hence, the Zener diode maintains a constant voltage for a range of variation in load current or supply voltage.

Example 3.3

(a) *A voltage stabilizer circuit uses a 10 V, 1.0 W Zener diode. The supply voltage is 20 V ±4 V. Determine the value and power rating of the series resistor R_s which is to be chosen from a range of preferred resistors with a tolerance of ±10%.*

(b) *Estimate the minimum value by which the load resistance can be decreased before stabilization fails to function correctly.*

Solution

(a)

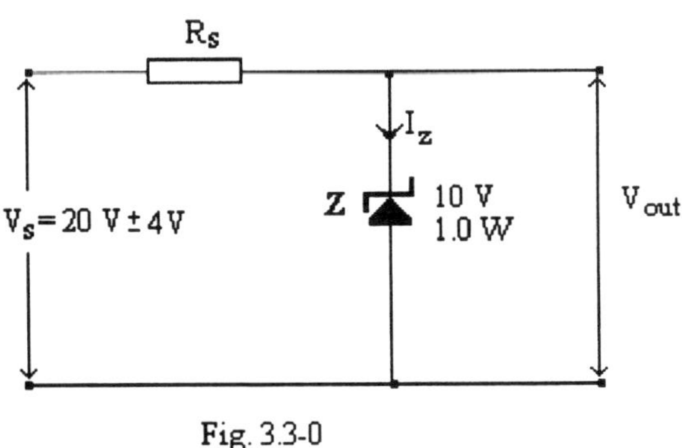

Fig. 3.3-0

<u>Refer to Fig. 3.3-0</u>

$$I_{zmax} = P_z/V_z = 1.0/10.0 = \underline{100\ mA}$$

The series resistance R_s must limit I_{zmax} to 100 mA when the supply voltage is maximum, that is, 20 V + 4 V

$$R_s = (V_{max} - V_z)/I_{zmax}$$

$$= \frac{24 - 10}{100 \times 10^{-3}} = \mathbf{\underline{140\ \Omega}}$$

Now R_s = 140Ω± 10%

(That is, R_s ranges from a minimum of 126 Ω to maximum of 154 Ω)

(contd)

When $R_s = 126\ \Omega$

$$I_{zmax} = (V_{max} - V_z)/R_s$$

$$= \frac{24 - 10}{126} \qquad = \underline{111.11\ mA}.$$

(Note: This current is greater than the current rating of the diode.)

When $R_s = 154\ \Omega$

$$I_{zmax} = (V_{max} - V_z)/R_s$$

$$= \frac{24 - 10}{154} \qquad = \underline{90.9\ mA}$$

Power dissipated in $R_s = I^2_{zmax} R_s$

$$= (90.9 \times 10^{-3})^2 \times 154 \qquad = \underline{1.27\ W}$$

(b) The circuit will cease to function as a voltage regulator when the Zener current and load current are

$$I_z = 0$$

$$I_L = I_{in}$$

The above condition occurs when the value of R_L is a minimum, giving

$$V_z/R_{Lmin} = (V_s - V_z)/R_s$$

$$\therefore \quad R_{Lmin} = V_z R_s/(V_s - V_z) \qquad \text{.....(Eq. 3.3-0)}$$

When $R_s = 126\ \Omega$

$I_{zmax} = 111.11$ mA (As mentioned earlier, this current is above the current rating of the Zener diode; hence $R_s = 126\ \Omega$ will not be used.)

To calculate R_{Lmin}, the component values to be used are

$$R_s = 154\ \Omega$$

$$V_{smin} = 16\ V$$

$$V_z = 10\ V$$

$\therefore$ From Eq. 3.3-0 $$R_{Lmin} = \frac{10 \times 154}{16 - 10} \qquad = \mathbf{\underline{256.67\ \Omega}}$$

Example 3.4

A Zener diode has a reverse breakdown voltage of 5.8 V with a slope resistance of 10 Ω measured at the approximate midpoint of the operating linear region. The Zener diode is used to stabilize 6 V across an output load of 2.5 kΩ from an input supply of 10 V. Calculate

(i) *the value and power rating of the series resistor*

(ii) *the change in load voltage when supply voltage increases by 10%*

(iii) *the minimum value of load resistance for which stabilization is still effective with 2.5 kΩ load.*

(iv) *the minimum value of the supply voltage*

Solution

(i)

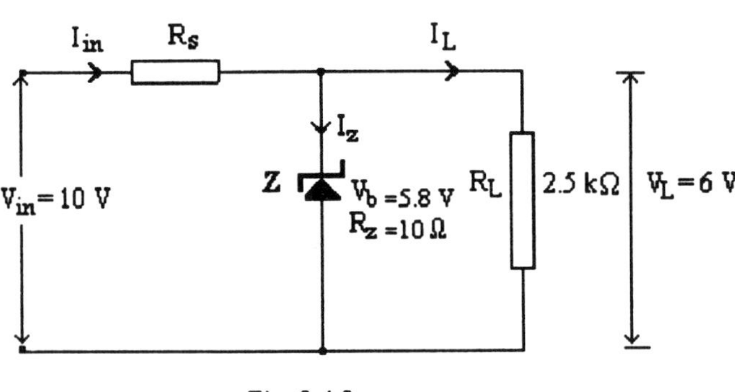

Fig. 3.4-0

Consider the simple circuit shown in Fig. 3.4-0.

$$I_L = V_L/R_L$$

$$= 6/(2.5 \times 10^3) \qquad = \underline{2.4\text{mA}}$$

$$V_L = V_z$$

$$= V_b + I_z R_z \quad (V_b = \text{breakdown voltage})$$

$$(R_z = \text{Zener slope resistance})$$

$$= 5.8 + 10 I_z$$

$$I_z = (V_L - 5.8)/10$$

$$= \frac{6 - 5.8}{10} \qquad = \underline{20\text{ mA}}$$

(contd)

$$I_{in} = I_L + I_z$$
$$= 2.4 + 20 \qquad = \underline{22.4 \text{ mA}}$$

$$R_s = (V_s - V_z)/I_{in}$$

$$= \frac{10 - 6}{22.4 \times 10^{-3}} \qquad = \underline{\mathbf{178.6\ \Omega}}$$

$$\text{Power rating of } R_s = I_{in}^2 \times R_s$$

$$= (22.4 \times 10^{-3})^2 \times 178.6$$
$$= \underline{\mathbf{89.6 mW}}$$

(The choice will be 1/8 W)

(ii) When the supply voltage increases by 10%

$$V_{in} = 10 + 1.0 \qquad = 11 \text{ V}$$

$$V_{in} = R_s (I_L + I_z) + I_L R_L \qquad \text{.........(Eq. 3.4-0)}$$

$$I_L R_L = V_b + I_z R_z$$

$$2500 I_L = 5.8 + 10 I_z$$

$$\therefore \quad I_z = (2500 I_L - 5.8)/10$$

$$= 250 I_L - 0.58$$

Substituting for I_z in Eq. 3.4-0

$$V_{in} = 178.6(I_L + 250 I_L - 0.58) + 2500 I_L$$

$$= 178.6 I_L + 44650 I_L - 103.59 + 2500 I_L$$

$$11.0 = 47328.6 I_L - 103.59$$

$$I_L = \frac{114.59}{47328.6} \qquad = \underline{2.42 \text{ mA}}$$

$$V_L = 2500 I_L$$

$$= 2500 \times 2.42 \times 10^{-3} = \underline{6.05 \text{ V}}$$

$$= 6.05 - 6.0 \qquad = \underline{\mathbf{0.05\ V}}$$

(contd)

(iii) *The minimum value of load resistance will be when the Zener diode just breaks down; that is*

when $I_z = 0$

and $V_L = V_z = V_b = 5.8$ V (Zener breakdown voltage)

$\therefore \quad I_L = I_{in}$

$= (V_s - V_z)/R_s$

$= \dfrac{10 - 5.8}{178.6} = \underline{23.52 \text{ mA}}$

$\therefore \quad R_{Lmin} = V_L/I_L$

$= 5.8/(23.52 \times 10^{-3}) = \underline{\mathbf{246.6\ \Omega}}$

(iv) *Similarly, the minimum value of supply voltage will be when*

$I_z = 0$

and $V_L = V_z = V_b = 5.8$ V (Zener breakdown voltage)

$\therefore \quad I_{in} = V_L/R_L \quad (V_z/R_L)$

$= 5.8/(2.5 \times 10^3) = \underline{2.32 \text{ mA}}$

$\therefore \quad V_{in(min)} = I_{in} \times R_s + V_L$

$= (2.32 \times 10^{-3} \times 178.6) + 5.8$

$= 0.414 + 5.8 = \underline{\mathbf{6.2\ V}}$

Example 3.5

(a) *Briefly describe two distinct uses for a Zener diode other than a simple stabilizer.*

(b) *A Zener diode has a reverse characteristic equation given by $V = 9.1 + 5.01\,I$ for values of I greater than 0.01 A. It is used as a simple stabilizer for a load which can vary between 0 and 150 mA from a d.c. supply of nominal voltage 25 V which may fluctuate by ±5 V. Estimate the maximum value of series resistor which may be used if the Zener diode current is not to fall below 0.01 A for all input and load conditions.*

(c) *For this value of series resistor, determine*

(i) *the minimum rating of the Zener diode*

(ii) *the maximum fluctuation in output voltage*

Solution

(a)

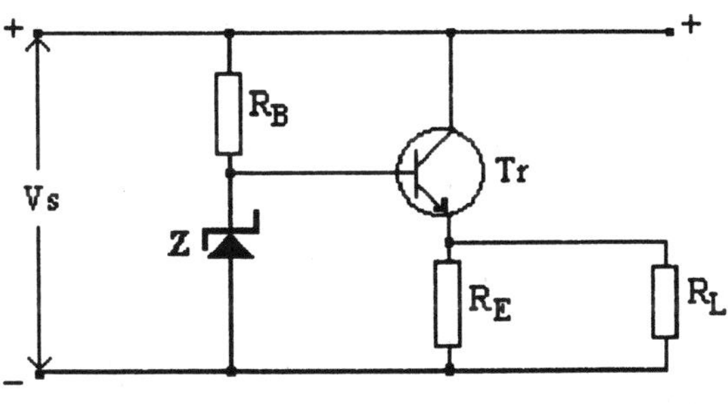

Fig. 3.5-0

Refer to Fig. 3.5-0

The Zener diode is commonly used as a voltage reference source in transistor-regulated power supplies. A sample of the output voltage is fed back and compared with the voltage reference source (the Zener voltage). The error voltage is then used to activate the transistor, which in turn restores the level of the output to the preset value.

(contd)

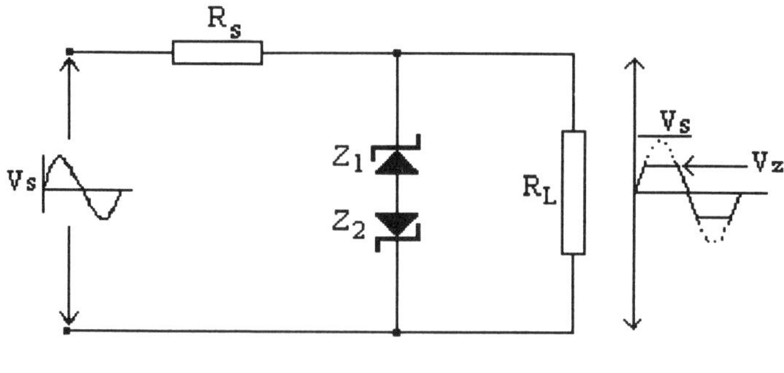

Fig. 3.5-1

Refer to Fig. 3.5-0

The Zener diode can be used in a signal-processing circuit. If two identical diodes are connected back-to-back as shown, a clipped output waveform approximating to a square wave can be obtained from a sine wave input.

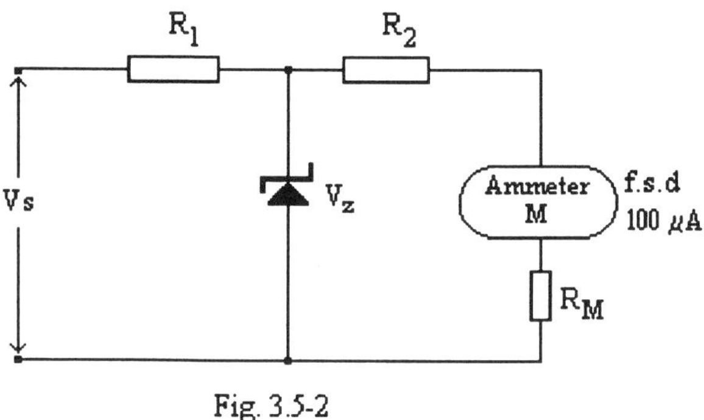

Fig. 3.5-2

Refer to Fig. 3.5-2

The Zener diode can be used as a protection device against overloading of a sensitive current-measuring instrument without affecting the linearity of the instrument.

When V_s is greater than V_z the Zener diode just breaks down and the voltage across it is V_z, but it will not be taking any current. At this instant the meter must indicate full-scale deflection (f.s.d). The current through R_1, R_2 and the meter will therefore be 100 μA . Hence, when V_s rises above V_z, the diode conducts and shunts away the overload current.

(contd)

(b) To determine the maximum value of the series resistor R_s, the worst case conditions of V_s and I_L must be considered. The worst case conditions will be when

V_s is minimum (V_{smin})

I_L is maximum (I_{Lmax})

I_z is passing a current of 0.01 A (10 mA)

Now $V_{smin} = 25 - 5 = 20V$

$I_{Lmax} = 150\text{ mA} = 0.15\text{ A}$

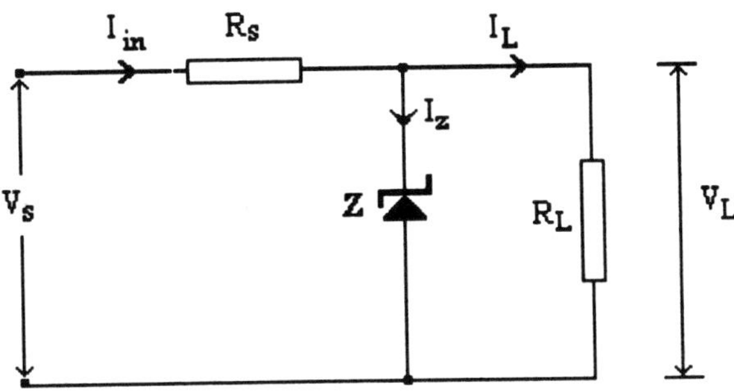

Fig. 3.5-3. Closed circuit

Consider Fig. 3.5-3.

$V_L = 9.1 + 5.01I$ (Equation given for $I \geq 0.01A$)

$= 9.1 + (5.01 \times 0.01)$

$= 9.1 + 0.05 = \underline{9.15\text{ V}}$

$R_s = (V_{smin} - V_L)/(I_L + I_z)$

$= \dfrac{20 - 9.15}{0.15 + 0.01}$

$= 10.85/0.16 = \underline{\mathbf{67.8\ \Omega}}$

(c) To calculate the minimum power rating of the Zener diode, the maximum current in the Zener diode must be calculated when the supply voltage is maximum.

The Zener current is at maximum when the load is open-circuited, so that

$I_{in} = I_z$ (when $I_L = 0$)

(contd)

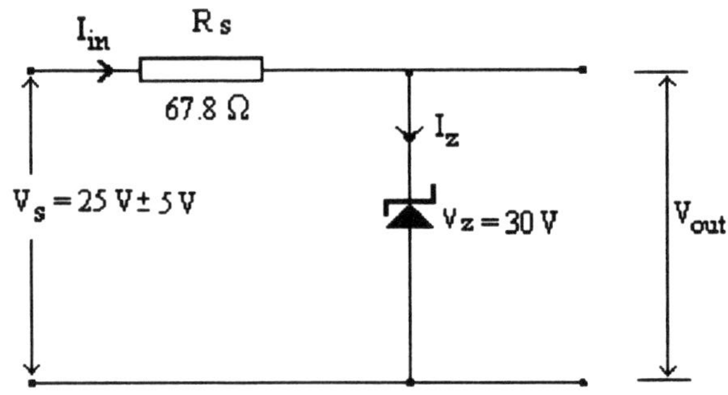

Fig. 3.5-4. Open circuit

Consider Fig. 3.5-4.

When the load is open-circuited, the supply voltage is at a maximum, that is 30 V (25 + 5V).

$$V_{out} = V_Z$$

and $$V_s = I_Z R_s + (9.1 + 5.01 I_Z)$$

that is $$30 = 67.8 I_Z + 9.1 + 5.01 I_Z$$

$$I_Z = \frac{30 - 9.1}{72.81} = \underline{0.287\ A}$$

$$\therefore \quad V_{out} = 9.1 + (5.01 \times 0.287)$$

$$= 9.1 + 1.44 = \underline{10.54\ V}$$

(i) Minimum power rating $= V_Z I_Z$

$$= 10.54 \times 0.287 = \mathbf{\underline{3.02\ W}}$$

(ii) Maximum fluctuation in output voltage

$$= V_{out(open\ circuit)} - V_{L(closed\ circuit)}$$

$$= 10.54 - 9.15 = \mathbf{\underline{1.39\ V}}$$

It should be noted that the maximum fluctuation in the output voltage occurred between two extreme circuit conditions, namely

$$V_s = \text{minimum with } I_L = \text{maximum}$$

and $$V_s = \text{maximum with } I_L = 0$$

Example 3.6

A galvanometer has an internal resistance of 620 Ω and requires 200 μA to give full-scale deflection (f.s.d). The galvanometer is to be used as a voltmeter to read 12 V full scale and is shown connected to the circuit in Fig. 3.6-0. The combined value of R_1 and R_2 equals 59.38 kΩ and the Zener break down voltage is 6.2 V. Determine the individual values of R_1 and R_2 so that when the supply voltage is 12 V the Zener diode will conduct, thus protecting the meter from the overload current.

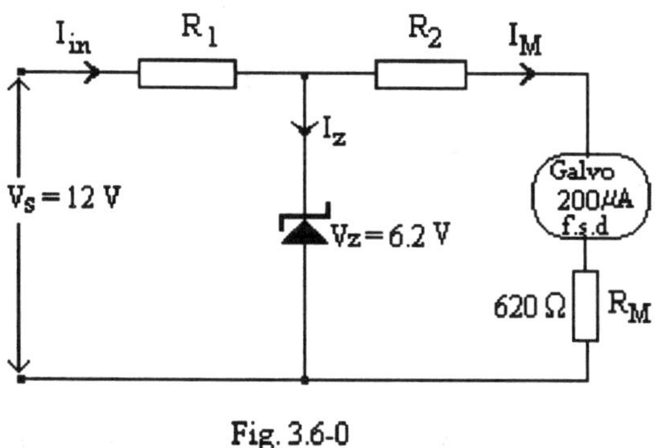

Fig. 3.6-0

Solution

When the input voltage is 12 V, the Zener diode just breaks down; that is

$$V_Z = 6.2 \text{ V}$$

and $$I_Z = 0$$

At this point the meter must give full-scale deflection; that is, the current through R_1, R_2 and the meter is 200 μA.

Thus $$I_{in} = I_Z + I_M$$

$$= 0 + (200 \times 10^{-6}) \text{ A}$$

$$R_1 = (V_s - V_Z)/(200 \times 10^{-6})$$

$$= (12 - 6.2)/(200 \times 10^{-6}) \qquad = \underline{\mathbf{29\ k\Omega}}$$

$$R_2 + R_1 = 59.38 \text{ k}\Omega$$

$$\therefore \quad R_2 = 59.38 - 29 \qquad = \underline{\mathbf{30.38 k\Omega}}$$

$$V_Z = I_M (R_2 + R_M)$$

$$= 200 \times 10^{-6} (30.38 + 0.620) \times 10^3$$

$$= \underline{6.2\text{V}}$$

Example 3.7

Describe the operation of the voltage doubler circuit shown in Fig.3.7-0.

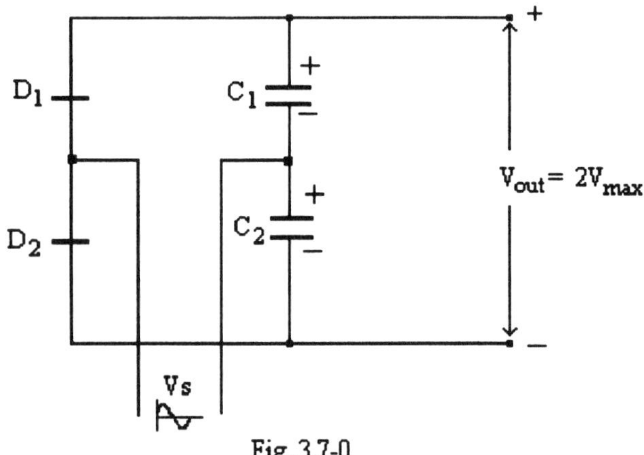

Fig. 3.7-0

Refer to Fig. 3.7-0

The ful-wave voltage doubler is obtained by replacing two diodes with capacitors.

When V_s is positive, D_1 charges C_1 to V_{max}.

When V_s is negative, D_2 charges C_2 to V_{max}.

The capacitors C_1 and C_2 are in series; hence, the voltage across them is twice the voltage in each, or is doubled. Thus

$$V_{out} = 2V_{max}$$

Example 3.8

Sketch the circuit diagram of a voltage quadrupler employing diodes and capacitors.

Describe the operation of the circuit and its limitations. Give an example where such a circuit would be practically employed.

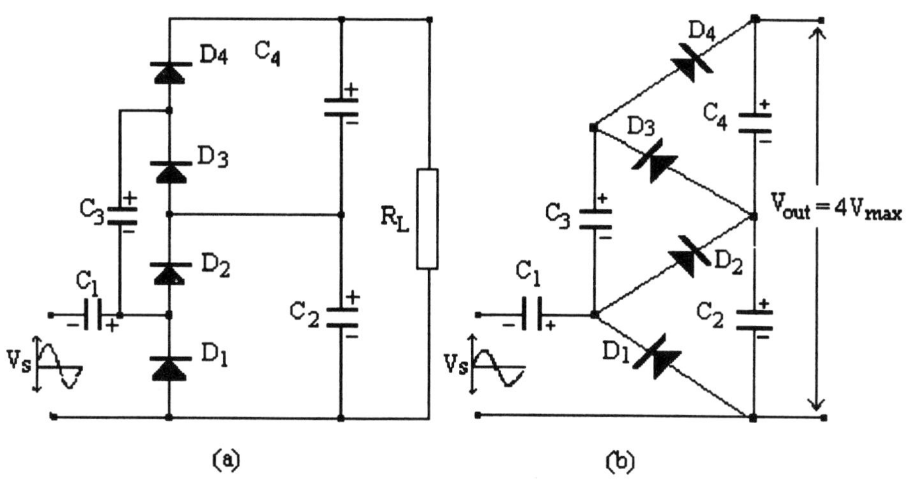

Fig. 3.8-0. A voltage quadrupler circuit arrangement

Consider the circuit arrangement shown in Fig. 3.8-0 (a) or 3.8-0(b). There are four stages. The output voltage V_{out} is the product of the number of stages and the maximum voltage.

For optimum performance, the capacitors would have different values integral with the number of stages.

If N = the number of stages

then $V_{out} = \sqrt{2}V_s \times N$

$= N \times V_{max}$

$\therefore$ For the quadrupler, $V_{out} = 4V_{max}$

The PIV of any diode $= 2V_{max}$

(contd)

For the operation of this circuit arrangement, it will be assumed that the load connected across the output has a very high resistance, mainly for two reasons:

(1) it will require a very small current

(2) it will prevent the capacitors from discharging through it in the time period T = CR seconds

In the following analysis, we will consider two complete cycles of the supply voltage V_s.

On the first half-cycle, C_1 is charged through D_1 to V_{max}.
On the second half-cycle, C_2 is charged through D_2 to $2V_{max}$ due to the supply voltage V_s and the voltage V_{max} across C_1 (V_{C1max}).
On the third half-cycle (that is, the first half-cycle of the second cycle), C_3 charges through D_3 to $2V_{max}$ due to the voltage V_s, and the voltage across C_2 (V_{C2max}) minus the voltage V_{C1max}.
On the fourth half-cycle (that is, the second half-cycle of the second cycle), C_4 charges through D_4 to $2V_{max}$ due to the voltage V_s, the voltage V_{C1max}, and the voltage V_{C3max} minus the voltage V_{C2max}.

Consequently, upon the completion of 2 cycles of the supply voltage V_s, the output voltage V_{out} equals $4V_{max}$.

The main limitations are

(1) the regulation is very poor (unless the capacitors are very large) and its ripple is very high when any appreciable load current is drawn

(2) the load resistance has to be very high, thus limiting the current to a very small value

Accordingly, its use is restricted to very high voltage, small current supplies. Some practical applications are

(i) the television EHT supply

(ii) the oscilloscope HT supply

(iii) the portable Geiger counter supply

In general, voltage multipliers can be extended to about 10 stages.

CHAPTER 4

TRANSISTOR h-PARAMETER EQUIVALENT CIRCUITS OF SMALL–SIGNAL AMPLIFIERS

Introduction

The conventional transistor is often referred to as a bipolar junction transistor (BJT). It is also the most widely used active device in electronic equipment. The parametric values of a transistor depend upon many factors, and the *h*-parameters are the most widely used and frequently quoted in transistor technical specifications. The transistor can be configured to operate in three different modes: Common-Base, Common-Emitter, and Common-Collector or Emitter-Follower. Each has its own advantages and disadvantages.

Overview of Basic Theory

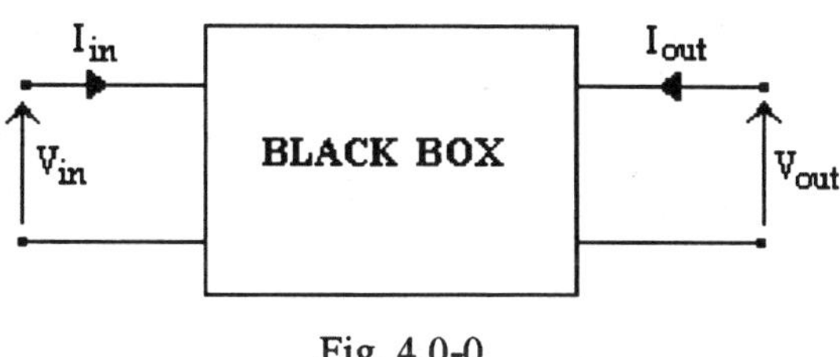

Fig. 4.0-0

A transistor operating in the small-signal range is conveniently regarded for the purpose of analysis as an active two-port **'black box'** having two input terminals and two output terminals. Consequently there are four external variables:

$$V_{in} \quad \text{and} \quad I_{in}$$

$$V_{out} \quad \text{and} \quad I_{out}$$

The four variables are related to each other by four constants called *h*-parameters as shown in Fig. 4.0-1.

(contd)

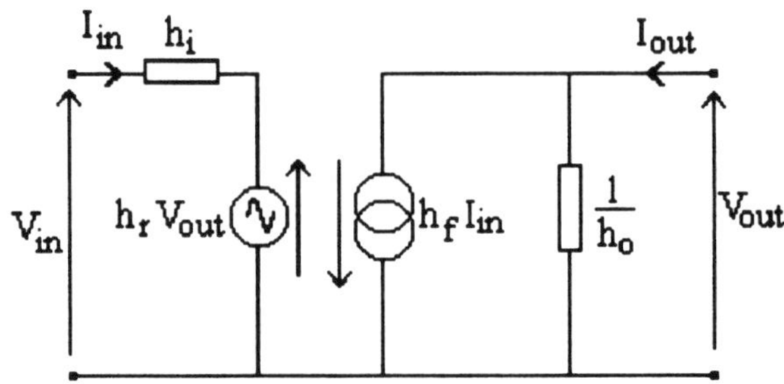

Fig. 4.0-1. *h*-parameter equivalent circuit

Consider the *h*-parameter equivalent circuit shown in Fig. 4.0-1. The relationships are

$$V_{in} = h_i I_{in} + h_r V_{out}$$
$$I_{out} = h_f I_{in} + h_o V_{out}$$

$$h_i = \left[\frac{\delta V_{in}}{\delta I_{in}}\right]_{V_{out}=0}$$: Input resistance with output terminals short-circuited to a.c.

$$h_f = \left[\frac{\delta I_{out}}{\delta I_{in}}\right]_{V_{out}=0}$$: Forward current gain with output terminals short-circuited to a.c.

$$h_r = \left[\frac{\delta V_{in}}{\delta V_{out}}\right]_{I_{in}=0}$$: Reverse voltage amplification with input terminals open-circuited to a.c.

$$h_o = \left[\frac{\delta I_{out}}{\delta V_{out}}\right]_{I_{in}=0}$$: Output admittance with input terminals open-circuited to a.c.

Subscripts and Symbols

Subscripts in uppercase letters are used to indicate quiescent d.c. quantities; for example

I_B = d.c. base bias current (See Fig. 4.0-2)

Subscripts in lowercase letters are used to indicate a.c. signal quantities; for example

I_b = a.c. base signal current (See Fig. 4.0-3)

(contd)

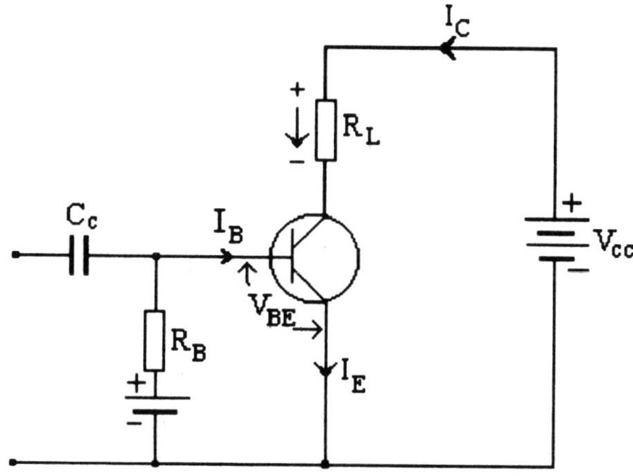

Fig. 4.0-2. No-signal d.c. bias

Refer to Fig. 4.0-2

$$I_E = I_C + I_B$$

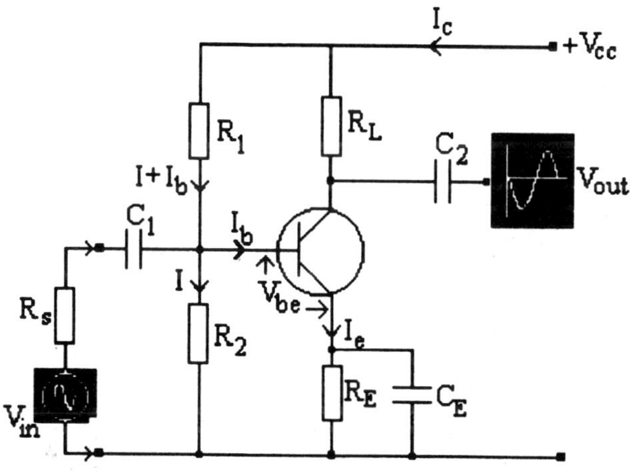

Fig. 4.0-3. a.c.signal input

Refer to Fig. 4.0-3

$$I_e = I_c + I_b$$

Example 4.1

Draw the h-parameter equivalent circuit for a transistor connected in common-emitter configuration and hence deduce the h-parameter equations for the following

(a) *input-output relationship*

(b) *input and output resistances*

(c) *current, voltage and power gains*

Solution

(a) **Input-Output Relationship**

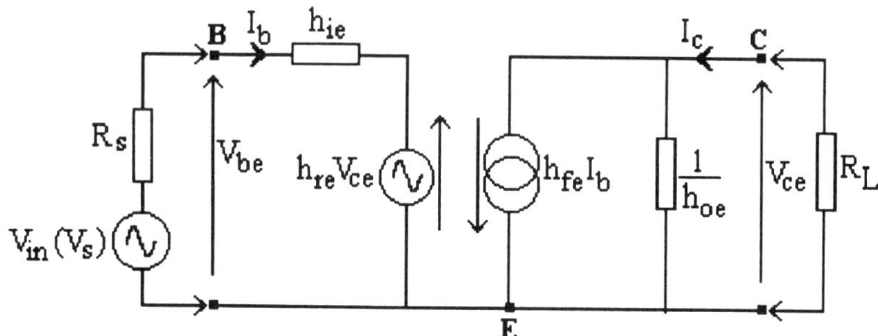

Fig. 4.1-0. Common-emitter *h*-parameter equivalent circuit

Comparing Fig. 4.0-1 with Fig. 4.1-0 we see that

$$I_b = I_{in}$$

$$I_c = I_{out}$$

$$V_{be} = V_{in}\ (V_s)$$

$$V_{ce} = V_{out}$$

From analysis of the equivalent circuit, we get

$$V_{be} = h_{ie}I_b + h_{re}V_{ce} \quad \text{(Eq. 4.1-0)}$$

$$I_c = h_{fe}I_b + h_{oe}V_{ce} \quad \text{(Eq. 4.1-1)}$$

(contd)

When R_s and R_L are taken into account, that is, when connected to the circuit, we get

$$V_{in} = I_b R_s + h_{ie} I_b + h_{re} V_{ce}$$
$$= I_b(R_s + h_{ie}) + h_{re} V_{ce} \qquad \text{(Eq. 4.1-2)}$$

$$I_c = h_{fe}\, I_b + h_{oe}(-\, I_c R_L)$$

(Note: $V_{ce} = -\, I_c R_L$. The minus sign indicates 180° phase shift from input to output.)

$$\therefore \quad I_c = h_{fe} I_b - h_{oe} I_c R_L$$

$$I_c + h_{oe} I_c R_L = h_{fe} I_b$$

from which

$$\mathbf{I_c} = \frac{\mathbf{h_{fe}\, I_b}}{\mathbf{1 + h_{oe}\, R_L}} \qquad \text{(Eq. 4.1-3)}$$

(b) **Input Resistance - R_{in}**

$$V_{be} = h_{ie} I_b + h_{re} V_{ce}$$
$$= h_{ie} I_b - h_{re} I_c R_L \qquad [V_{ce} = -\, I_c R_L]$$

Substituting for I_c from (Eq. 4.1-3), we get

$$V_{be} = h_{ie} I_b - h_{re} \left[\frac{h_{fe}\, I_b}{1 + h_{oe} R_L} \right] R_L$$

$$= I_b \left[h_{ie} - \frac{h_{re} h_{fe}\, R_L}{1 + h_{oe} R_L} \right]$$

$$\mathbf{R_{in}} = \frac{\mathbf{V_{be}}}{\mathbf{I_b}}$$
$$= \mathbf{h_{ie}} - \frac{\mathbf{h_{re} h_{fe}\, R_L}}{\mathbf{1 + h_{oe} R_L}} \qquad \text{(Eq. 4.1-4)}$$

(contd)

Output Resistance - R_{out}

$$V_{be} = h_{ie}I_b + h_{re}V_{ce} \qquad \text{(Refer to Eq. 4.1-0)}$$

$$I_c = h_{fe}I_b + h_{oe}V_{ce} \qquad \text{(Refer to Eq. 4.1-1)}$$

Also $$V_{be} = I_bR_s$$

Then $$I_bR_s = h_{ie}I_b - h_{re}I_cR_L \qquad (V_{ce} = -I_cR_L)$$

$$h_{ie}I_b - I_bR_s = h_{re}I_cR_L$$

$$-I_b(h_{ie} + R_s) = h_{re}V_{ce}$$

$$I_b = \frac{-h_{re}V_{ce}}{h_{ie}+R_s}$$

Now $$I_c = h_{fe}I_b + h_{oe}V_{ce}$$

Substituting for I_b, we get

$$I_c = -h_{fe}\frac{h_{re}V_{ce}}{h_{ie}+R_s} + h_{oe}V_{ce}$$

$$= V_{ce}\left[h_{oe} - \frac{h_{fe}h_{re}}{h_{ie}+R_s}\right]$$

Output Admittance

$$Y_{out} = \frac{I_c}{V_{ce}}$$

$$= h_{oe} - \frac{h_{fe}h_{re}}{h_{ie}+R_s} \qquad \text{(Eq. 4.1-5)}$$

$$\mathbf{R_{out}} = \frac{\mathbf{V_{ce}}}{\mathbf{I_c}}$$

$$= \frac{1}{\left[\mathbf{h_{oe}} - \dfrac{\mathbf{h_{fe}h_{re}}}{\mathbf{h_{ie}+R_s}}\right]} \qquad \text{(Eq. 4.1-6)}$$

(contd)

(c) **Current Gain - A_i**

$$I_c = h_{fe}I_b + h_{oe}V_{ce}$$

$$= h_{fe}I_b + h_{oe}(-I_cR_L)$$

$$= h_{fe}I_b - h_{oe}I_cR_L$$

$$-I_c(1 + h_{oe}R_L) = h_{fe}I_b$$

$$I_c = \frac{-h_{fe}I_b}{1 + h_{oe}R_L}$$

$$A_i = \frac{I_c}{I_b}$$

$$= \frac{-h_{fe}}{1 + h_{oe}R_L} \qquad \text{(Eq. 4.1 -7)}$$

Note: The minus sign indicates phase inversion, that is, 180° phase shift from input to output.

The equation may be written as

$$A_i = \frac{h_{fe}}{1 + h_{oe}R_L}\angle 180^\circ$$

Alternatively

Current Gain A_i $= -I_{out}/I_{in} = -h_{fe}/(1 + h_{oe}R_L)$

Voltage Gain - A_v

$$V_{be} = h_{ie}I_b + h_{re}V_{ce} \qquad \text{(i)}$$

$$\frac{I_c}{I_b} = \frac{-h_{fe}}{1 + h_{oe}R_L} \qquad \text{(Refer to Eq. 4.1-7)}$$

$$I_b = \frac{I_c(1 + h_{oe}R_L)}{-h_{fe}}$$

Substituting for I_b in (i) above, we get

$$V_{be} = h_{ie}I_c\left[\frac{1 + h_{oe}R_L}{-h_{fe}}\right] + h_{re}V_{ce} \qquad \text{(ii)}$$

(contd)

The load current $I_c = -\frac{V_{ce}}{R_L}$

Substituting for I_c in (ii) we get

$$V_{be} = -h_{ie}\frac{V_{ce}}{R_L}\left[\frac{1+h_{oe}R_L}{-h_{fe}}\right] + h_{re}V_{ce}$$

$$= -V_{ce}\left[h_{ie}\left(\frac{1+h_{oe}R_L}{h_{fe}R_L}\right) - h_{re}\right]$$

$$= -V_{ce}\left[\frac{h_{ie}(1+h_{oe}R_L) - h_{re}h_{fe}R_L}{h_{fe}R_L}\right]$$

$$A_v = -\frac{V_{ce}}{V_{be}}$$

$$= \frac{-h_{fe}R_L}{h_{ie}(1+h_{oe}R_L) - h_{re}h_{fe}R_L} \qquad \text{(iii)}$$

$$= \frac{-h_{fe}R_L}{h_{ie} - \dfrac{h_{re}h_{fe}R_L}{1+h_{oe}R_L}} \qquad \text{(iv)}$$

But $R_{in} = h_{ie} - \frac{h_{re}h_{fe}R_L}{1+h_{oe}R_L}$

Substituting for R_{in} in (iv), we get

$$\mathbf{A_v} = -\frac{\mathbf{h_{fe}R_L}}{\mathbf{R_{in}}} \qquad \text{(Eq. 4.1-8)}$$

Note:

Equation 4.1-8 gives the *approximate value* of the voltage gain; it is assumed that the transistor gain, h_{fe} is the same as the stage gain, A_i.

However, the stage gain $A_i = (-h_{fe})/(1 + h_{oe}R_L)$.

An alternative derivation of $\mathbf{A_v}$ follows.

(contd)

Alternatively

$$A_v = \frac{V_{out}}{V_{in}}$$

$$= \frac{-I_{out}R_L}{I_{in}R_{in}}$$

$$= -A_i\frac{R_L}{R_{in}}$$

$$= \frac{-h_{fe}}{1+h_{oe}R_L}\ \frac{R_L}{R_{in}} \qquad \text{(Eq. 4.1-9)}$$

Equation (4.1-9) gives the true value of the voltage gain A_v compared with the approximate value given by Eq. 4.1-8.

Power Gain - A_p

$$A_p = \text{current gain x voltage gain}$$

$$= \frac{-h_{fe}}{1+h_{oe}R_L} \text{ x } A_i\frac{R_L}{R_{in}}$$

$$= A_i \text{ x } \frac{-h_{fe}}{1+h_{oe}R_L}\ \frac{R_L}{R_{in}}$$

$$= A_i \text{ x } A_v \qquad \text{(Eq. 4.1-10)}$$

Alternatively (approximate value)

$$A_p = P_{out}/P_{in}$$

$$P_{out} = (h_{fe}I_b)^2 \text{ x } R_L \qquad \equiv [I_c^2\, R_L]$$

$$P_{in} = I_b^2R_{in} \qquad \equiv [\, I_b^2\, R_{in}]$$

$$P_{out}/P_{in} = h_{fe}^2I_b^2R_L/I_b^2R_{in} \qquad \equiv [\, A_i^2\, R_L/R_{in}]$$

$$= h_{fe}^2R_L/R_{in} \qquad \text{(Eq. 4.1-11)}$$

Example 4.2

(a) *A transistor connected in a common-emitter circuit produces changes in the emitter and collector currents of 1.0 mA and 0.98 mA, respectively. Determine both the common-base short-circuit current gain and the common-emitter short-circuit current gain.*

(b) *If a resistor of 4.0 kΩ is connected to the collector and the transistor input resistance is 1.0 kΩ, calculate the voltage and power gains of the transistor.*
Express the power gain in decibel units.

Solution

(a) **Common-Base Short-Circuit Current Gain**

$$h_{fb} = I_c/I_e$$
$$= {}^{0.98}/_{1.0} \qquad = \underline{\mathbf{0.98}}$$

Common-Emitter Short-Circuit Current Gain

$$h_{fe} = I_c/I_b$$
$$= -h_{fb}/(1 + h_{fb})$$
$$= h_{fb}/(1 - h_{fb})$$
$$= {}^{0.98}/_{(1 - 0.98)} \qquad = \underline{\mathbf{49}}$$

(b) **Voltage Gain**

$$A_v = h_{fe}R_L/R_{in}$$
$$= 49 \times {}^{4000}/_{1000} \qquad = \underline{\mathbf{196}}$$

(contd)

Power Gain

$$A_p = h_{fe}^2 R_L / R_{in}$$

$$= 49^2 \times {}^{4000}/1000 = \underline{\mathbf{9604}}$$

Alternatively

$$A_p = A_i \times A_v$$

$$= 49 \times 196 = \underline{\mathbf{9604}}$$

In decibel units $A_p = 10 \log_{10} 9604 = \underline{\mathbf{39.82}}\ \mathbf{dB}$

Example 4.3

The data given in Tables 4.3-0 and 4.3-1 refer to a transistor in the common-emitter configuration.

Table 4.3-0

Collector Voltage V_{ce} (volts)		2	4	6	8	10
Collector Current (mA)	Base current 120 μA	6.4	7.0	7.6	8.2	8.8
	Base current 80 μA	4.4	4.8	5.2	5.6	6.0
	Base current 40 μA	2.2	2.4	2.6	2.8	3.0

Use the data in Table 4.3-0 to plot the collector-voltage/collector-current characteristics for I_b = 40 μA, 80 μA, and 120 μA and from the characteristic for I_b = 80 μA deduce the output resistance of the transistor. Use the data in Table 4.3-1 to plot the collector-current/base-current characteristic for V_{ce} = 4.5 V and from the graph deduce the current gain.

Table 4.3-1

Base Current I_b (μA)	0	5	10	15	20
Collector Current (mA)	0.15	0.45	0.75	1.1	1.4

(contd)

Solution

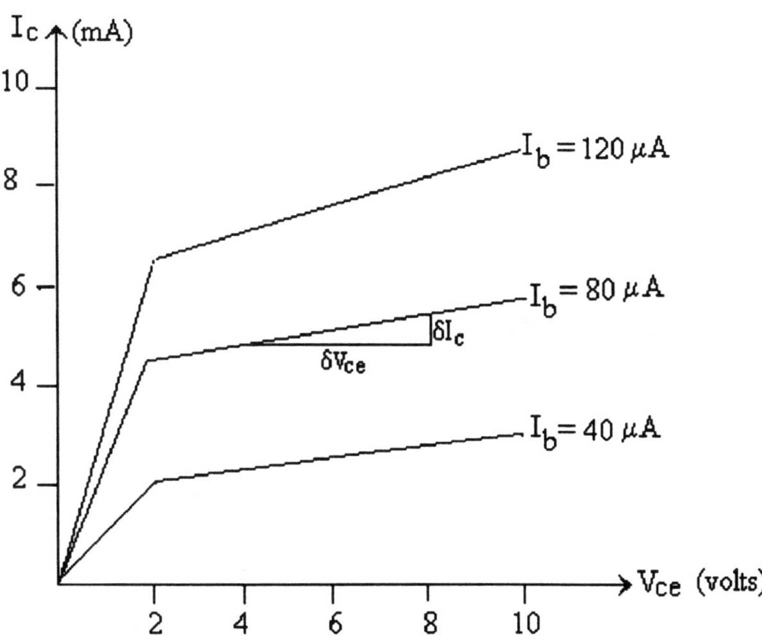

Fig. 4.3-0. Collector-voltage/collector-current characteristics

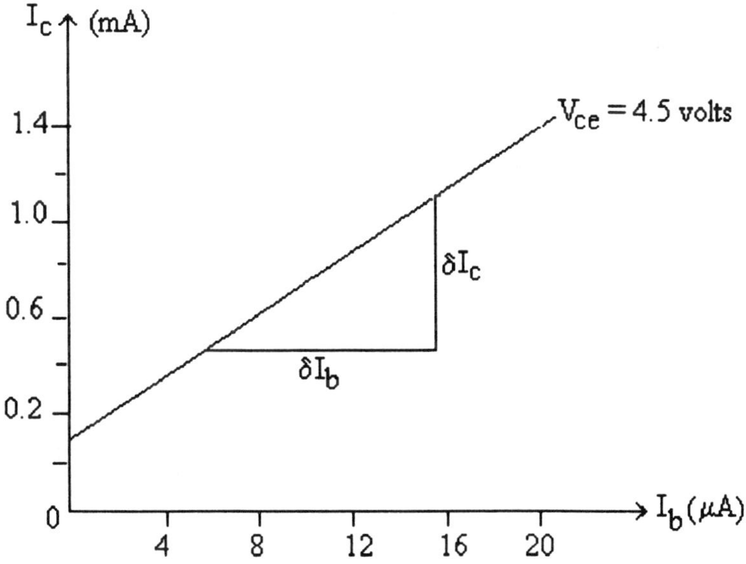

Fig. 4.3-1. Collector-current/base-current characteristics

(contd)

Refer to Fig. 4.3-0

Output resistance, R_{out} $= \delta V_{ce}/\delta I_c$

$= \dfrac{8 - 4}{(5.6 - 4.8) \times 10^{-3}}$

$= \dfrac{4 \times 10^3}{0.8}$ $=$ **5.0 kΩ**

Refer to Fig. 4.3-1

Current gain, h_{fe} $= \delta I_c/\delta I_b$

$= \dfrac{(1.1 - 0.45) \times 10^{-3}}{(15 - 5.0) \times 10^{-6}}$

$= \dfrac{0.65 \times 10^3}{10}$ $=$ **65**

Example 4.4

(a) *Tabulate the h-parameters for a transistor connected in*

common-base configuration

common-emitter configuration

common-collector (emitter-follower) configuration

(b) *Manufacturers do not normally publish data for the above configurations; frequently only common-emitter h-parameters are quoted. Prepare a table in comparative form showing the relationship between the h-parameters.*

(c) *Summarize the common-emitter and common-base h-parameter equations for the following:*

input resistance

output resistance

current gain

voltage gain

power gain

(d) *Values of a typical silicon n.p.n transistor are given below:*

$h_{ib} = 30\ \Omega$

$h_{rb} = 1.5 \times 10^{-4}$

$h_{fb} = -0.98$

$h_{ob} = 0.5 \times 10^{-6}$

Calculate the corresponding values for the common-emitter h-parameters.

Solution

(a) The *h*-parameters for the three configurations are shown in Tables 4.4-0, 4.4-1 and 4.4-2.

(contd)

Table 4.4-0

h-parameter			Conditions S.C = Short-Circuited O.C = Open-Circuited
Common-Base (C.B)	Common-Emitter (C.E)	Common-Collector (C.C)	
$h_{ib} = \frac{V_{eb}}{I_e}$	$h_{ie} = \frac{V_{be}}{I_b}$	$h_{ic} = \frac{V_{bc}}{I_b}$	Input resistance with S.C output
$h_{rb} = \frac{V_{eb}}{V_{cb}}$	$h_{re} = \frac{V_{be}}{V_{ce}}$	$h_{rc} = \frac{V_{bc}}{V_{ec}}$	Reverse voltage amplification with O.C input
$h_{fb} = \frac{I_c}{I_e}$	$h_{fe} = \frac{I_c}{I_b}$	$h_{fc} = \frac{I_e}{I_b}$	Forward current gain with S.C output
$h_{ob} = \frac{I_c}{V_{cb}}$	$h_{oe} = \frac{I_c}{V_{ce}}$	$h_{oc} = \frac{I_e}{V_{ec}}$	Output admittance with O.C input

(b)

Table 4.4-1

Common-Base (C.B)	Common-Emitter (C.E)	Common-Collector (C.C)
$h_{ib} = \frac{h_{ie}}{1 + h_{fe}}$	$h_{ie} = \frac{h_{ib}}{1 + h_{fb}}$	$h_{ic} = \frac{h_{ib}}{1 + h_{fb}}$
$h_{rb} = \frac{h_{ie}h_{oe}}{1 + h_{fe}} - h_{re}$	$h_{re} = \frac{h_{ib}h_{ob}}{1 + h_{fb}} - h_{rb}$	$h_{rc} \approx 1$
$h_{fb} = \frac{-h_{fe}}{1 + h_{fe}}$	$h_{fb} = \frac{-h_{fe}}{1 + h_{fe}}$	$h_{fc} = \frac{-1}{1 + h_{fb}}$
$h_{ob} = \frac{h_{oe}}{1 + h_{fe}}$	$h_{oe} = \frac{h_{ob}}{1 + h_{fb}}$	$h_{oc} = \frac{h_{ob}}{1 + h_{fb}}$

(contd)

(c) Summary of C.B and C.E *h*-parameter equations

Table 4.4-2

	C.E ↓	Common-Emitter True Equations	Common-Emitter Approximate Equations	Common-Base
R_{in}	$\frac{V_{be}}{I_b}$	$h_{ie} - \frac{h_{re} h_{fe} R_L}{1 + h_{oe} R_L}$	h_{ie}	$h_{ib} - \frac{h_{rb} h_{fb} R_L}{1 + h_{oe} R_L}$
R_{out}	$\frac{V_{ce}}{I_c}$	$\frac{1}{h_{oe} - \frac{h_{fe} h_{re}}{h_{ie} + R_s}}$	$\frac{1}{h_{oe}}$	$\frac{1}{h_{ob} - \frac{h_{fb} h_{rb}}{h_{ib} + R_s}}$
A_i	$\frac{I_c}{I_b}$	$\frac{-h_{fe}}{1 + h_{oe} R_L}$	h_{fe}	$\frac{-h_{fb}}{1 + h_{ob} R_L}$
A_v	$\frac{V_{ce}}{V_{be}}$	$-A_i \frac{R_L}{R_{in}}$	$-h_{fe} \frac{R_L}{R_{in}}$	$-A_i \frac{R_L}{R_{in}}$
A_p	$\frac{P_{out}}{P_{in}}$	$A_i^2 \frac{R_L}{R_{in}}$	$h_{fe}^2 \frac{R_L}{R_{in}}$	$A_i^2 \frac{R_L}{R_{in}}$

Corresponding C.E and C.C (emitter-follower) h-parameters

$$h_{fc} = -(h_{fe} + 1)$$

$$h_{ic} = h_{ie}$$

$$h_{oc} = h_{oe}$$

$$h_{rc} = 1 - h_{re}$$

(contd)

(d)

$$h_{ie} = h_{ib}/(1 + h_{fb})$$
$$= {}^{30}/(1 - 0.98) \qquad = \underline{\mathbf{1500}}$$

$$h_{re} = [\,(h_{ib}\,h_{ob})/(1 + h_{fb})] - h_{rb}$$

$$= \frac{30 \times 0.5 \times 10^{-6}}{1 - 0.98} - (1.5 \times 10^{-4})$$

$$= \frac{15 \times 10^{-6}}{0.02} - (\,1.5 \times 10^{-4})$$

$$= (7.5 \times 10^{-4}) - (1.5 \times 10^{-4})$$

$$= \underline{\mathbf{6.0 \times 10^{-4}}}$$

$$h_{fe} = -\,h_{fb}/(1 + h_{fb})$$
$$= {}^{0.98}/(1 - 0.98) \qquad = \underline{\mathbf{49}}$$

$$h_{oe} = h_{ob}/(1 + h_{fb})$$

$$= \frac{0.5 \times 10^{-6}}{1 - 0.98} \qquad = \mathbf{25 \times 10^{-6}\ S}$$
$$= \underline{\mathbf{25\ \mu S}}$$

Example 4.5

For the common-emitter amplifier stage shown in Fig. 4.5-0, determine from the labeled components the following:

(a) *the a.c. input resistance*

(b) *the voltage gain*

The data given for the transistor h-parameters are $h_{ie} = 1.5\ k\Omega$ *and* $h_{fe} = 50$ *(*h_{oe} *and* h_{re} *may be neglected).*

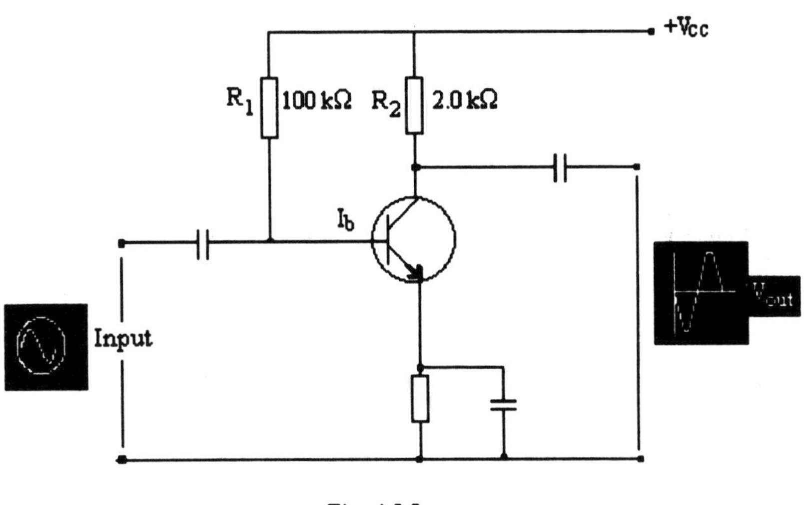

Fig. 4.5-0

Solution

(a)

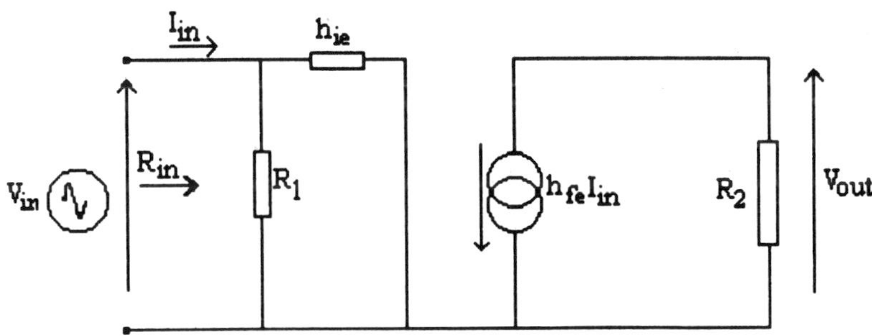

Fig. 4.5-1. Equivalent circuit for C.E amplifier stage in Fig. 4.5-0

(contd)

Input Resistance - R_{in}

Now $V_{in} = I_{in} \times (R_1 h_{ie})/(R_1 + h_{ie})$

and $R_{in} = V_{in}/I_{in}$

$$= (R_1 h_{ie})/(R_1 + h_{ie})$$

$$= \frac{(100 \times 10^3) \times (1.5 \times 10^3)}{(100 \times 10^3) + (1.5 \times 10^3)}$$

$$= \frac{150 \times 10^3}{101.5} = \mathbf{1.48\ k\Omega}$$

(b) *Voltage Gain - A_v*

Now $V_{in} = I_{in} \times R_{in}$

and $V_{out} = -h_{fe} I_{in} R_2$

$$A_v = V_{out}/V_{in}$$

$$= (-h_{fe} I_{in} R_2)/(I_{in} \times R_{in})$$

$$= (-h_{fe} R_2)/R_{in}$$

$$= -\frac{50 \times 2000}{1.48 \times 10^3} = \mathbf{-67.57}$$

The minus sign indicates phase inversion, that is, 180° phase shift between the input and the output voltages.

Example 4.6

The common-collector, or emitter-follower as it is usually called, is shown in Fig. 4.6-0. For this arrangement

(a) *draw the common-emitter equivalent circuit and determine approximate values for the input and output resistances of the stage*

(b) *draw the common-collector (emitter-follower) equivalent circuit and determine approximate values for the input and output resistances of the stage*

From the results obtained, explain where a common-collector or emitter-follower amplifier stage would be practically employed. The transistor h-parameters are given in Table 4.6-0.

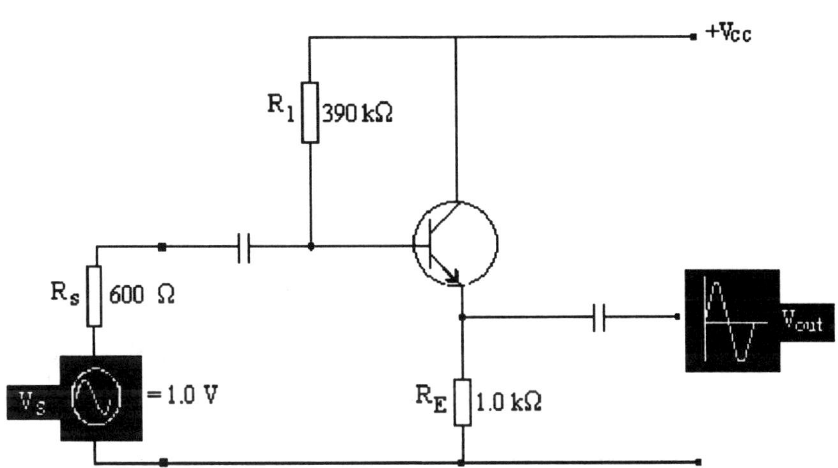

Fig. 4.6-0. Common-collector/emitter-follower

Table 4.6-0

COMMON–EMITTER	COMMON–COLLECTOR
h_{ie} = 600 Ω	h_{ic} = 600 Ω
h_{re} = 0	h_{rc} = 1
h_{fe} = 99	h_{fc} = −100
h_{oe} = 0	h_{oc} = 0

(contd)

Solution *Common-Emitter Equivalent Circuit*

(a) Since $h_{re} = 0$ and $h_{oe} = 0$, these two parameters are omitted from the equivalent circuit.

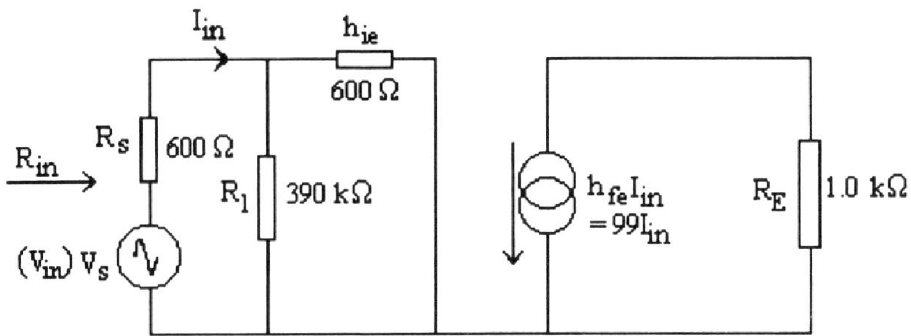

Fig. 4.6-1. C.E equivalent circuit

Input Resistance - R_{in}

$$R_{in} = R_s + \frac{R_1 h_{ie}}{R_1 + h_{ie}}$$

The shunting effect of R_1 compared with h_{ie} is negligible; hence R_1 can be safely neglected.

$$\therefore \quad R_{in} = R_s + h_{ie}$$

$$= (600 + 600) \times 10^{-3} = \underline{\mathbf{1.2\ k\Omega}}$$

Output Resistance - R_{out} (Refer to Fig. 4.6-2)

The source V_s (V_{in}) is removed and replaced with its internal resistance, while a source V_{out} volts of zero internal resistance is injected across the output terminals.

[Note: From Table 4.6-0 there will be an error with the assumed values $h_{re} = 0$ in the input circuit and $h_{oe} = 0$ in the output circuit. However, the assumed values considerably simplify the circuit calculations.]

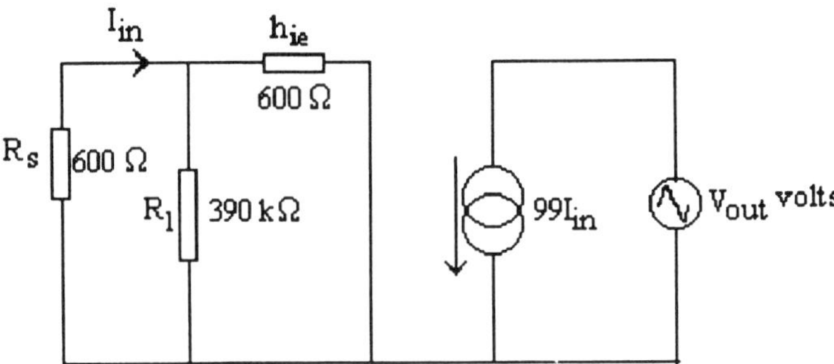

Fig. 4.6-2

(contd) Refer to Fig. 4.6-2

$$\text{Iin} = 0$$

$$\therefore \quad R_{out} = V_{out} / 99I_{in}$$

$$= V_{out}/0 \qquad = \infty \text{ (infinity)}$$

[Note: $R_{out} = {}^{1}/h_{oe}$ and since $h_{oe} = 0$, $R_{out} = {}^{1}/0 = \infty$.]

The output resistance R_{out} of an amplifier may be defined as the ratio of V_{out} to I_{out} when an a.c. source is applied to the output terminals, with the input source removed and replaced by its internal resistance.

For all practical purposes,

R_{out} ranges from 25 kΩ for germanium to 70 kΩ for silicon.

(b) *Common-Collector or Emitter-Follower Equivalent Circuit*

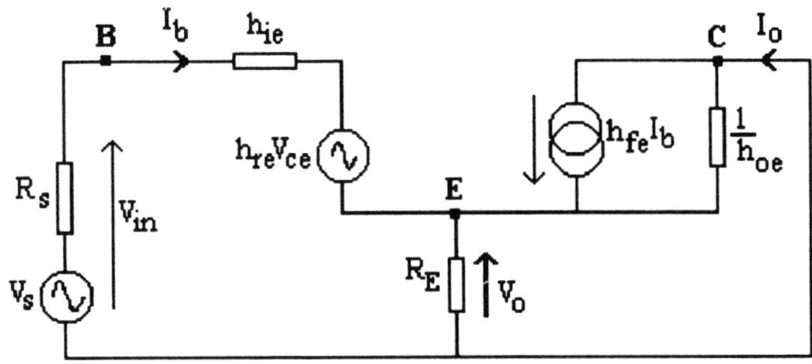

Fig. 4.6-3. Common-collector/emitter-follower equivalent circuit

Approximate Input Resistance - R_{in} (Refer to Fig. 4.6-4)

For the approximate input resistance, the equivalent circuit is shown in Fig. 4.6-4; again R_1 and $h_{oe} = 0$ have been omitted from the equivalent circuit because of their insignificance.

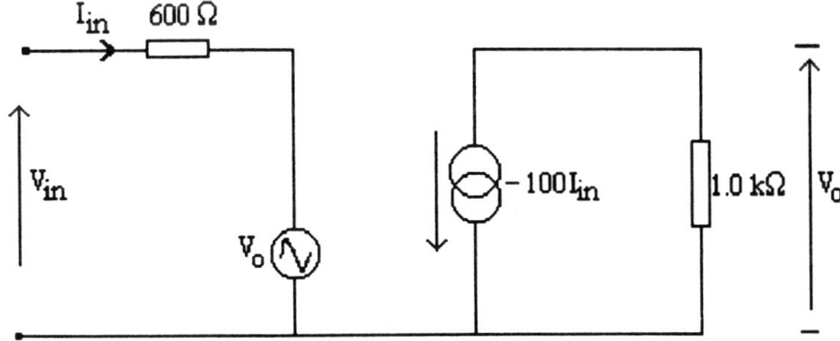

Fig. 4.6-4. Common-collector/emitter-follower approximate equivalent circuit

(contd)

$$V_{in} = V_o + 600I_{in} \qquad \text{..........(Eq. 4.6-0)}$$

$$V_o = -(-100I_{in}) \times R_E$$

$$= 100I_{in} \times 1.0 \times 10^3 = 100I_{in} \times 10^3$$

Substituting for V_o in (Eq. 4.6-0); $V_{in} = 100I_{in} + 0.6I_{in}$

$$R_{in} = Vin/Iin = (100 + 0.6)\text{k}\Omega = \underline{\mathbf{100.6k\Omega}}$$

Note: When I_o is low, $V_o \approx I_oR_E = (1 + h_{fe})\, I_bR_E$

Now, in the emitter-follower $V_o = h_{ie}\, I_b + V_o$

$\therefore\ V_{in} = h_{ie}I_b + (1 + h_{fe})\, I_bR_E = I_b[h_{ie} + (1 + h_{fe})\, R_E]$

Hence, Rin $= V_{in}/I_b = h_{ie} + (1 + h_{fe})\, R_E \approx h_{ie} + h_{fe}\, R_E$

$\therefore\ R_{in} = h_{ie} + (1 + h_{fe})\, R_E = 0.6 + (1 + 100) \times 1.0 = \underline{\mathbf{101.6\ k\Omega}}$

or $R_{in} \approx h_{fe}\, R_E \approx 100 \times 1.0 = \underline{\mathbf{100\ k\Omega}}$

(c) *To determine R_{out}* (Refer to Fig. 4.6-5)

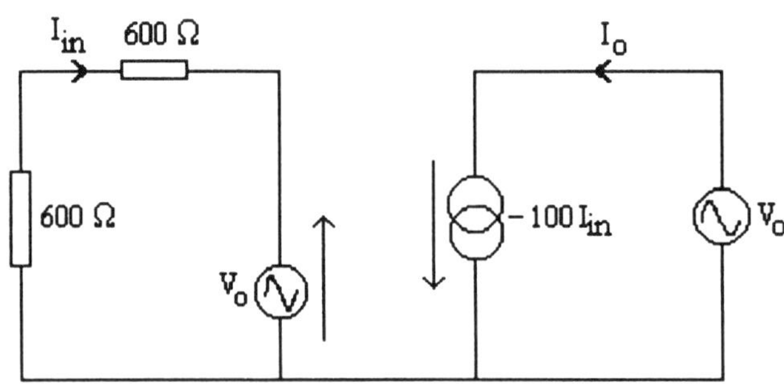

Fig. 4.6-5

$$I_{in} = -V_o/R_{in} = V_o/(1.2 \times 10^3)$$

$$I_o = -100I_{in}$$

$$R_{out} = V_o/I_o = \frac{1.2 \times 10^3}{-100I_{in}} = \underline{\mathbf{12.0\ \Omega}}$$

Summary

$$\mathbf{R_{in} = 100.6\ k\Omega\ (high);\ R_{out} = 12\ \Omega\ (low)}$$

From the results of the input and output resistances, it can reasonably be concluded that the common-collector/emitter-follower amplifier may be practically employed as a buffer amplifier stage, conveniently connected between a high-impedance source and a low-impedance load.

Example 4.7

A transistor employed in a common-base amplifier has the following h-parameters:

$$h_{ib} = 42\ \Omega$$
$$h_{rb} = 4 \times 10^{-4}$$
$$h_{fb} = -0.98$$
$$h_{ob} = 0.4\ \mu S$$

Determine

(i) *the voltage gain*

(ii) *the input and output resistances of the amplifier if the load resistance is 2.0 kΩ and the signal source has an internal resistance of 400 Ω.*

Solution

(i) The equivalent circuit of the amplifier is shown in Fig. 4.7-0.

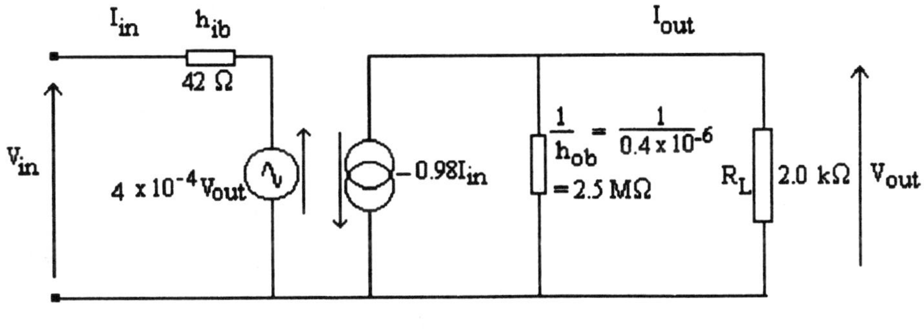

Fig. 4.7-0

To determine the voltage gain -A_v (Refer to Fig. 4.7-0)

$$V_{in} = h_{ib}I_{in} + 4 \times 10^{-4}\, V_{out} \quad \text{......(Eq. 4.7-0)}$$

$$V_{out} = -(-0.98 I_{in})\ 2.0 \times 10^{3} \quad \text{.....(Eq. 4.7-1)}$$

$$= 1960\ I_{in}$$

[Note: The shunting effect of $1/h_{ob}$ = 2.5 MΩ on the load is negligible and can safely be omitted in the calculations.]

(contd)

$$I_{in} = V_{out}/1960$$

Substituting for I_{in} in Eq. 4.7-0

$$V_{in} = 42 \times (V_{out}/1960) + 4 \times 10^{-4} V_{out}$$

$$= 214.3 \times 10^{-4} V_{out} + 4 \times 10^{-4} V_{out}$$

$$= 218.3 \times 10^{-4} V_{out}$$

$\therefore$ Voltage gain, $A_v = V_{out}/V_{in} = 10^4/218.3 = \underline{\mathbf{45.8}}$

(ii) *To Determine R_{in}*

$$V_{in} = 218.3 V_{out}/10^4$$

$$V_{out} = (V_{in} \times 10^4)/218.3$$

Substituting for V_{out} in Eq. 4.7-0

$$V_{in} = 42 I_{in} + (4 \times 10^{-4} \times 45.8) V_{in}$$

$$= 42 I_{in} + 0.0183 V_{in}$$

$$V_{in} - 0.0183 V_{in} = 42 I_{in}$$

$$V_{in}(1 - 0.0183) = 42 I_{in}$$

$$R_{in} = V_{in}/I_{in}$$

$$= 42/0.982 \qquad = \underline{\mathbf{42.77\ \Omega}}$$

The equivalent circuit to determine R_{out} is shown in Fig. 4.7-1.

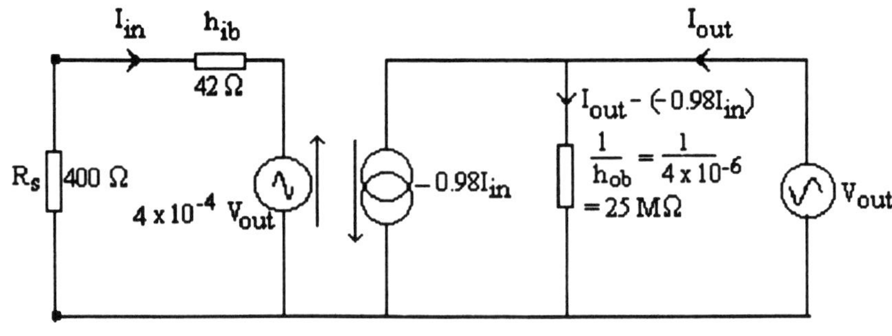

Fig. 4.7-1

(contd) Refer to Fig. 4.7-1

$$V_{out} = 2.5 \times 10^6 [\, I_{out} - (-0.98 I_{in})] \quad \text{...(Eq. 4.7-2)}$$

(From the input)

$$4 \times 10^{-4} V_{out} = -(h_{ib} + R_s)\, I_{in}$$

$$= -442 I_{in}$$

$$\therefore \quad I_{in} = \frac{-4 \times 10^{-4} V_{out}}{442}$$

Substituting for I_{in} in Eq. 4.7-2

$$V_{out} = 2.5 \times 10^6 \left[I_{out} - \left(-0.98 \times \frac{-4 \times 10^{-4} V_{out}}{442}\right)\right]$$

$$= 2.5 \times 10^6\, I_{out} - \frac{(0.98 \times 4 \times 10^{-4} \times 2.5 \times 10^6)}{442} V_{out}$$

$$= 2.5 \times 10^6\, I_{out} - \frac{9.8 \times 10^2}{442} V_{out}$$

$$V_{out} + \frac{980}{442} V_{out} = 2.5 \times 10^6\, I_{out}$$

$$V_{out}\left(1 + \frac{980}{442}\right) = 2.5 \times 10^6\, I_{out}$$

$$V_{out}\left(\frac{1422}{442}\right) = 2.5 \times 10^6\, I_{out}$$

$$\therefore \quad R_{out} = V_{out}/I_{out}$$

$$= \frac{2.5 \times 10^6 \times 442}{1422} \qquad = \mathbf{\underline{777.1\ k\Omega}}$$

Example 4.8

Sketch typical input and output characteristics of a transistor connected in the common-emitter configuration for various values of I_B , I_C , V_{BE} and V_{CE} .

Indicate on the characteristics how the h_{ie}, h_{re}, h_{fe}, and h_{oe} parameters can be estimated, giving typical values.

Solution

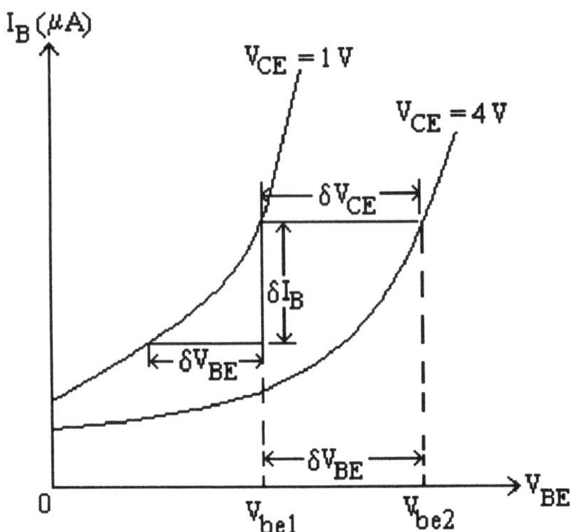

Fig. 4.8-0. Common-emitter input characteristics

h_{ie} = $\delta V_{BE}/\delta I_B$: Typical value = 1.4 kΩ

h_{re} = $\dfrac{V_{be2} - V_{be1}}{\delta V_{CE}}$ that is , $\delta V_{BE}/\delta V_{CE}$:

Typical value = 6 x 10^{-4}

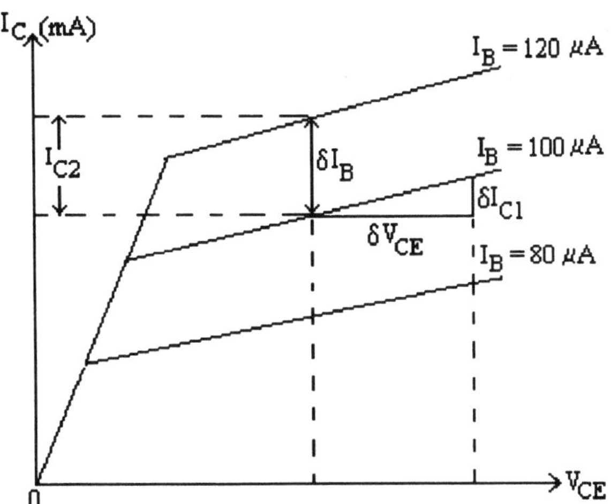

Fig. 4.8-1. Common-emitter output characteristics

h_{fe} = $\delta I_{C2}/\delta I_B$: Typical value = 49

h_{oe} = $\delta I_{C1}/\delta V_{CE}$: Typical value = 25 μS

Example 4.9

The common-emitter amplifier shown in Fig. 4.9-0 has a silicon transistor with the following h-parameters:

$h_{ie} = 1.4\ k\Omega;\ h_{re} = 6 \times 10^{-4};\ h_{fe} = 50;\ h_{oe} = 25 \times 10^{-6}$.

Using the values of the labeled components and neglecting the leakage current, determine

(i) *the approximate d.c. conditions*

(ii) *the small-signal mid-frequency current and power gains.*

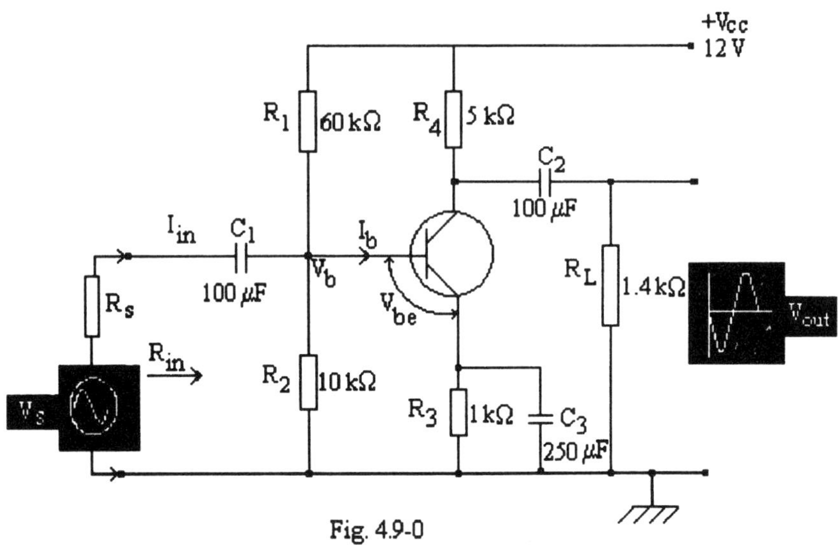

Fig. 4.9-0

Solution

Note: The coupling capacitors C_1 and C_2 isolate the source and the load from the direct voltages on the transistor.

C_3 is the emitter bypass capacitor and effectively connects the emitter to ground on the a.c. signal.

It is assumed that the reactances of the coupling capacitors are negligibly small and the d.c. supply is of negligible impedance

As a result of the preceding assumptions, the equivalent circuit is drawn as shown in Fig. 4.9-1.

(contd)

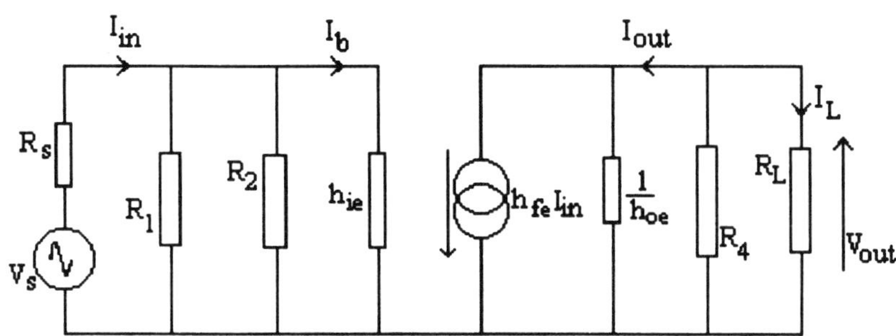

Fig. 4.9-1. Equivalent circuit

$$R_{in} = R' h_{ie}/(R' + h_{ie}) \qquad \text{where } R' = \frac{R_1 R_2}{R_1 + R_2}$$

$$R' = \frac{60 \times 10}{60 + 10} = \underline{8.57\ k\Omega}$$

$$R_{in} = \frac{8.57 \times 1.4}{8.57 + 1.4} = \underline{1.2\ k\Omega}$$

(i) Approximate d.c. Conditions

To obtain the d.c. operating conditions, it is assumed that the base bias current I_B is very much smaller than the d.c. current through R_2; hence

$$V_B = V_{cc}R_2/(R_1 + R_2)$$

$$= \frac{12 \times 10 \times 10^3}{(60 + 10) \times 10^3} = \underline{\mathbf{1.7\ V}}$$

Since the amplifier is a silicon transistor, it can be assumed that the voltage drop across the base-emitter V_{BE} is 0.7 V.

$\therefore$ Emitter bias voltage, $V_E = V_B - V_{BE}$

$$= 1.7 - 0.7 = \underline{\mathbf{1.0\ V}}$$

Emitter current, $I_E = V_E/R_3$

$$= 1.0/1000 = \underline{1.0\ mA}$$

(contd)

$$I_E \approx I_C$$

$\therefore$ Collector voltage, $V_C = V_{cc} - I_c R_4$

$= 12 - (1.0 \times 10^{-3} \times 5 \times 10^3)$

$= 12 - 5 \qquad = \underline{\mathbf{7\ V}}$

$I_B \approx I_C/h_{FE}$ (h_{FE} is the same as h_{fe})

$= 1/50 \qquad = \underline{\mathbf{0.02\ mA\ (20\ \mu A)}}$

Current through $R_2 = I_{R2} = V_B/R_2$

$= 1.7/(10 \times 10^3) \qquad = \underline{\mathbf{0.17\ mA}}$

Alternatively $I_{R2} = (I_E \times 1000) + V_{be})/R_2$

$= [1 \times 10^{-3} \times 10^3 + 0.7]/R_2$

$= 1.7/R_2$

$= 1.7/(10 \times 10^3) \qquad = \underline{\mathbf{0.17\ mA}}$

[Note: $I_{R2} = 0.17$ mA substantiate the earlier assumption that $I_B < I_{R2}$.]

(ii) *Small-Signal Mid-frequency Current and Power Gains*

Mid-frequency gain $= I_{out}/I_b$

$= -h_{fe}/(1 + h_{oe}R_{eq})$

where $R_{eq} = R_4 R_L/(R_4 + R_L)$

$= (5 \times 1.4)/(5 + 1.4)$

$= 7/6.4 \qquad \approx \underline{\mathbf{1.1\ k\Omega}}$

The actual current gain $A_i = I_{out}/I_{in}$ is less than the transistor current gain because

(a) I_{in} is partially shunted by R_1 and R_2 and

(b) I_{out} is shared between R_4 and R_L

R' = parallel combination R_1 and R_2

$\therefore \quad R' = (60 \times 10)/(60 + 10) = \underline{\mathbf{8.57\ k\Omega}}$

(contd)

$$I_b = I_{in} \times \frac{R'}{R' + h_{ie}}$$

$$I_{in} = \frac{I_b (R' + h_{ie})}{R'}$$

$$I_L = \frac{I_{out} R_4}{R_4 + R_L}$$

Current gain of stage, $A_i = \dfrac{I_L}{I_{in}}$

$$= \frac{-I_{out} R_4}{R_4 + R_L} \times \frac{R'}{I_b (R' + h_{ie})} \qquad \text{(Eq. 4.9-0)}$$

Now $\dfrac{I_{out}}{I_b} = \dfrac{-h_{fe}}{(1 + h_{oe} R_{eq})}$

Substituting for $\dfrac{I_{out}}{I_b}$ in Eq. 4.9-0

$$A_i = \frac{-h_{fe}\, R_4\, R'}{(1 + h_{oe} R_{eq})(R_4 + R_L)(R' + h_{ie})}$$

$$= \frac{-50 \times 5 \times 10^3 \times 8.57 \times 10^3}{[1 + (25 \times 10^{-6} \times 1.1 \times 10^3)]\,[(5 + 1.4) \times 10^3)]\,[(8.57 + 1.4) \times 10^3}$$

$= \dfrac{-2142.5}{65.56}$ $= \mathbf{\underline{-32.68}}$

The minus sign indicates phase inversion, i.e., 180° phase shift from input to output, and may be written as **32.68/180°**

Power gain, $A_p = A_i^2\, R_L/R_{in}$

$= (32.68)^2 \times 1.4/1.2 =$ **<u>1246</u>** (31 dB)

Example 4.10

The common-emitter amplifier shown in Fig. 4.10-0 has a power supply of 10 V. The no-signal base bias current and voltages are I_B = 20 μA and V_{BE} = 0.5 V. The emitter current is 1.0 mA. Suggest suitable values for R_1, R_2, R_E and R_L.

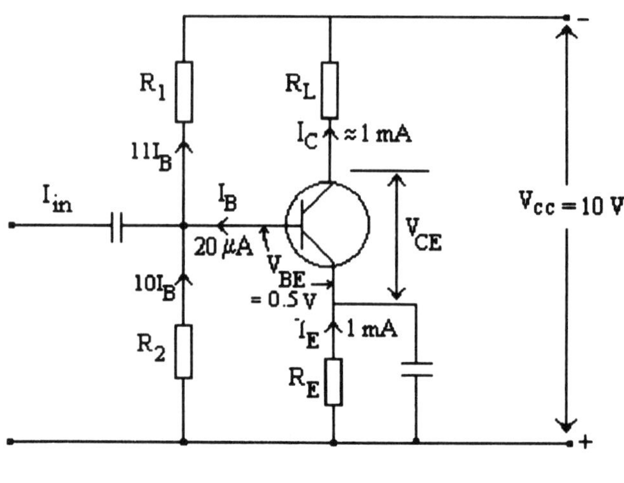

Fig. 4.10-0

Solution

Refer to the circuit of Fig. 4.10-0

Emitter current , I_E = 1 mA

Let the voltage drop across R_E = V_{RE} = 1 V (a normal value in this kind of circuit)

Resistance of R_E = V_{RE}/I_E

= $1/(1 \times 10^{-3})$ = **1.0 kΩ**

The current through voltage divider bias resistors R_1 and R_2 is assumed to be very large compared with I_B.

Let the current in R_2 (I_{R2}) = $10I_B$ (a conventional value)

= $10 \times 20 \times 10^{-6}$ = 200μA (0.2mA)

Voltage across R_2 (V_{R2}) = $V_{BE} + I_ER_E$

= $0.5 + (1 \times 10^{-3} \times 1000)$

= 1.5 V

(contd)

Resistance of **R_2** $= V_{R2}/I_{R2}$

$= 1.5/(0.2 \times 10^{-3}) = \underline{\mathbf{7.5\ k\Omega}}$

Current in $R_1(I_{R1}) = 10I_B + (20 \times 10^{-6}) = \underline{0.22\ mA}$

Voltage across $R_1(V_{R1}) = V_{CC} - V_{R2}$

$= 10 - 1.5 = \underline{8.5\ V}$

Resistance of **R_1** $= V_{R1}/I_{R1}$

$= 8.5/(0.22 \times 10^{-3}) = \underline{\mathbf{38.6\ k\Omega}}$

Since $V_{RE} = 1.0\ V$

Voltage across V_{CE} and $R_L = V_{CC} - V_{RE}$

$= 10 - 1.0 = \underline{9.0\ V}$

This voltage (9 V) can be divided equally between V_{CE} and R_L, making the quiescent collector voltage 4.5 V

Thus $V_{CE} = 4.5\ V$

and $V_{RL} = 4.5\ V$

Neglecting $I_B = 20\ \mu A$,

$I_C = I_E$

$= 1.0\ mA$

$\therefore$ **R_L** $= V_{RL}/I_C$

$= 4.5/(1.0 \times 10^{-3}) = \underline{\mathbf{4.5\ k\Omega}}$

Preferred suggested values would be resistors of ±10% tolerance of nearest standard values.

Example 4.11

An amplifier consists of two transistor stages connected in common-emitter configuration. The collector resistors of the first and second stages are 6 kΩ and 2 kΩ, respectively. The transistor hybrid parameters at the operating point are

Stage 1:	$h_{ie} = 1.0\ k\Omega$:	$h_{re} = 5 \times 10^{-4}$
	$h_{fe} = 50$;	$h_{oe} = 20\ \mu S$
Stage 2:	$h_{ie} = 1.2\ k\Omega$;	$h_{re} = 5 \times 10^{-4}$
	$h_{fe} = 80$;	$h_{oe} = 60\ \mu S$

Assuming that the effect of biasing components and signal source impedance can be neglected, calculate the overall

(i) *voltage gain*

(ii) *current gain*

(iii) *power gain*

Solutions

The approach to this problem (or to any problem) is to follow a systematic procedure, focusing on details.

Procedure 1: Draw the equivalent circuit - (Fig.4.11-0)

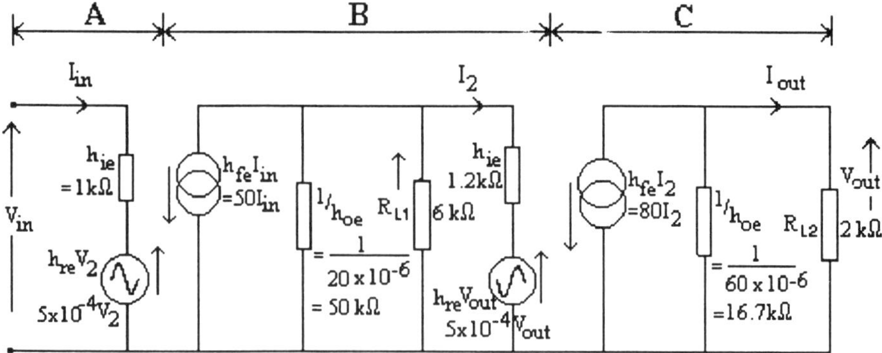

Fig. 14.11-0. Equivalent circuit of a two-stage amplifier

(contd)

Procedure 2: Lump all parallel resistors with their equivalent value

Refer to Fig. 4.11-0

In section **B**, h_{oe} is in parallel with R_{L1} so that

$$\mathbf{R'_{L1}} = (R_{L1} \times h_{oe})/(R_{L1} + h_{oe})$$
$$= (6 \times 50)/(6 + 50) \qquad = \underline{\mathbf{5.4\ k\Omega}}$$

In section **C**, h_{oe} is in parallel with R_{L2} so that

$$\mathbf{R'_{L2}} = (R_{L2} \times h_{oe})/(R_{L2} + h_{oe})$$
$$= (2 \times 16.7)/(2 + 16.7) \qquad = \underline{\mathbf{1.8\ k\Omega}}$$

The equivalent circuit of Fig. 4.11-0 is now reduced to the circuit shown in Fig. 4.11-1.

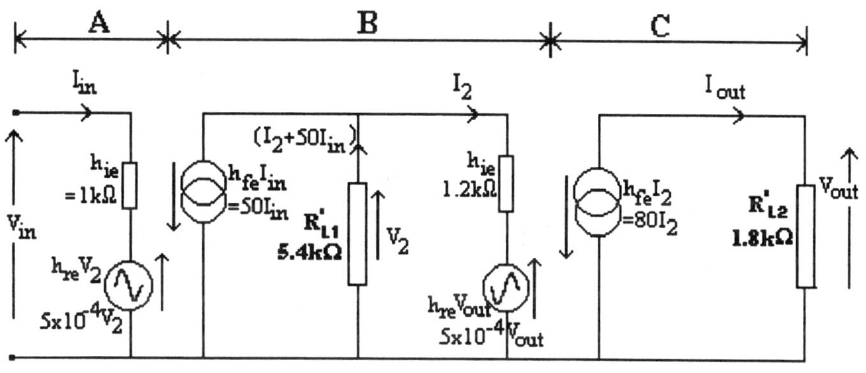

Fig. 4.11-1

Procedure 3: Obtain equations for I_2 and V_2 and for I_{in} and V_{in}

Refer to Fig. 4.11-1

(Section **A**) $\quad V_{in} = 1.0I_{in} + (5 \times 10^{-4}V_2)$ .....(Eq. 4.11-0)

(Section **B**) $\quad V_2 = -(I_2 + 50I_{in})5.4$ ....(Eq. 4.11-1)

$-(I_2 + 50I_{in})5.4 = 5 \times 10^{-4}V_{out} + 1.2I_2$ .... (Eq. 4.11-2)

(Section **C**) $\quad V_{out} = -80I_2 \times 1.8$

$$\mathbf{I_2} = -\mathbf{V_{out}/144}$$

(contd) - *Procedure (3)*

Substituting for I_2 in Eq. 4.11-1

$$V_2 = -(-V_{out}/144 + 50I_{in})5.4$$

$$= (5.4V_{out}/144) - 270I_{in}$$

$$= 375 \times 10^{-4}V_{out} - 270I_{in} \quad \text{..(Eq. 4.11-3)}$$

Substituting for I_2 also in Eq. 4.11-2

$$-[(-V_{out}/144) + 50I_{in}]5.4 = (5 \times 10^{-4}V_{out}) - (1.2\ V_{out}/144)$$

$$375 \times 10^{-4}V_{out} - 270I_{in} = (5 \times 10^{-4}V_{out}) - (83 \times 10^{-4}\ V_{out})$$

$$\mathbf{I_{in}} = (453 \times 10^{-4}V_{out})/270$$

$$= \mathbf{1.678 \times 10^{-4}\ V_{out}} \quad \text{...(Eq. 4.11-4)}$$

Substituting for I_{in} in Eq. 4.11-3

$$\mathbf{V_2} = (375 \times 10^{-4}V_{out}) - [270(1.678 \times 10^{-4}V_{out})]$$

$$= \mathbf{-78 \times 10^{-4}V_{out}} \quad \text{...(Eq. 4.11-5)}$$

Procedure 4: Use the equations to determine A_v, A_i, and A_p

Substituting Eqs .4.11-4 and 4.11-5 into Eq. 4.11-0

$$V_{in} = 1.0I_{in} + (5 \times 10^{-4}V_2) \quad \textbf{(Eq. 4.11-0 recalled)}$$

$$= (1.0 \times 1.678 \times 10^{-4}V_{out}) + [5 \times 10^{-4} \times \ldots (-78 \times 10^{-4})V_{out}]$$

$$= 1.678 \times 10^{-4}V_{out} - 3.9 \times 10^{-6}V_{out}$$

$$= 1.678 \times 10^{-4}V_{out} - 0.039 \times 10^{-4}V_{out}$$

$$\therefore \quad \mathbf{V_{in}} = \mathbf{1.639 \times 10^{-4}V_{out}}$$

(i) *Voltage gain - A_v*

$$\mathbf{A_v} = \mathbf{V_{out}/V_{in}}$$

$$= 1/(1.639 \times 10^{-4}) = \underline{\mathbf{6101}}\ \mathbf{(\approx 76dB)}$$

(contd)

(ii) *Current gain - A_i*

Refer to Fig. 4.11-0 Section **C**

$$V_{out} = I_{out} \times R_{L2}$$

$$\therefore \quad I_{out} = V_{out}/R_{L2}$$

$$= V_{out}/2.0 \quad \ldots\ldots\ldots\ldots\ldots\ldots \text{(Eq. 4.11-6)}$$

$$A_i = I_{out}/I_{in} \quad \text{i.e., (Eq .4.11-6)/(Eq. 4.11-4)}$$

$$= (V_{out}/2.0) \div (453 \times 10^{-4} V_{out}/270)$$

$$= (V_{out}/2.0) \times \frac{270 \times 10^4}{453 V_{out}}$$

$$\therefore \quad \mathbf{A_i} = \frac{135 \times 10^4}{453} = \underline{\mathbf{2980}} \ (\approx \mathbf{69dB})$$

(iii) *Power gain - A_p*

$$\mathbf{A_p} = A_i \times A_v$$

$$= 2980 \times 6101 \approx \underline{\mathbf{18 \times 10^6}} \ (\approx \mathbf{73dB})$$

CHAPTER 5

SMALL-SIGNAL AMPLIFIERS AND STABILIZATION

Introduction

The primary function of an amplifier is to increase the output level by amplifying the low-level input signal voltage or current. The amplified output is referred to as the amplifier ***gain***. The gain of the amplifier is the ratio of the amplified signal to the low-level input signal.

Small-signal amplifiers are restricted to signals sufficiently small to ensure that the region of operation lies on the linear portion of the transistor parametric value. Hence, **Class A** is assigned, under Class A operation the bias at the **Q** point (operating point) is so set that collector current flows at all times. The output waveform is a magnified mirror image of the input waveform.

The transistor, being an active device, has been popularized by its attributes to amplify small-signal-voltage or current, especially in low-level circuits. However, it has its disadvantages, which can only be overriden by suitable circuit designs. Its main disadvantage is its inability to handle high operating temperature efficiently without going haywire. Temperature influences the operation of a transistor and must be a determinant in the design of a transistor amplifier.

The **stabilization** of the operating point is vital with transistors. The variation of leakage current with temperature will result in alteration of input and output impedances, shifting of the operating point, and imminent thermal runaway.

Consequently, the transistor circuit must be biased by reliable circuitry which will prevent excessive shifting of the operating point. Freedom from distortion is therefore, the design criterion.

Generally, the main function of stabilization is to prevent or control the variation of leakage current with rise in temperature.

Example 5.1

(a) *The basic bipolar transistor bias circuit is shown in Fig.5.1-0 in which the quiescent bias current I_B is derived from the supply voltage through resistor R_B. Assuming Class A operation, estimate the value of R_B given that V_{cc} = 20 V, the quiescent collector voltage and current are 10 V and 10 mA respectively and V_{BE} = 0.5 V. h_{FE} = 50.*

(b) *What would be the value of R_B if the quiescent collector voltage is 12 V? Assume that the value of R_L is unchanged.*

(c) *What is the main disadvantage of this type of biasing?*

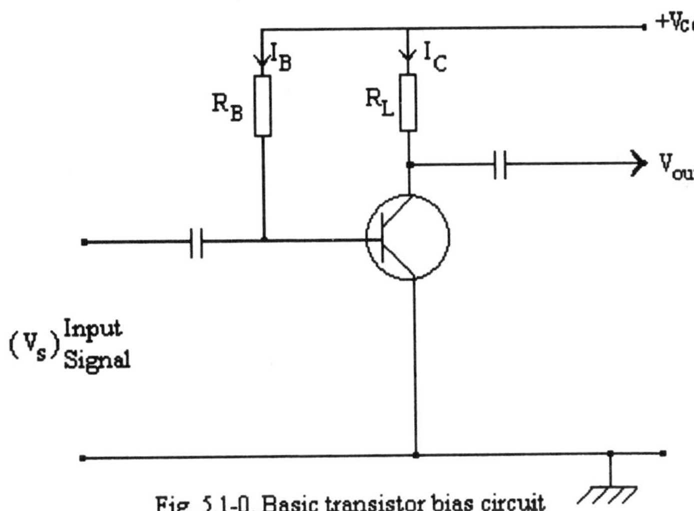

Fig. 5.1-0. Basic transistor bias circuit

Solution

Refer to Fig. 5.1-0

(a) *Estimated Value of R_B*

$$I_B = (V_{CC} - V_{BE})/R_B$$

$$I_C = h_{FE}\, I_B$$
$$= h_{FE}(V_{CC} - V_{BE})/R_B \qquad \text{(Eq.5.1-0)}$$

With $V_s = 0$

Collector Q voltage, $V_C = V_{CC}/2$ (for Class A operation)

$\therefore \quad I_C R_L = V_{CC}/2$

$I_C = V_{CC}/2R_L$ .... (i)

But from Eq.5.1-0 $\quad I_C = h_{FE}(V_{CC} - V_{BE})/R_B$ ...... (ii)

Equating (i) and (ii), $V_{CC}/2R_L = h_{FE}(V_{CC} - V_{BE})/R_B$ .... (iii)

$$\therefore \quad R_B = [h_{FE}(V_{CC} - V_{BE})/V_{CC}] \times 2R_L \qquad \text{(Eq.5.1-1)}$$

Given that $V_C = 10$ V

and $I_C = 10$ mA

then $R_L = V_C/I_C$

$= 10/(10 \times 10^{-3}) \quad = \mathbf{\underline{1.0\ k\Omega}}$

Substituting for R_L in Eq.5.1-1

$$R_B = \frac{50\,(20 - 0.5) \times 2 \times 10^3}{20}$$

$= 97.5 \times 10^3 \quad = \mathbf{\underline{97.5\ k\Omega}}$

(contd)

Note: If it is assumed that $V_{BE} << V_{CC}$ and on the basis of this assumption V_{BE} is neglected, then from Eq.5.1-1

$$R_B = 2h_{FE} R_L$$

$$= 2 \times 50 \times 10^3 \qquad = \underline{\mathbf{100\ k\Omega}}$$

The difference between $R_B = 100k\Omega$ and $R_B = 97.5\ k\Omega$ is $+2.5\ k\Omega$ which is about 2.56%.
This percentage error can be safely tolerated and hence V_{BE} may be neglected with no adverse effect depending upon the design limitation.

Alternative Method to Calculate the Value R_B

$$h_{FE} = I_c/I_B = 50$$

$$\therefore \quad I_B = I_c/h_{FE} = 10/50 = \underline{0.2mA}$$

(Omit V_{BE}) $\quad R_B = V_{CC}/I_B = 20\ V/0.2\ mA$

$$= \underline{100k\Omega}$$

(Admit V_{BE}) $\quad R_B = (V_{CC} - V_{BE})/I_B$

$$= (20 - 0.5)V/0.2mA = \underline{97.5\ k\Omega}$$

(b) *Q point - Collector Voltage* = *12 V*
New Value of R_B

$$V_c = 12V$$

$$R_L = 1.0\ k\Omega \text{ (unchanged)}$$

$$\therefore \quad I_c = (V_{CC} - V_c)/R_L$$

$$= (20 - 12)/10^3 = 8mA$$

Base current, $I_B = I_c/h_{FE}$

$$= (8 \times 10^{-3})/50 = 0.16\ mA$$

(contd) ***(b)***

(V_{BE} ignored)	R_B	$= V_{cc}/I_B$
		$= 20/(0.16 \times 10^{-3}) \quad = \underline{\mathbf{125k\Omega}}$
(V_{BE} included)	R_B	$= (V_{CC} - V_{BE})/I_B$
		$= (20 - 0.5)\ V/0.16\ mA$
		$= \underline{\mathbf{121.9\ k\Omega}}$

(c) The main disadvantage of this simple bias circuit is that it does not compensate for temperature changes in the transistor parameters.

The net effect of temperature change on the collector voltage is the shifting of the operating point (Q point) due to slow variation of the collector voltage, known as *"drift"*.

Shifting of the quiescent point leads to changes in the amplifier constants.

In a power amplifier the cumulative effect of temperature rise is known as "*thermal runaway*" and may result in the damage of the transistor.

Example 5.2

A transistor connected in the common-emitter mode has the data given in Table 5.2-0.

(a) *Plot the output characteristics of the transistor and draw the load line for a collector load resistance of 2.5 kΩ*

Using the load line, estimate

(b) *the total power dissipated in the circuit under steady collector voltage and current (quiescent condition) if the base bias current is 50μA and the collector supply voltage is 10 V*

(c) *the amplifier current gain*

(d) *the a.c. input resistance if the voltage gain is 100*

(e) *the effect on the power and current gain when the load resistance is reduced to 1.0 kΩ.*

Table 5.2-0

V_{CE}	I_C = Collector current (mA)		
	$I_B = -20\ \mu A$	$I_B = -40\ \mu A$	$I_B = -60\ \mu A$
-2 V	-0.9	-1.8	-2.8
-8 V	-1.5	-2.55	-3.85

Solution

(a) *The load line*

The Load Line equation is

$$V_{CC} = V_{CE} + I_C R_L$$

This equation is in the form of y = mx + c and is therefore a straight line.

The load line is superimposed on the output characteristics.

(contd)

(a) *The Load Line*

The two points for the load line on the output characteristics can best be found in the following manner.

Refer to Fig. 5.2-0

Point. 1			**Point. 2**		
Let	I_C	$= 0$	Let	V_{CE}	$= 0$
Then	V_{CC}	$= V_{CE}$	Then	I_C	$= V_{CC}/R_L$
		$= 10$ V			$= 10/2.5 \times 10^3$
					$= 4.0$ mA

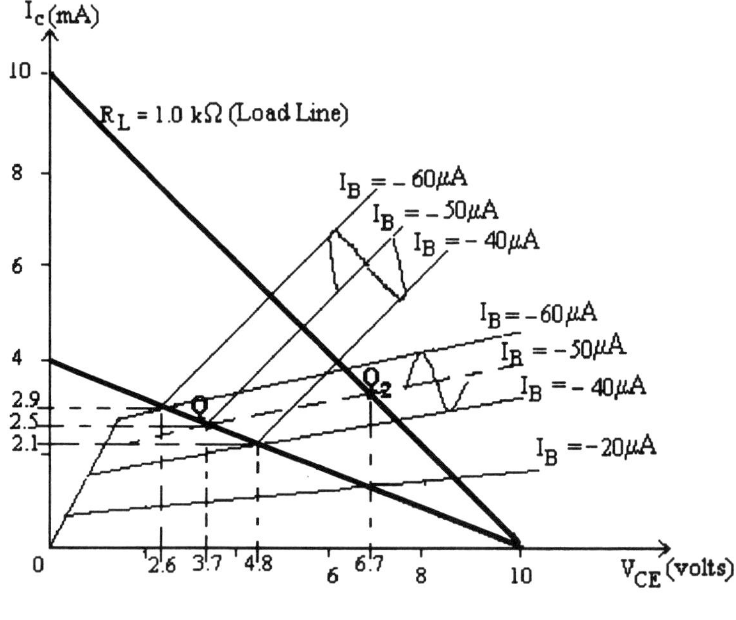

Fig. 5.2-0

The characteristic for a base current of -50 μA is drawn by taking the average value for $I_B = -40$ μA and $I_B = -60$ μA.

This characteristic value is shown in Table 5.2-1.

Table 5.2-1

$I_B = -50$ μA	
V_{CE}	I_c
- 2 V	2.5 mA
- 10 V	3.4 mA

(contd) *Total power dissipated*

(b) Refer to Fig. 5.2-0

The quiescent(Q) collector current is

$$I_Q = 2.5 \text{ mA}$$

$\therefore$ Power dissipated $= V_{CC} \times I_Q = 10 \times 2.5 \times 10^{-3} =$ **25 mW**

(c) *Current gain, A_i*

Refer to the Load Line on Fig. 5.2-0

The base current swings from 60 μA to 40 μA. That is, $I_B = 20$ μA.
For the same base current swing, the collector current swings from 2.9 mA to 2.1 mA.

$$I_c = 0.8 \text{ mA}$$

$$A_i = I_{out}/I_{in}$$

$$= \frac{0.8 \times 10^{-3}}{20 \times 10^{-6}} = \mathbf{40}$$

(d) *Input resistance - R_{in}*

Given $A_v = 100$

$$A_v = V_{out}/V_{in} = I_{out} R_L / I_{in} R_{in}$$

$$= A_i \times R_L/R_{in}$$

$$\therefore \quad R_{in} = (A_i R_L)/A_v = (40 \times 2.5 \times 10^3)/100$$

$$= 100 \times 10^3/100 = \mathbf{1.0\ k\Omega}$$

(e) *Load resistance reduced to 1.0 kΩ: effect on A_p and A_i*

It can be clearly seen in Fig. 5.2-0 that the load line for $R_L = 1.0$ kΩ has shifted the operating point (Q point) to the position Q_2 on the -50 μA base bias characteristic. As a result the power dissipation and current gain will increase.

$$I_{Q2} = 3.0 \text{ mA (i.e., collector current } I_c \text{ at } Q_2)$$

Power dissipation, $A_p = V_{cc} I_{Q2}$

$$= 10 \times 3 \times 10^{-3} = \mathbf{30\ mW}$$

The collector current swings between 3.7 mA and 2.3 mA

$$\therefore \quad I_{out} = I_c = 1.4 \text{ mA [i.e., } (3.7 - 2.3) \text{ mA]}$$

and $A_i = I_{out}/I_{in} = 1.4 \times 10^{-3}/20 \times 10^{-6} =$ **70**

Example 5.3

(a) *An elementary n.p.n transistor current amplifier is shown in Fig. 5.3-0 with its associated output characteristics and load line in Fig. 5.3-1. The Q point is at $I_B = 0.2$ mA and $V_{CE} = 4.8$ V; the supply voltage $V_{cc} = 10$ V and $V_{BB} = 3$ V. Determine the value of R_L and R_B assuming that V_{BE} is approximately 0.7 V.*

(b) *What would be the current gain when the signal current is 0.1 mA?*

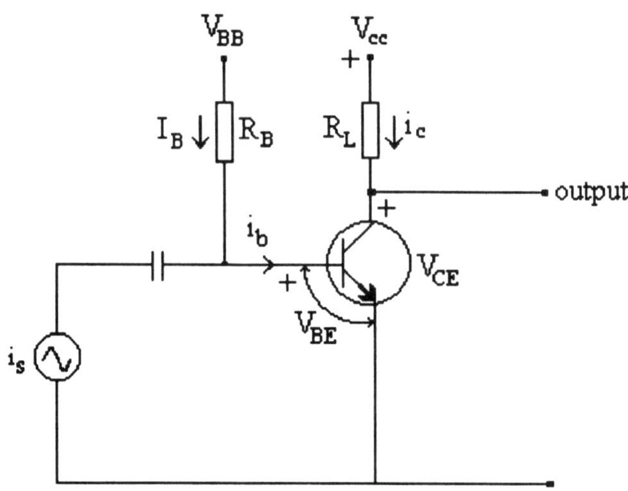

Fig. 5.3-0. Elementary transistor amplifier

(a) *For no-signal input*:

$$i_s = 0$$

$$i_b = I_B \text{ (the quiescent value)}$$

$$V_{BB} = I_B R_B + V_{BE}$$

$$V_{BE} \approx 0.7\text{V}$$

$$\therefore \quad R_B = (V_{BB} - V_{BE})/I_B$$

$$= \frac{3 - 0.7}{0.2 \times 10^{-3}} = \underline{\mathbf{11.5\ k\Omega}}$$

$$R_L = (V_{CC} - V_{CE})/\ I_c$$

$$= \frac{10 - 4.8}{10.4 \times 10^{-3}} = \underline{\mathbf{500\ \Omega}}$$

(contd)

(b) *Signal of 0.1 mA applied*

When signal is applied the base current becomes

$$i_b = I_B + i_s$$

That is, an a.c. component superimposes on a d.c. component.

<u>Refer to Fig. 5.3-1</u>

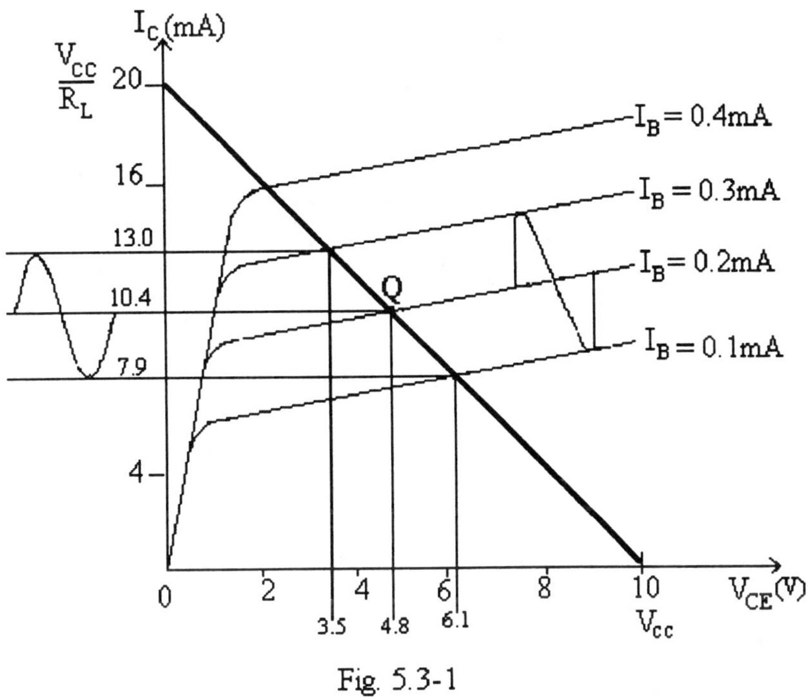

Fig. 5.3-1

A signal current of 0.1 mA causes the base current to swing from 0.2 mA to 0.3 mA, that is, the base current swings 0.1 mA. Simultaneously, the collector current swings from 10.4 mA to 13.0 mA, that is, 2.6 mA.

$\therefore$ $i_b = 0.1$ mA

and $i_c = 2.6$ mA

The current gain, $A_i = {}^{i_c}/_{i_b}$ $({}^{I_{out}}/_{I_{in}})$

$= {}^{2.6}/_{0.1}$ $=$ **<u>26</u>**

Example 5.4

(a) *The circuit shown in Fig. 5.4-0 is that of a transistor amplifier with a fixed current bias. Estimate the values of R_c and R_B if the mean (quiescent) d.c. collector current and voltage are 1.0 mA (I_Q) and 10 V(V_Q) respectively.*
The transistor has an h_{FE} of 100 and V_{BE} = 0.7V.

(b) *To improve the d.c. stabilization of the circuit, the bias to the base is to be obtained by returning the bias resistor to the collector. Draw a circuit diagram showing how this can be accomplished without introducing unwanted a.c. feedback and calculate suitable values for the components required.*

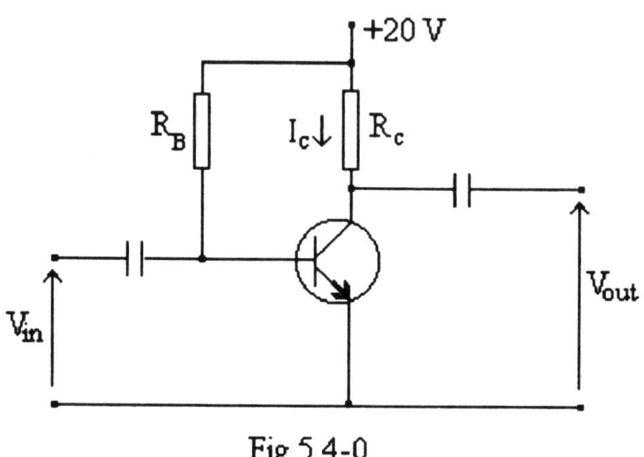

Fig.5.4-0

Solution

(a)

Voltage drop across R_c $= V_{CC} - V_Q$

$= 20 - 10 = \underline{10\ V}$

Resistance of R_c $= (V_{CC} - V_Q)/I_c$

$= (20 - 10)/(1.0 \times 10^{-3})$

$= 10 \times 10^3 = \underline{\mathbf{10\ k\Omega}}$

Voltage drop across R_B $= V_{CC} - V_{BE}$

$= 20 - 0.7 = \underline{19.3\ V}$

Resistance of R_B $= 19.3/I_B$

$I_B = I_c/h_{FE} = (1.0 \times 10^{-3})/100$

$= 10 \times 10^{-6}$A **or** 10 μA

$\therefore$ $R_B = 19.3/(10 \times 10^{-6})$

$= \underline{\mathbf{1.93\ M\Omega}}$

(contd) *(a)*

If for simplicity it is assumed that V_{BE} is negligible and can be omitted, then

$$R_B = V_{cc}/I_B = \mathbf{2\ M\Omega}$$

The fixed current bias resistor in Fig. 5.4-0 does not compensate for the effects of temperature change in the transistor parameters. In particular, the parameter h_{FE} or β changes with temperature.

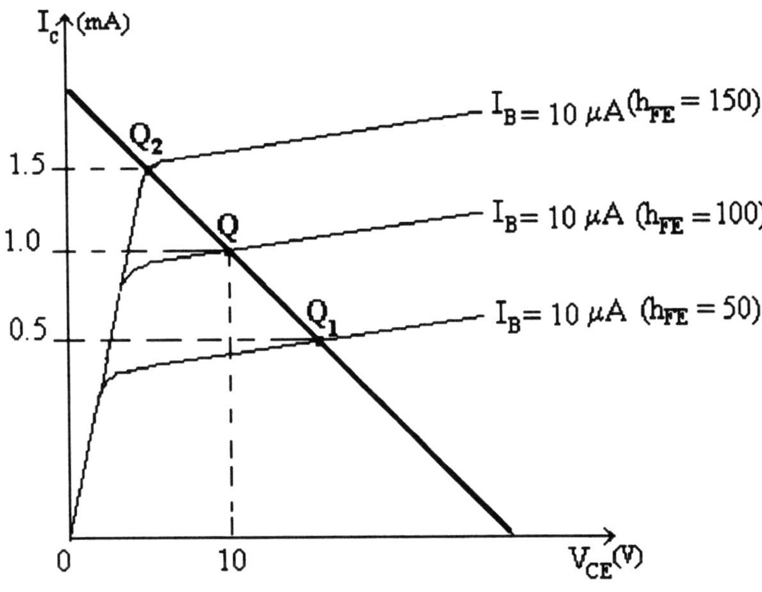

Fig. 5.4-1

Refer to Fig. 5.4-1

It is assumed that h_{FE} varies from 50 to 150 due to temperature variation. The fixed quiescent point (**Q**) is set at

$$I_c = 1.0\ \text{mA}$$
$$V_{CE} = 10\ \text{V}$$
and
$$I_B = 10\ \mu\text{A}$$

When the transistor h_{FE} increases, the operating point of the amplifier shifts upward along the load line, from **Q** to **Q_2**. As a result the collector current shifts from 1.0 mA to 1.5 mA; that is

$$I_c = h_{FE} \times I_B$$
$$= 150 \times 10 \times 10^{-6} = 1.5\ \text{mA}$$

(contd) *(a)*

When h_{FE} of the transistor decreases, the operating point of the amplifier shifts downward along the load line, from **Q** to **Q_1**.
As a result, the collector current shifts from 1.0 mA to 0.5 mA.

Consequently, because of temperature variation, what would be an acceptable signal swing at **Q** could possibly produce a cut-off at **Q_1** and saturation at **Q_2**.

(b) *Improving d.c. stabilization*

Thermal stabilization is improved with the circuit of Fig. 5.4-2. This circuit uses collector-voltage feedback.

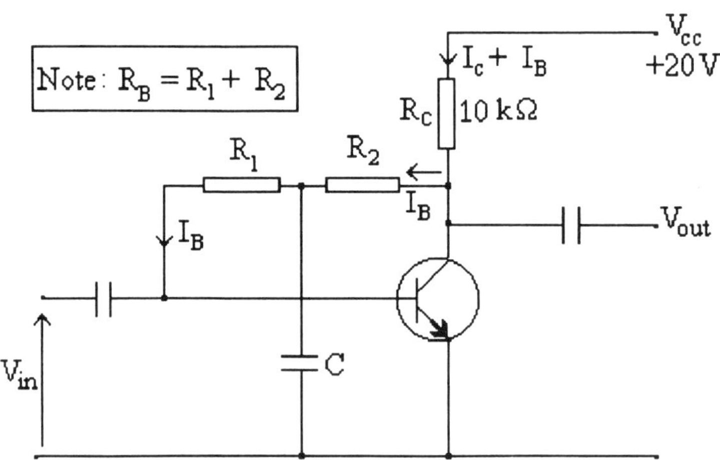

Fig. 5.4-2. Collector-voltage feedback bias

The values of I_B and I_c are the same as those in Fig. 5.4-0; that is,

$$I_B = 10\ \mu A$$
$$I_c = 1.0\ mA$$
$$V_{BE} = 0.7\ V$$

Now
$$V_{CE} = V_{CC} - \text{voltage drop across } R_c$$
$$= V_{CC} - [(I_c + I_B) \times 10 \times 10^3]$$
$$= 20 - (1.0 + 0.01) \times 10$$
$$= 20 - 10.1 \qquad = \underline{9.9\ V}$$

(contd) *(b)*

Voltage drop across $R_B = V_{CE} - V_{BE}$
$= 9.9 - 0.7 = \underline{9.2\text{ V}}$

$\therefore$ Resistance of $R_B = 9.2/I_B$
$= 9.2/(10 \times 10^{-6}) = \underline{920\text{ k}\Omega}$

In this example R_1 and R_2 will be made equal; that is

$R_1 = 920/2 = \underline{460\text{ k}\Omega}$

$R_2 = 920/2 = \underline{460\text{ k}\Omega}$

The reason for making $R_1 = R_2$ may be summarized as follows:
If R_1 is made small, it will reduce the input resistance of the amplifier. The decoupling causes R_1 to shunt the amplifier input.

Similarly, if R_2 is made small, it will reduce the collector load resistance which it effectively shunts by the decoupling.

The capacitor C decouples the junction of R_1 and R_2, thus limiting the effect of a.c. feedback by providing a low reactance path to ground. Capacitor C is chosen so that its reactance at the lowest signal frequency (i.e., 50 Hz) is small compared with the resistance of R_1 and R_2.

Further improvement in thermal stability is achieved by the use of the potential divider network and emitter resistor bias circuit.

Note: In the preceding calculation of R_B, if it is assumed that

$$V_{BE} \ll V_{CE}$$

Then the voltage drop across $R_B = V_{CE} \approx 9.9\text{ V}$

and $R_1 = R_2 = \underline{495\text{ k}\Omega}$

This value is acceptable depending upon the design limitation.

Example 5.4

Compute the values of the quiescent collector current and voltage of the transistor amplifier shown in Fig. 5.5-0 given that

$$R_C = 4.7k\,\Omega$$
$$R_1 + R_2 = 560k\,\Omega$$
$$V_{CC} = 10\ V$$
$$h_{FE} = 100$$

Derive any formula used.

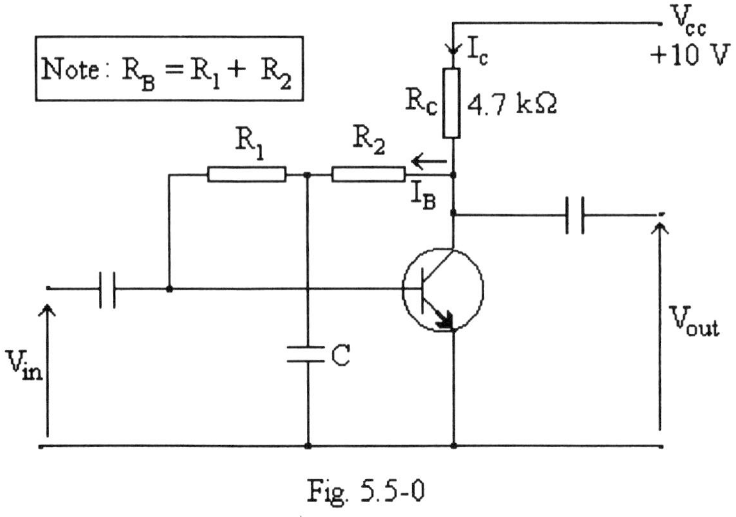

Fig. 5.5-0

Solution

[Note the use of the term $\mathbf{V_c}$ (collector voltage) which is the same as $\mathbf{V_{CE}}$.]

<u>Refer to the circuit of Fig. 5.5-0</u>

$$I_C = h_{FE}\, I_B$$
$$= h_{FE} V_C / R_B$$
$$V_C = V_{CC} - I_C R_L$$
$$= V_{CC} - (h_{FE} V_C R_L / R_B)$$

Now $\quad V_C\,[\,1 + (h_{FE} R_L / R_B)] = V_{CC}$

$\therefore \quad (V_{CC}/V_C) - 1 = h_{FE} R_L / R_B$

$$= 100 \times [4.7/560] \qquad = 0.84$$

$\therefore \quad V_{CC} = 1.84 V_C$

and $\quad V_C = 10/1.84 \qquad = \mathbf{5.43V}$

Now $\quad I_C = (V_{CC} - V_C)/R_C$

$$= (10 - 5.43)/(4.7 \times 10^3) \qquad = \mathbf{0.97mA}$$

Example 5.6

(a) *Briefly explain the effect of leakage current in a transistor amplifier and define stability* ***S*** *and temperature stability* ***k****.*

(b) *A silicon n.p.n transistor with an* $h_{FE} = 99$ *and a leakage current* $I_{CBO} = 10^{-11}$ *A is connected as shown in Fig. 5.6-3, estimate the values of* I_C, I_E *and* V_{CE}.

(c) *Determine the stability factor of the circuit shown in Fig. 5.6-4 in which* $R_E = 1.0\ k\Omega$, $R_1 = 82\ k\Omega$, $R_2 = 16\ k\Omega$ *and* $h_{FE} = 99$.

Solution

(a) If the temperature of a transistor increases, the leakage current increases, thereby changing the values of I_C, V_C and I_B.
In the common-base (CB) circuit, I_{CBO} is defined as collector-to-base leakage current that flows in the transistor with the emitter open-circuited (collector-to-base cut-off current).

In the common-emitter (CE) circuit, I_{CBO} is now a current in the base; this leakage current is amplified by h_{FE} to give a value of $h_{FE}I_{CBO}$ in both the collector and the base, known as I_{CEO}.

The total leakage current in the collector is the amplified leakage current $h_{FE}I_{CBO}$ plus the initial current I_{CBO}.

The two leakage currents I_{CEO} and I_{CBO} are related as follows:

$$I_{CEO} = I_{CBO} + h_{FE}I_{CBO}$$

$$= (1 + h_{FE})I_{CBO} \qquad \text{(Eq. 5.6-0)}$$

The *stability* is defined in terms of the in the leakage current I_{CBO} and the corresponding change in I_C.
The relationship is given as

$$S = \delta I_C / \delta I_{CBO}$$

where S is the stability factor.

(contd) *(a)*

Refer to Fig. 5.6-0

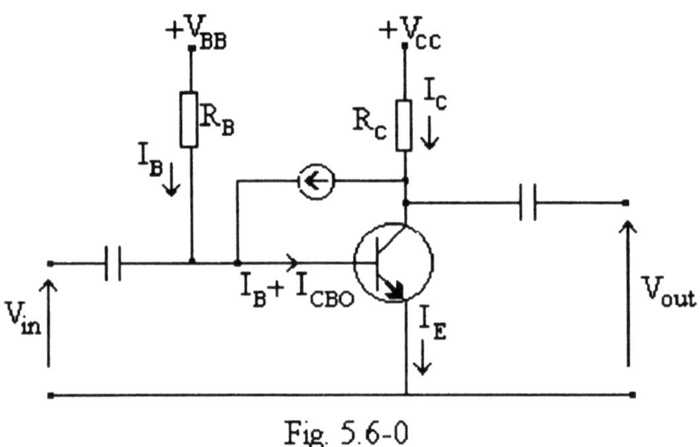

Fig. 5.6-0

The effective base current may be expressed as

$$I_{B\,Effective} = I_B + I_{CBO}$$

$$= \frac{V_{BB} - V_{BE}}{R_B} + I_{CBO} \quad \cdots \quad \text{(Eq. 5.6-1)}$$

$$\therefore \quad I_C = h_{FE}\left[\frac{V_{BB} - V_{BE}}{R_B} + I_{CBO}\right]$$

$$= h_{FE}I_B + I_{CEO} \qquad \text{(Eq. 5.6-2)}$$

A close examination of the collector current, I_C shows that it has two separate components; that is

$$I_C = \text{a d.c component + the leakage current}$$

$$= h_{FE}\left[\frac{V_{BB} - V_{BE}}{R_B} + I_{CBO}\right] + I_{CEO}$$

From Eq. 5.6-0 $I_{CEO} = (1 + h_{FE})\, I_{CBO}$

(contd) *(a)*

$$I_C = h_{FE}\left[\frac{V_{BB}-V_{BE}}{R_B} + I_{CBO}\right] + [1 + h_{FE}]\, I_{CBO} \quad \cdots \text{(Eq. 5.6-3)}$$

When I_{CBO} increases by δI_{CBO}, I_C will increase by δI_C

$$\therefore I_C + \delta I_C = h_{FE}\left[\frac{V_{BB}-V_{BE}}{R_B} + I_{CBO}\right] + (1 + h_{FE})(I_{CBO} + \delta I_{CBO})$$

Subtracting Eq.5.6-3 from the above expression, we get

$$\delta I_C = (1 + h_{FE})\, \delta I_{CBO} \quad \ldots\ldots \quad \text{(Eq. 5.6-4)}$$

The stability factor S is given as

$$S = \delta I_C / \delta I_{CBO} \qquad [\text{where } 1 \le S \le (1 + h_{FE})]$$

Then by definition (refer to Eq. 5.6-4).

$$S = \delta I_C / \delta I_{CBO}$$

$$= 1 + h_{FE} \quad \ldots\ldots\ldots\ldots\ldots\ldots \quad \text{(Eq. 5.6-5)}$$

The general formula to evaluate S is

$$S = \frac{\text{sum of resistors in which } I_B \text{ flows (say } R_B)}{\text{same sum of resistors } R_B + (1 + h_{FE})}$$

$$= \frac{R_B}{R_B/(1 + h_{FE})}$$

$$= 1 + h_{FE}$$ (This substantiates the results of Eq. 5.6-5)

(contd) *(a)*

The thermal stability factor k relates the change in I_C to the change in I_{CEO}, given as

$$k = \delta I_C / \delta I_{CEO}$$
$$= S/(1 + h_{FE})$$

The effectiveness by which the change in I_C due to temperature rise can be controlled may be expressed by the stability factor S or the thermal stability factor k. Summarizing

$$S = \delta I_C / \delta I_{CBO} \quad \text{and} \quad k = \delta I_C / \delta I_{CEO}$$

The worst case Thermal Stability occurs when the emitter resistor $R_E = 0$ (that is, no emitter resistor is used).

Refer to Fig. 5.6-1

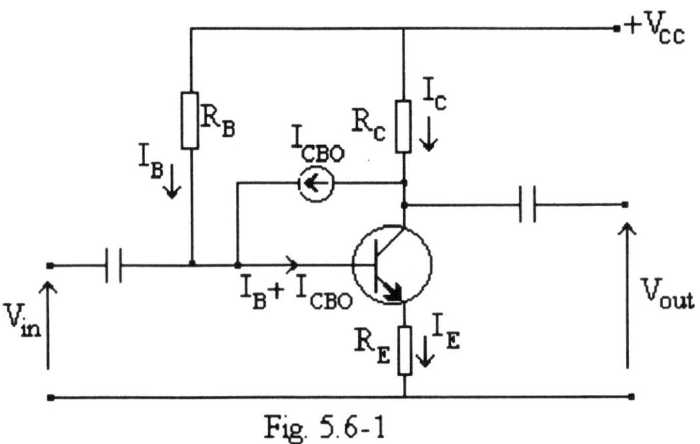

Fig. 5.6-1

The circuit has an emitter resistor R_E added so that

$$S = \frac{R_E + R_B}{R_E + \dfrac{R_B}{1 + h_{FE}}} \quad \cdots\cdots\cdots \quad \text{(Eq. 5.6-6)}$$

(contd) *(a)*

<u>Refer to Fig. 5.6-2</u>

The voltage divider base bias circuit is employed.

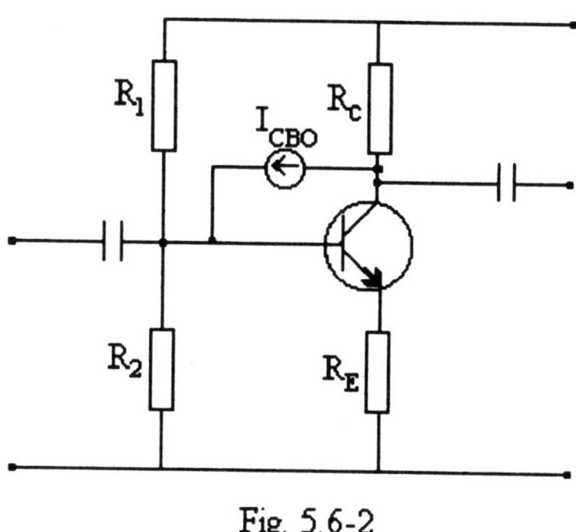

Fig. 5.6-2

With this type of bias circuit, the thermal stability is

$$S = \frac{R_E + R_B'}{R_E + \dfrac{R_B'}{1 + h_{FE}}}$$

where $$R_B' = \frac{R_1 R_2}{R_1 + R_2}$$

(contd)

(b) Refer to Fig. 5.6-3

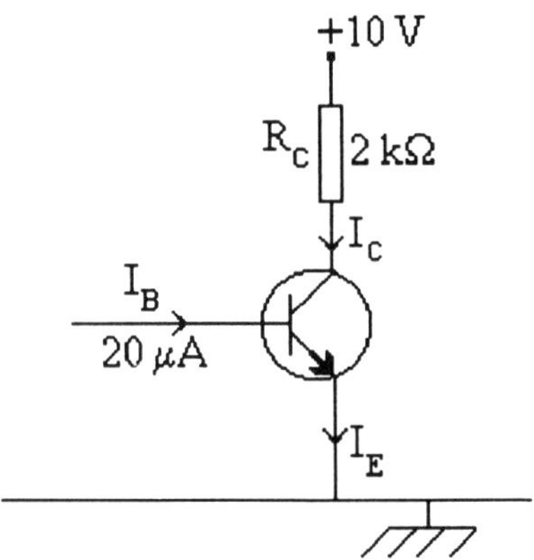

Fig. 5.6-3

The collector cut-off current is

$$I_{CEO} = (1 + h_{FE})\, I_{CBO} \quad \text{(refer to Eq. 5.6-0)}$$

$$= (1 + 99) \times 10^{-11}$$

$$= 100 \times 10^{-11} \text{ A}$$

The collector current, $I_C = h_{FE}\, I_B + I_{CEO}$

$$= 99 \times 20 \times 10^{-6} + 100 \times 10^{-11}$$

Note that I_{CEO} is a negligibly small part of I_C.

$\therefore \quad I_C = 99 \times 20 \times 10^{-6}$ = **1.98 mA**

The emitter current, $I_E = I_B + I_C$

$= (20 + 1980) \times 10^{-6}$ = **2.0 mA**

Alternatively $I_E = h_{FE} I_B + I_B$

$$= (h_{FE} + 1)\, I_B$$

$= (99 + 1) \times 20 \times 10^{-6}$ = 2.0 mA

The collector-emitter voltage is

$$V_{CE} = V_{CC} - I_C R_C$$

Assuming $I_C = I_E$ $\quad V_{CE} = 10 - (2 \times 10^{-3} \times 2 \times 10^{3})$ = **6 V**

(contd)

(c)

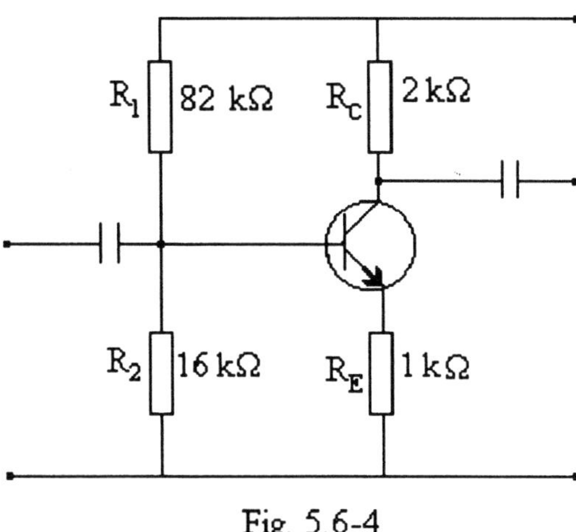

Fig. 5.6-4

This circuit employs the voltage divider base bias; therefore the stability factor of the circuit is as follows:

$$S = \frac{R_E + R_B'}{R_E + R_B'/(1 + h_{FE})}$$

$$R_B' = \frac{82 \text{ x } 16}{82 + 16} = 13.39 \text{ k}\Omega$$

$$S = \frac{1.0 + 13.39}{1.0 + \dfrac{13.39}{1 + 99}} = \underline{\mathbf{12.69}}$$

Example 5.7

(a) *The circuit shown in Fig. 5.7-0 employs a voltage divider circuit commonly used in common-emitter amplifiers for base current bias and operating point stabilization. Briefly explain how stabilization is achieved in this circuit.*

(b) *The common-emitter amplifier in Fig. 5.7-0 has $R_1 = 33\ k\Omega$, $R_2 = 5\ k\Omega$, $R_E = 2.2\ k\Omega$ and $h_{FE} = 99$. Determine the stability factor S of the circuit.*

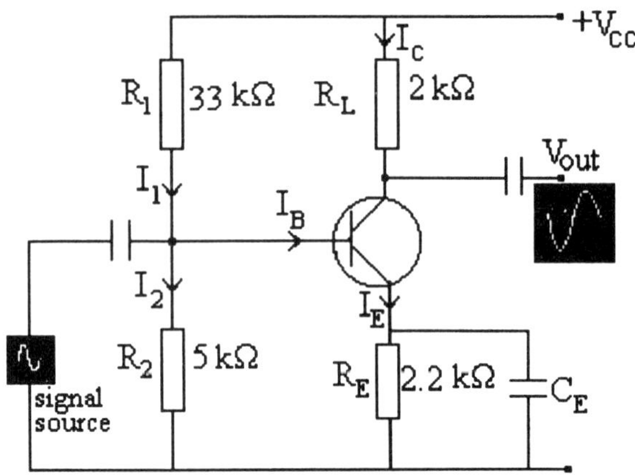

Fig. 5.7-0. Voltage divider base bias

Solution

(a) The resistors R_1 and R_2 provide and fix the base potential.

The capacitor C_E provides a low-reactance path for the a.c. signal (that is, short-circuits the emitter resistor R_E to a.c.) to ground so as to prevent reduction of the signal gain.
Any increase in I_{CBO} will cause an increase in I_c and consequently in I_E, and as a result a chain reaction will follow. Thus

An increase in I_E will cause the emitter potential to increase.
An increase in emitter potential follows a decrease in base-emitter voltage, V_{BE}. A decrease in V_{BE} will reduce the base current, I_B; a reduced I_B will cause I_c to fall toward its original value. Hence by maintaining the base potential constant or nearly constant, stabilization is achieved against any increase in I_{CBO}.

(contd)

(b)

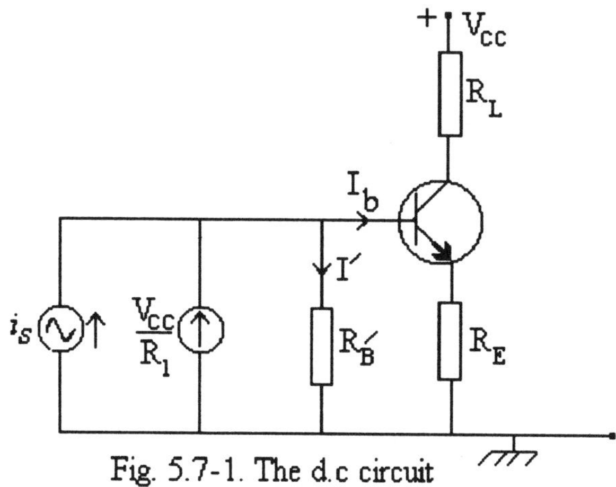

Fig. 5.7-1. The d.c circuit

Refer to Fig. 5.7-1

Stability factor, S $= \delta I_c / \delta I_{CBO}$

$$= \frac{R_E + R'_B}{R_E + \dfrac{R'_B}{1 + h_{FE}}}$$

$$R'_B = \frac{R_1 R_2}{R_1 + R_2}$$

$$= \frac{33 \times 5}{33 + 5} \qquad = \underline{4.3\ k\Omega}$$

$$\therefore \qquad S = \frac{(2.2 + 4.3) \times 10^3}{[2.2 + \dfrac{4.3}{1 + 99}] \times 10^3}$$

$$= \frac{6.5}{(2.2 + 0.043)} \qquad = \underline{\mathbf{2.90}}$$

Note: The stability factor 'S' is a measure of the d.c. stability of the circuit when leakage current changes.
In this example, if the leakage current increases by 1.0 μA, the collector current would increase by 2.90 μA. Hence the lower the value of S the better the stability.

Example 5.8

(a) *Sketch the gain/frequency characteristic for a typical R.C-coupled C.E amplifier and hence explain the term bandwidth.*

(b) *The transistor amplifier shown in Fig. 5.8-0 has an a.c. current gain h_{fe} of 99 and an a.c. input resistance of 800 Ω. It is connected to a second stage whose input resistance is 1.2 kΩ. Determine the current flowing into the second stage assuming negligible reactance of all capacitors at the operating frequency. If the stage-coupling capacitor C_C is the main component responsible for low-frequency cut-off, calculate its value such that the lower 3 dB point is 50 Hz.*

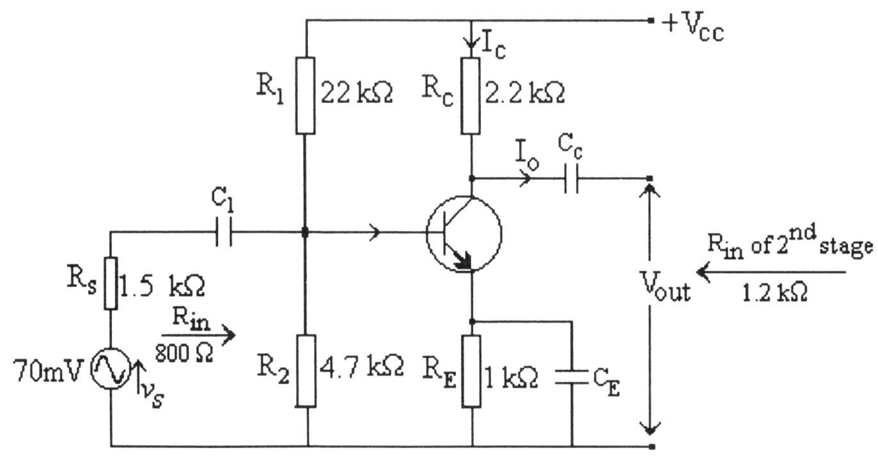

Fig. 5.8-0

Note: The input resistance of the second stage is connected to the output of the first stage; that is, R_{in} of the second stage becomes the load resistance R_L of the first stage. Figure 5.8-1 is a simulated diagram.

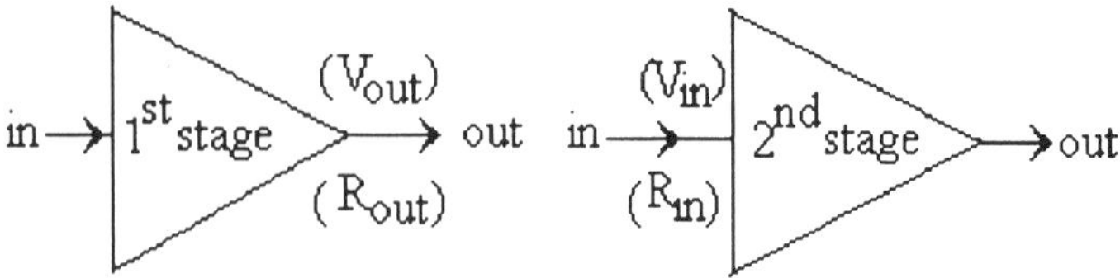

Fig. 5.8-1

(contd)

Solution

(a)

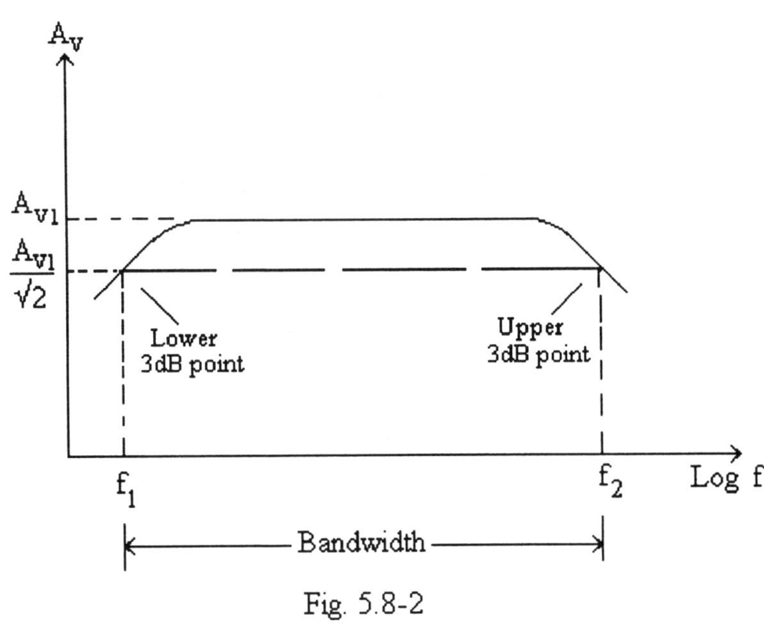

Fig. 5.8-2

Refer to Fig. 5.8-2

This is the gain-frequency response curve of an R.C-coupled C.E amplifier. The amplifier gain, A_V vs. frequency curve is drawn, where frequency f has a logarithmic scale.

The 3 dB points on the response curve are obtained as follows:

Lower 3 dB point

At low frequencies the reactance (X_C) of the coupling capacitor is high, causing a higher voltage drop of the signal voltage across it, thus reducing the actual signal strength to the transistor and thereby decreasing the voltage gain A_V.

Mid-frequency points

Over the medium frequency range, the voltage gain remains reasonably constant, since over this range of frequencies the reactance of the coupling capacitor is negligible.

(contd) *(a)*

Upper 3 dB point

At high frequencies the reactance of the transistor's inherent internal capacitances plus external stray capacitances decreases, thus effectively shunting the load and reducing the overall load impedance. As a result the output voltage decreases and the voltage gain A_v falls.

[Note: Capacitive reactance (X_c) is inversely proportional to frequency ($X_c = 1/2\pi fC$). Thus as the frequency increases, the capacitive reactance decreases, hence the shunting effect at high frequencies]

The 3dB points are defined as the points at which the gain falls to $^1/\sqrt{2}$ of its maximum value and are expressed in decibel units (dB) as follows:

$$\text{dB} = 20 \log A_{v1} - 20 \log A_{v1}/\sqrt{2}$$

$$= 20 \log \frac{A_{v1}}{A_{v1}/\sqrt{2}}$$

$$= 20 \log \sqrt{2}$$

$$= 20 \times 0.150 \qquad = 3 \text{ dB}$$

Refer to Fig. 5.8-2

Bandwidth is defined as the frequencies corresponding to the upper and lower 3dB points.

$$\text{Bandwidth} = (f_2 - f_1) \text{ Hz}$$

(b) The a.c. input equivalent circuit of Fig. 5.8-0 is shown below in Fig. 5.8-3.

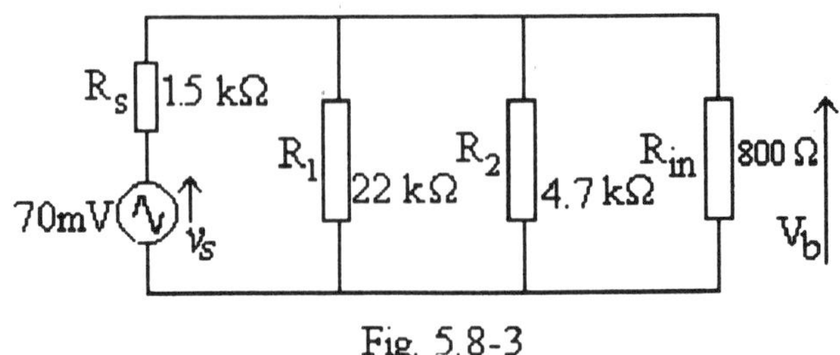

Fig. 5.8-3

(contd) *(b)*

Refer to Fig. 5.8-3

R_1 and R_2 are in parallel.

Let R' represent this parallel combination.

$$\therefore \quad R' = R_1R_2/(R_1 + R_2)$$

$$= 22 \times 4.7/(22 + 4.7) = \underline{3.87\ k\Omega}$$

The equivalent circuit of Fig. 5.8-3 is now redrawn in Fig. 5.8-4 with R' replacing the parallel combination of R_1 and R_2.

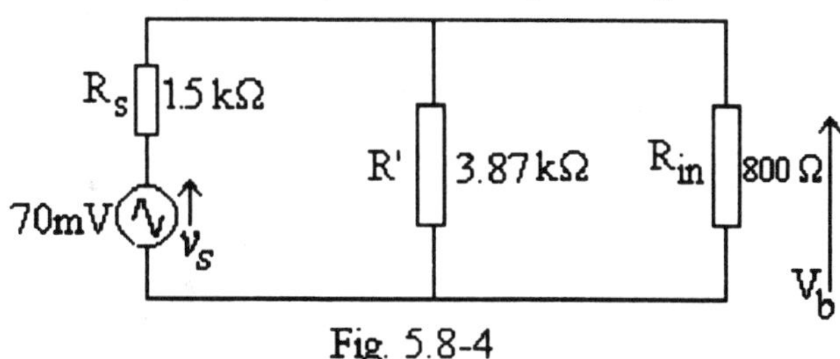

Fig. 5.8-4

Refer to Fig. 5.8-4

It can be seen that R' and R_{in} are in parallel.

Let $R_{eq.}$ represent the parallel combination of R' and R_{in}.

$$\therefore \quad R_{eq.} = R' R_{in}/(R' + R_{in})$$

$$= 3.87 \times 0.8/(3.87 + 0.8) \qquad = \underline{0.663\ k\Omega}$$

Refer to Fig. 5.8-5

Rs 1.5kΩ

70mV Vs

Req. 0.663kΩ Vb

Fig.5.74

$$V_b = \frac{V_S}{R_s + R_{eq}} \times R_{eq}$$

$$= \frac{70 \times 10^{-3}}{(1.5 + 0.663) \times 10^3} \times 0.663 \times 10^3$$

$$= \frac{46.41}{2.163} \times 10^{-3} \qquad = \underline{\mathbf{21.46\ mV}}$$

(contd) *(b)*

The a.c. base current, I_b = V_b/R_{in}

$= \frac{21.46 \times 10^{-3}}{800} = \underline{26.83\ \mu A}$

The a.c. collector current, I_c = $h_{fe}I_b$

$= 99 \times 26.83 \times 10^{-6} = \underline{2.656\ mA}$

Part of I_c, the collector current in the first stage is the input current (I_o) to the second stage as shown in Figs. 5.8-0 and 5.8-6.

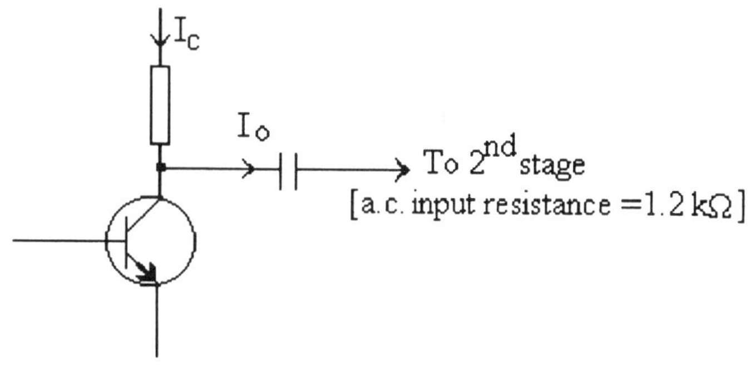

Fig. 5.8-6

The a.c. equivalent circuit to the second stage is shown in Fig. 5.8-7.

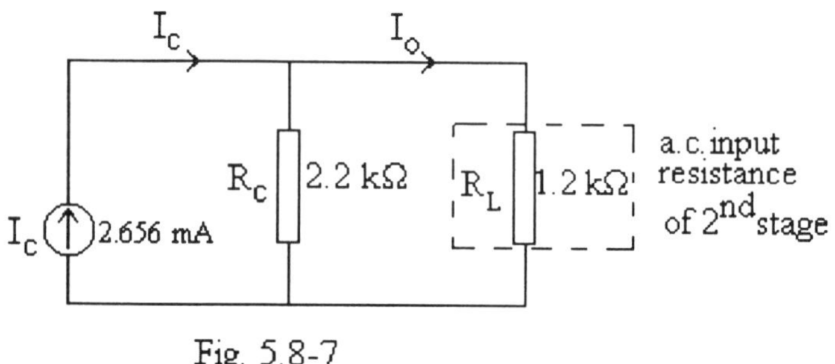

Fig. 5.8-7

Note: Refer to Fig. 5.8-7. $R_L = R_{in}$ of the second stage; that is, the input resistance of the second stage is the load resistance to the first stage.

$$I_o = \frac{R_C}{R_C + R_L} \times I_C$$

$$= \frac{2.2}{2.2 + 1.2} \times 2.656 \times 10^{-3} = \underline{\mathbf{1.72\ mA}}$$

(contd) *(b)*

At low frequencies the output circuit of Fig. 5.8-7 is replaced with its Thévenin equivalent circuit, shown in Fig. 5.8-8.

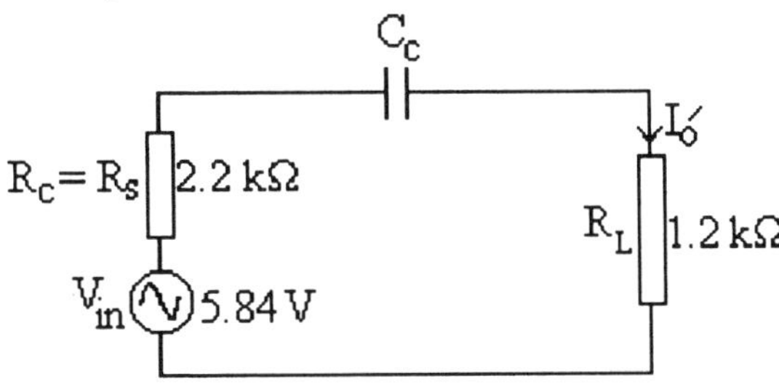

Fig. 5.8-8

Refer to Fig. 5.8-8

The equivalent voltage generator is V_{in}.

$$V_{in} = I_c R_c$$

$$= 2.656 \times 10^{-3} \times 2.2 \times 10^3 \qquad = \underline{\mathbf{5.84\ V}}$$

[*Note: The equivalent generator internal resistance is $R_C = R_S$.*]

Let the current in the 1.2 kΩ resistor be denoted by I_o'

At the lower 3 dB point, $I_o' = {}^{I_o}/_{\sqrt{2}} \qquad = {}^{1.72}/_{\sqrt{2}}$

Also $\qquad I_o' = {}^{V_{in}}/_{Z}$

$$= V_{in}/\sqrt{(R^2 + X_c^2)}$$

$$= 5.84/\sqrt{(3.4^2 + X_c^2)}$$

$$\therefore \quad 5.84/\sqrt{(3.4^2 + X_c^2)} = 1.72/\sqrt{2}$$

$$\sqrt{2} \times 5.84/1.72 = \sqrt{(3.4^2 + X_c^2)}$$

$$4.8 = \sqrt{(3.4^2 + X_c^2)}$$

$$X_c^2 = 4.8^2 - 3.4^2$$

$$X_c = \sqrt{(23.04 - 11.56)} \qquad = \underline{3.388\ \text{k}\Omega}$$

At a frequency of 50 Hz, $X_c = 1/(2\pi f C_c)$

$$\therefore \quad C_c = 1/(2\pi f X_c)$$

$$= 1/(2\pi \times 50 \times 3.388 \times 10^3) = \underline{\mathbf{0.94\ \mu F}}$$

Example 5.9

The transistor used in Fig. 5.9-0 has the data given in Table 5.9-0. Plot the output characteristics of the transistor and establish a suitable operating point on the d.c. load line.

Draw the a.c. load line and use it to find the alternating current which flows in the load resistor R_L when an input signal of 30 μA is superimposed on the d.c. base bias current. The reactances of all capacitors are negligible at the operating signal frequencies.

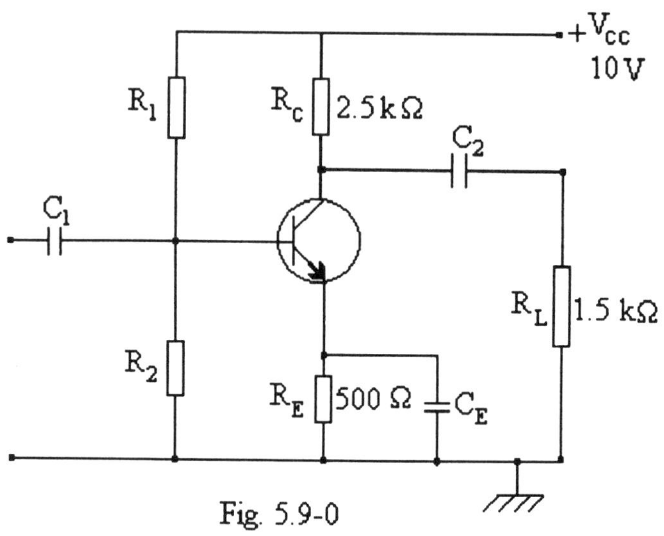

Fig. 5.9-0

Table 5.9-0

V_{CE}	I_C (mA)			
	I_b = - 20 μA	I_b = - 40 μA	I_b = - 60 μA	I_b = - 80 μA
-2 V	-0.8	-1.5	-2.3	-3.1
-9 V	-1.30	-2.1	-3.0	-3.9

(contd)

Solution

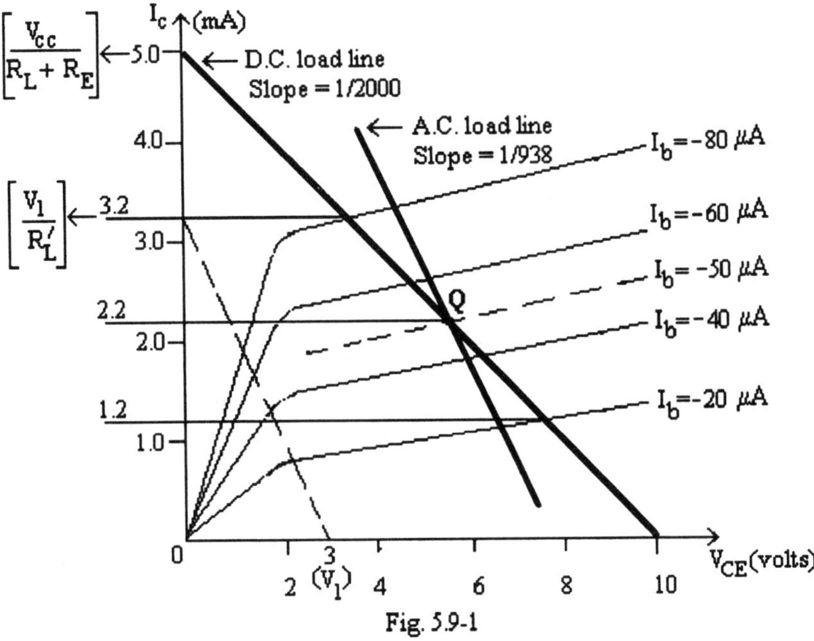

Fig. 5.9-1

Figure 5.9-1 shows the output characteristics, with the d.c. and a.c. load lines.

The load lines are constructed as follows:

D.C. load line.

At zero frequency (that is, d.c.) the reactance C_E is infinitely high and does not shunt R_E. Hence the d.c load on the transistor is

$$\mathbf{R_E + R_L} = 500 + 1500 = \underline{2000\ \Omega}$$

The d.c. load line is drawn on the output characteristics between two points (I_c = 5mA and V_{CE} =10 V).

The points are obtained as indicated hereunder.

When $I_c = 0$, $V_{CE} = V_{cc} = 10V$ (Point 1)

When $V_{CE} = 0$, $I_c = V_{cc}/(R_E + R_L)$

$= 10/2000$

$= \underline{5\ mA}$ (Point 2)

(contd) Refer to Fig. 5.9-1

A suitable operating point (quiescent point) for a base current swing of 30 μA will be **Q** at $I_b = -50\ \mu A$.

[Note: $I_b = -50\ \mu A$ is drawn by taking the mean value for I_b between $I_b = -40\ \mu A$ and $I_b = -60\ \mu A$.]

A.C load line

[*Note:* *With a.c. signal input the emitter Resistor* R_E *is shunted to ground by the decoupling capacitor* C_E. *Hence the a.c. load line is independent of* R_E.]

The a.c. load line is drawn passing through **Q** with a slope equal to $^1/R_L$.

To establish the a.c. load line proceed as follows:

(i) Assume that the a.c. load is actually a d.c. load.

Let the a.c. load $= R'_L$

(where R'_L is the parallel combination of R_L and R_c)

$$\therefore \quad R'_L = \frac{2.5 \times 1.5}{2.5 + 1.5} = \underline{0.938\ k\Omega}$$

(ii) Take any convenient value of V_{CE}.

Use this value of V_{CE} to determine I_c.

Now draw a broken line joining the points V_{CE} and I_c.

In this example the two points are at V_1 and $I_c = {}^{V}cc/R'_L$.

$$V_1 = 3\ V$$

$$I_c = V_1/R'_L = {}^3/938 = \underline{3.20\ mA}$$

(iii) Finally, the a.c. load line is drawn parallel to this broken line passing through **Q** (the operating point). When the signal of 30 μA is applied to the input, the base current swings between -80 μA and -20 μA [that is, between (50 + 30) μA and (50 – 30) μA], resulting in a collector current swing between 3.2 mA and 1.2 mA The peak collector signal current is

$$I_c = (3.2 - 1.2)/2 = 1.0\ mA$$

(contd)

The collector resistor R_C is in parallel with R_L so that the proportion of the collector current that flows through R_L is determined in a similar manner as when two resistors are in parallel sharing a common current. Thus:

Refer to Fig. 5.9-2

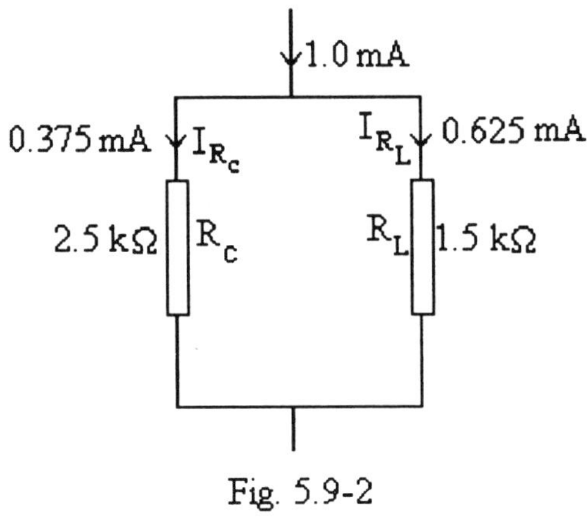

Fig. 5.9-2

$$I_{R_c} = \frac{R_L}{R_c + R_L} \times 1.0 \times 10^{-3}$$

$$= \frac{1.5}{2.5 + 1.5} \times 1.0 \times 10^{-3} = \underline{0.375\ \text{mA}}$$

$$I_{R_L} = (1.0 - 0.375)\ \text{mA} = \underline{0.625\ \text{mA}}$$

Therefore the current in the load resistor R_L = **0.625mA**

Example 5.10

(a) *Briefly explain the term drift as applied to a direct coupled amplifier.*

(b) *The circuit of Fig. 5.10-0 has the following data:*

R_C	$= 10\ k\Omega$	I_C	$= 1.0\ mA$
R_E	$= 2\ k\Omega$	V_{BE}	$= 0.7\ V$
I_B	$= 10\ \mu A$	h_{FE}	$= 100$

If the range of variation in h_{FE} is from 100 to 150 determine the change in I_C.

(c) *Assume that V_{BB} is derived from a voltage divider circuit that gives it a value of 3.5 V. Determine from this assumption the percentage change in the value of I_C.*

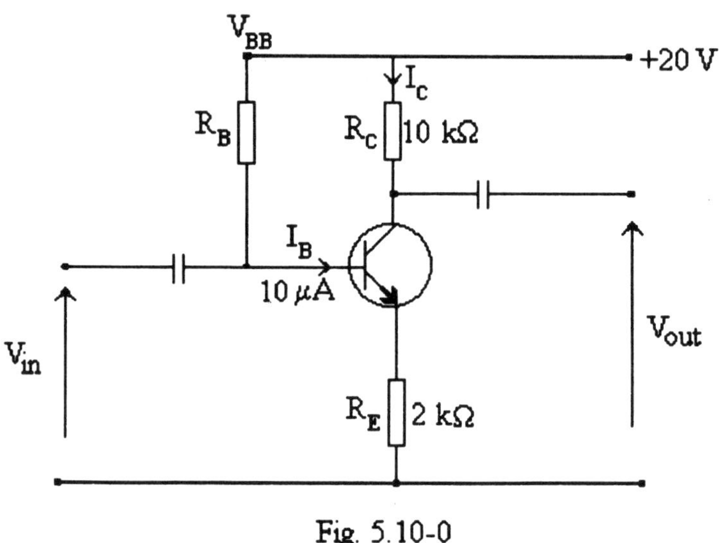

Fig. 5.10-0

Solution

(a) *Drift* is a gradual change in the output voltage or current of a d.c amplifier when the input signal is maintained at a constant level.

Drift may also be defined as the movement of holes (+charges) or free electrons (-charges) due to the influence of an electric field or temperature.

(contd) *(a)*

The primary effects of temperature increase are

(i) change in base-emitter voltage, V_{BE}

(ii) change in leakage current, I_{CBO}

(iii) change in current gain, h_{FE}

The net result is the shifting of the operating point (Q point) and hence the amplifier constants.

(b) In Fig. 5.10-0, the value of R_B is evaluated to determine the thermal stability factor *k*.

$$V_{BB} - V_{BE} = I_B R_B + (I_B + I_C)R_E$$

$$20 - 0.7 = 10 \times 10^{-6} R_B + (10 \times 10^{-6} + 1.0 \times 10^{-3}) \times 2 \times 10^3$$

$$= [\,0.010 \times 10^{-3} R_B + (0.010 \times 10^{-3} + 1.0 \times 10^{-3}) \times 2 \times 10^3]$$

$$= 0.010 \times 10^{-3} R_B + (1.010 \times 10^{-3}) \times 2 \times 10^3$$

$$= 0.010 \times 10^{-3} R_B + 2.020$$

$$\therefore \quad R_B = \frac{20 - 0.7 - 2.020}{0.010 \times 10^{-3}}$$

$$= \frac{20 - 0.7 - 2.020}{10} \times 10^6 \qquad = \underline{1.728\ \text{M}\Omega}$$

For a change of I_C over a variation of h_{FE}, the thermal stability factor is

$$k = \frac{1}{1 + (h_{FE} + \delta h_{FE})\dfrac{R_E}{R_E + R_B}}$$

$$= \frac{1}{1 + (100 + 50)\dfrac{2}{2 + 1728}}$$

$$= \frac{1}{1 + 0.17} \qquad = \underline{\mathbf{0.85}}$$

(contd) *(b)*

The current gain h_{FE} changes from 100 to 150, that is, an increase of 50%.

$\therefore$ Change in I_C = 50% x k

= 50% x 0.85 = <u>43%</u>

= 1.0 mA + (43% of 1.0 mA)

= 1.0 mA + 0.43 mA = **<u>1.43 mA</u>**

<u>That is, I_C changes from 1.0 mA to 1.43 mA</u>

(c) Since V_{BB} is derived from a voltage divider circuit and is now given a new value, the result will be a new value for R_B and hence *k*.

$$V_{BB} - V_{BE} = I_B R_B + (I_B + I_C)\,R_E$$

$$3.5 - 0.7 = 0.010 \times 10^{-3} R_B + [(0.010 + 1.0) \times 10^{-3}] \times 2 \times 10^3$$

$$2.8 = 0.010 R_B + 2.020$$

$$\therefore \quad R_B = \frac{2.8 - 2.02}{0.01} \times 10^3 = \underline{78\ \text{k}\Omega}$$

$$k = \frac{1}{1 + 150 \times \dfrac{2}{2 + 78}} = \underline{\mathbf{0.21}}$$

A transistor in which its h_{FE} is increased by 50% causes the collector current I_C to change by only 10.5%, obtained as follows:

I_C = 50% x k

= 50% x 0.21 = **<u>10.5%</u>**

Example 5.11

(a) *The direct coupled amplifier shown in Fig. 5.11-0 has the following d.c. current gains:*

h_{FE} *of Tr1* = *50*

h_{FE} *of Tr2* = *49*

and the base potential of Tr2 = *-2 V.*

Determine the collector and emitter currents and voltages of Tr2. The base potential of Tr1 can be neglected.

(b) *Briefly explain why the d.c. stability of the second stage is superior to that of the first stage.*

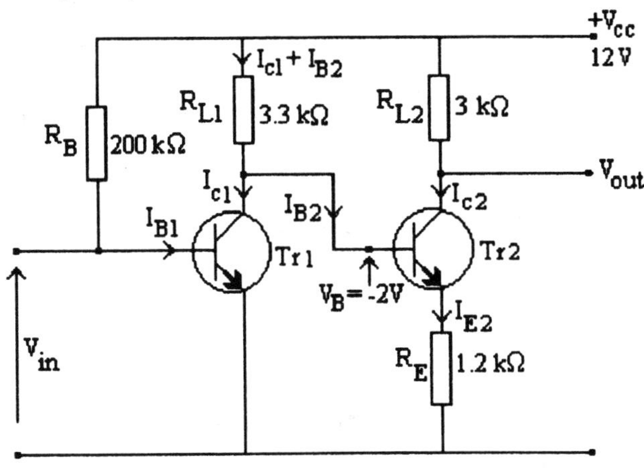

Fig. 5.11-0

Solution

(a) Since V_{BE} of Tr1 can be neglected, then

$$I_{B1} = V_{BB}/R_B$$

$$= 12/(200 \times 10^3) \qquad = \underline{60\ \mu A}$$

$$\therefore \quad I_{C1} = h_{FE1} \times I_{B1}$$

$$= 50 \times 60 \times 10^{-6} \qquad = \underline{3\ mA}$$

Current through $R_{L1} = I_{C1} + I_{B2}$

Tr1 collector voltage, $V_{RL1} = V_{CC} - V_B$

$$= 12 - 2 \qquad = \underline{10\ V}$$

(contd) *(a)*

Also $V_{RL1} = R_{L1}(I_{C1} + I_{B2})$

That is $10 = 3.3\,(3 + I_{B2})$

$\therefore \quad I_{B2} = \dfrac{10 - 9.9}{3.3}$ $= \underline{0.03\ \text{mA}}$ $\underline{(30\ \mu\text{A})}$

Analyzing Tr2

$I_{C2} = h_{FE2}I_{B2}$

$= 49 \times 0.03$ $= \underline{\mathbf{1.47\ mA}}$

$I_{E2} = I_{C2} + I_{B2}$

$= 1.47 + 0.03$ $= \underline{\mathbf{1.5\ mA}}$

Voltage drop across $R_{E2} = R_E \times I_E$

$= 1.2 \times 1.5$ $= \underline{\mathbf{1.8\ V}}$

Tr2 collector voltage, $V_{RL2} = V_{CC} - I_{C2}\,R_{L2}$

$= 12 - (1.47 \times 3)$ $= \underline{\mathbf{7.59\ V}}$

(b) The d.c. stability of the second stage is superior to that of the first stage because of the inclusion and the effect of the 1.2 kΩ emitter resistor R_E on the stability factor.

Example 5.12

(a) *For nearly all FETs (JFET: DE or E MOSFET) operating in the constant-current region (saturation region of their output characteristics), the drain current is given by the following equations:*

$$I_D = I_{DSS}(1 - V_{GS}/V_p)^2 \quad \text{(Eq. 5.12-0)}$$

For an enhancement-only MOSFET, the corresponding relation is $I_D = K(V_{GS} - V_T)^2$ *(Eq. 5.12-1)*

Define each term in the two equations.

(b) *The n-channel enhancement-only MOSFET shown in Fig. 5.12-0 has the following specifications:*

$$V_T = 4\ V$$

$$I_D = 5.0\ mA \text{ at } V_{GS} = 8\ V$$

Determine the value of R_D *for operation at* $V_{DS} = 6\ V$, $V_{DD} = 12\ V$ *and* $R_G = 100\ M\Omega$.

(c) *If the circuit of Fig. 5.12-0 is to be used as an amplifier with* $I_{DS} = 2\ mA$ *and* $V_{GS}(off) \approx -4\ V$, *estimate the voltage gain with a load resistance,* $R_L = 4.8\ k\Omega$.

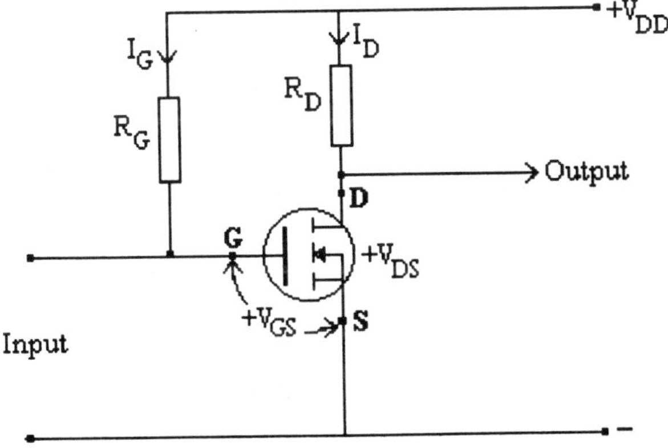

Fig. 5.12-0. An n-channel enhancement MOSFET

Solution

(a) $$\mathbf{I_D = I_{DSS}\,(1 - V_{GS}/V_p)^2}$$ (Refer to Eq. 5.12-0)

This equation describes the **D**epletion or **E**nhancement MOSFET, and both positive (+ve) and negative (–ve) values of V_{GS} can be used.

For the **E**nhancement-only MOSFET, the corresponding relation is

$$\mathbf{I_D = K\,(V_{GS} - V_T)^2}$$

Equation 5.12-0: *Definition of Terms*

I_D = drain current (sometimes denoted by I_{DS})

I_{DSS} = drain current with gate shorted to source

V_{GS} = gate-source voltage

V_p = pinch-off voltage

Equation 5.12-1: *Definition of Terms*

I_D = drain current (sometimes denoted by I_{DS})

V_{GS} = gate-source voltage

K = constant of proportionality

V_T = threshold voltage (that is, turn-on voltage)

(b) Substituting the given specifications in Eq. 5.12-1 , we get

$$I_D = K\,(V_{GS} - V_T)^2$$ (Refer to Eq. 5.12-1)

$$5 \times 10^{-3} = K\,(8 - 4)^2$$

$$\therefore \quad K = \frac{5 \times 10^{-3}}{16} = \underline{0.3125 \times 10^{-3}}$$

I_G is not specified and hence it is taken to be zero; that is, the voltage drop across R_G = 0 V.

(contd) *(b)*

Since the voltage drop across R_G is considered to be 0 V

$\therefore \quad V_{GS} = V_{DS} = 6\text{ V}$

$I_D = K\,(V_{GS} - V_T)^2$ (Refer to Eq. 5.12-1)

$= 0.3125 \times 10^{-3}(8 - 4)^2$

$= 0.3125 \times 10^{-3} \times 16$ $= \underline{5.0\text{ mA}}$ (given)

and $R_D = [V_{DD} - V_{DS}]/I_D$

$= \dfrac{12 - 6}{5 \times 10^{-3}}$ $= \underline{\mathbf{1.2\ k\Omega}}$

(c) $I_D = I_{DSS}\,(1 - V_{GS}/V_p)^2$ (Refer to Eq.5.12-0)

$i_D = 2 \times 10^{-3}\,(1 - V_{GS}/4)^2$

$= 2 \times 10^{-3}\,(1 - 0.25V_{GS})^2$

For a signal of $V_{GS} = V_m \sin \omega t$

Output voltage $V_{RL} = i_D R_L$

$= 2 \times 10^{-3} \times 4.8 \times 10^3(1 + 0.25V_m \sin \omega t)^2$

$= 9.6\,(1 + 0.5V_m \sin \omega t + 0.0625V_m^2 \sin^2 \omega t)$

$= \mathbf{9.6 + 4.8V_m \sin \omega t + 0.6V_m^2 \sin^2 \omega t}$

The output voltage consists of three components, namely

9.6 V = d.c. component

$4.8V_m \sin \omega t$ = signal component at input frequency. (this is an amplifier with a voltage gain of 4.8 for small values of V_m)

$0.6V_m^2 \sin^2 \omega t$ = distortion component due to the non linear transfer characteristic

(contd) *(c)*

Refer to Fig. 5.12-1

This is the transfer characteristic of the MOSFET amplifier in Fig. 5.12-0.

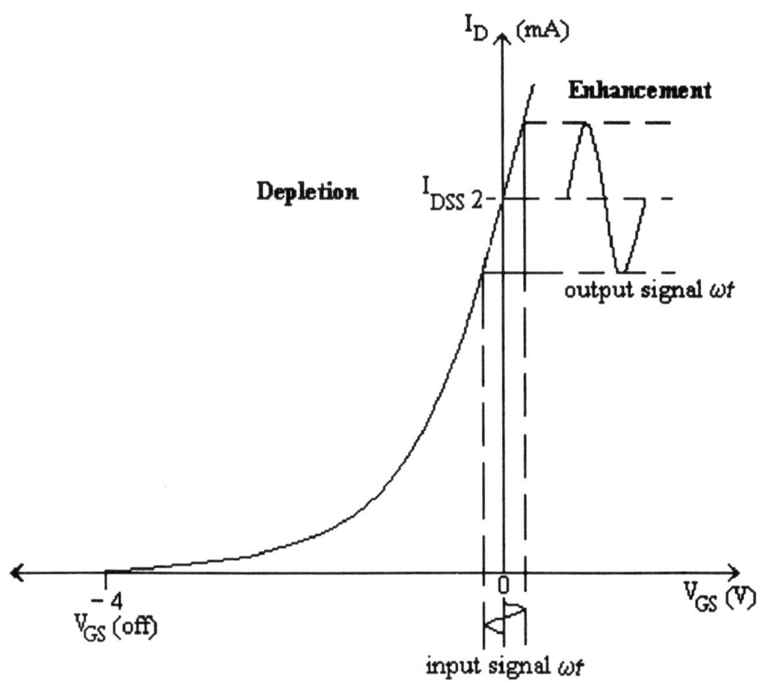

Fig. 5.12-1

Example 5.13

(a) *Sketch the output characteristics and corresponding transfer characteristics for*

(1) *JFET*

(2) *E MOSFET*

(3) *DE/E MOSFET*

and briefly explain the uses of FETs in comparison to bipolar transistors.

(b) *Define the mutual conductance of an FET.*

(c) *The circuit of Fig. 5.13-0 shows an amplifier using an n-channel FET, for which $V_p = -3$ V and $I_{DSS} = 2.5$ mA. With V_{DD} at 20 V, it is desired to bias the circuit at $I_D = 1.2$ mA. Assume that the resistance from drain to source, r_{DS} is very much greater than R_D. Determine*

(i) *the gate-source voltage, V_{GS}*

(ii) *the transconductance, g_{fS}*

(iii) *the source resistance, R_S*

(iv) *the voltage gain, A_v*

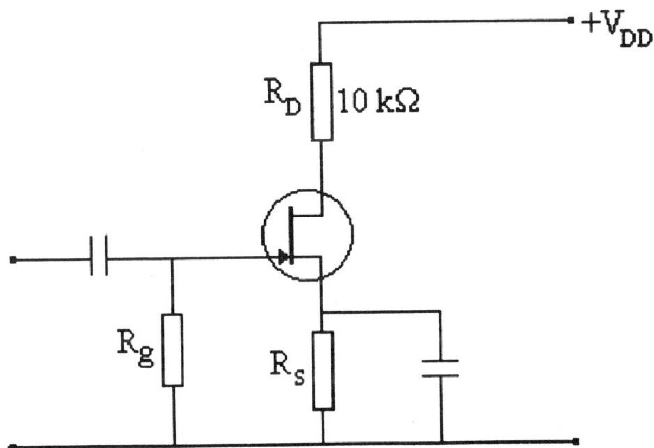

Fig. 5.13-0. An n-channel FET amplifier

Solution

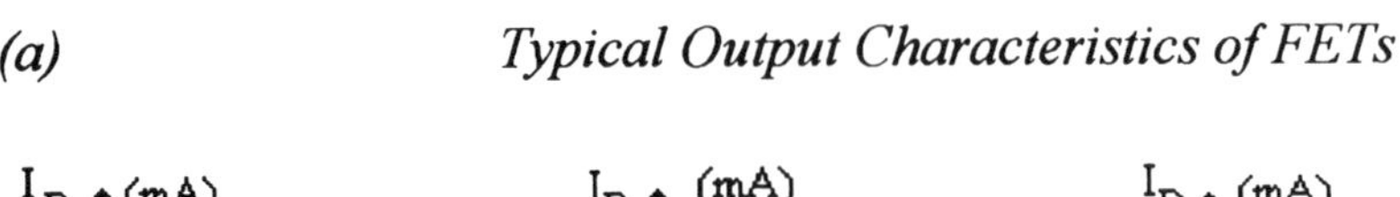

(a) *Typical Output Characteristics of FETs*

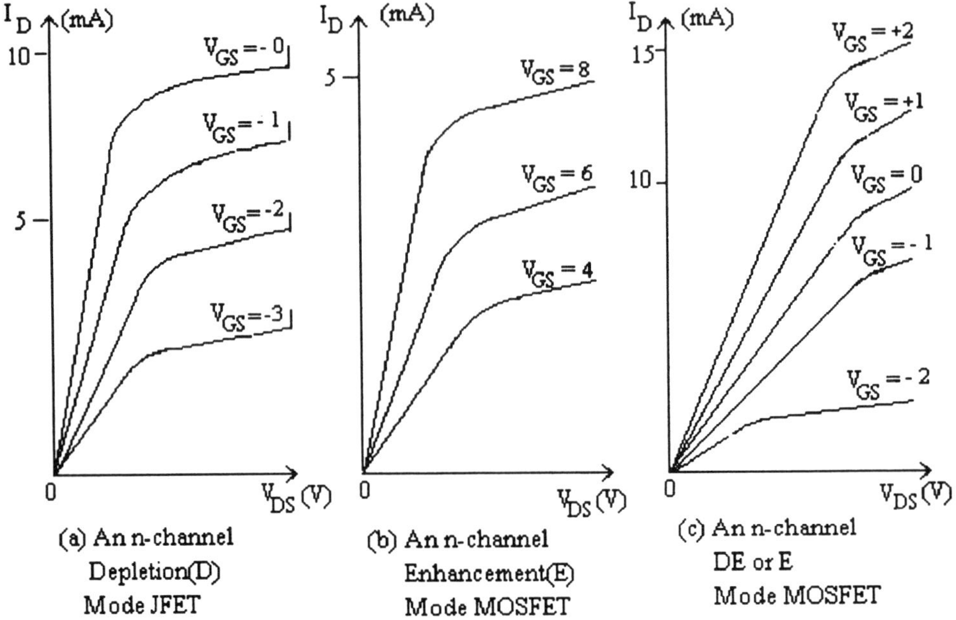

Typical Transfer Characteristics of FETs

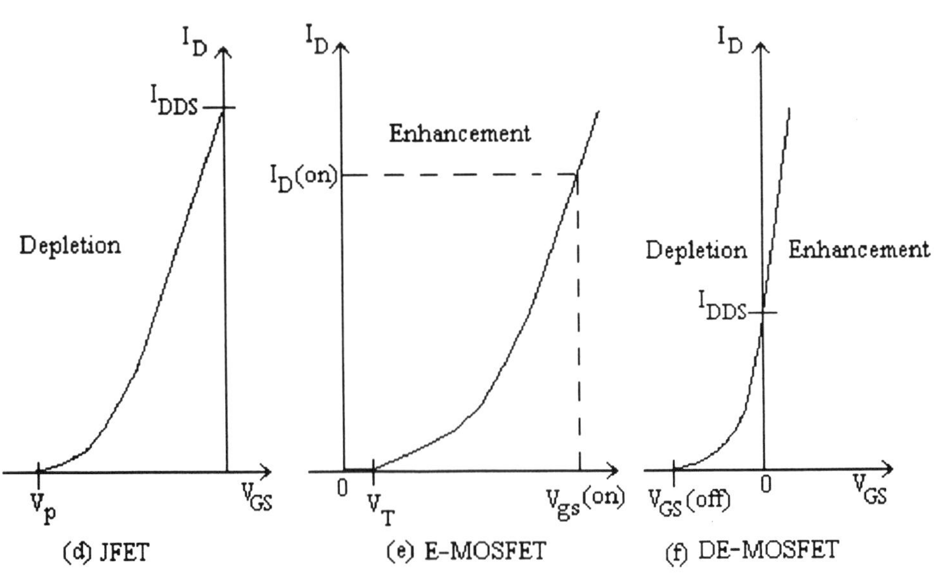

Fig. 5.13-1

(contd) *(a)*

FETs are extensively used in digital circuit systems, more so, because their structure permits the fabrication of units acting as passive components (e.g., resistors and capacitors). Thousands of units can be fabricated on a single silicon chip at reduced cost.

The disadvantages of FETs are their high operating voltages, lower voltage gain and poor high-frequency response in comparison to bipolar transistors. However, with improved quality of material and better design control, these disadvantages can be minimized.

(b) The mutual conductance (g_{fs}) of an FET is defined as the small change in Drain current (δI_D) produced by a small change in gate-source voltage (δV_{GS}). That is

$$g_{fs} = \delta I_D / \delta V_{GS}$$

For all FETs operating in the constant-current region, the following equation is true.

$$I_D = I_{DSS}(1 - V_{GS}/V_p)^2 \qquad \text{(Refer to Eq. 5.12-0)}$$

$$\therefore \quad g_{fs} = \delta I_{DSS}(1 - V_{GS}/V_p)^2$$

Differentiating, $$g_{fs} = -\frac{2 I_{DSS}}{V_p}(1 - V_{GS}/V_p) \qquad \text{(Eq. 5.13-0)}$$

When $V_{GS} = 0$

$$g_{fso} = -2\, I_{DSS}/V_p$$

This is the value of the mutual conductance of an FET when the gate-source voltage is zero. I_{DSS} and V_p can be easily found by direct measurement or from the gate transfer characteristic.

Refer to Fig. 5.13-1(d) for typical transfer characteristic.

(contd)

(c) (i) Gate-Source Voltage: V_{GS}

Drain current, $I_D = I_{DSS}(1 - {}^{V_{GS}}/_{V_p})^2$

$\therefore \quad 1.2 = 2.5\,[1 - (-{}^{V_{GS}}/_{-3})^2$

$= 2.5\,(1 + {}^{V_{GS}}/_{3})^2$

$(1 + {}^{V_{GS}}/_{3})^2 = {}^{1.2}/_{2.5}$

$\therefore \quad V_{GS} = 3 \times \sqrt{({}^{1.2}/_{2.5})} - 1$

$= 3 \times \sqrt{0.48} - 1$

$= 3 \times (-0.307)$ $= \underline{\mathbf{-0.922V}}$

(ii) Transconductance: g_{fs}

$g_{fs} = {}^{\delta I_D}/_{\delta V_{GS}}$

$g_{fs} = -\dfrac{2\,I_{DSS}}{Vp}(1 - {}^{V_{GS}}/_{V_p})$ (Refer to Eq. 5.13-0)

From Eq. 5.12-0 $I_D = I_{DSS}(1 - {}^{V_{GS}}/_{V_p})^2$

$\therefore \quad \sqrt{({}^{I_D}/_{I_{DSS}})} = (1 - {}^{V_{GS}}/_{V_p})$

Substituting in Eq. 5.13-0, $g_{fs} = -\dfrac{2\,I_{DSS}}{V_p}({}^{I_D}/_{I_{DSS}})^{1/2}$

$= -2\,({}^{2.5}/_{-3})\,({}^{1.2}/_{2.5})^{1/2}$

$= 1.667 \times 0.693$ $= \underline{\mathbf{1.15\ mS}}$

(iii) <u>Source Resistance</u>: R_S

$R_S = {}^{V_{GS}}/_{I_D}$

$= {}^{0.922}/_{(1.2 \times 10^{-3})} = \underline{\mathbf{768\ \Omega}}$

Refer to Figs. 5.13-2(a) and 5.13-2(b) for graphical representations of the operating and bias conditions.

(contd) *(c)* Graphical representations of the operating and bias conditions

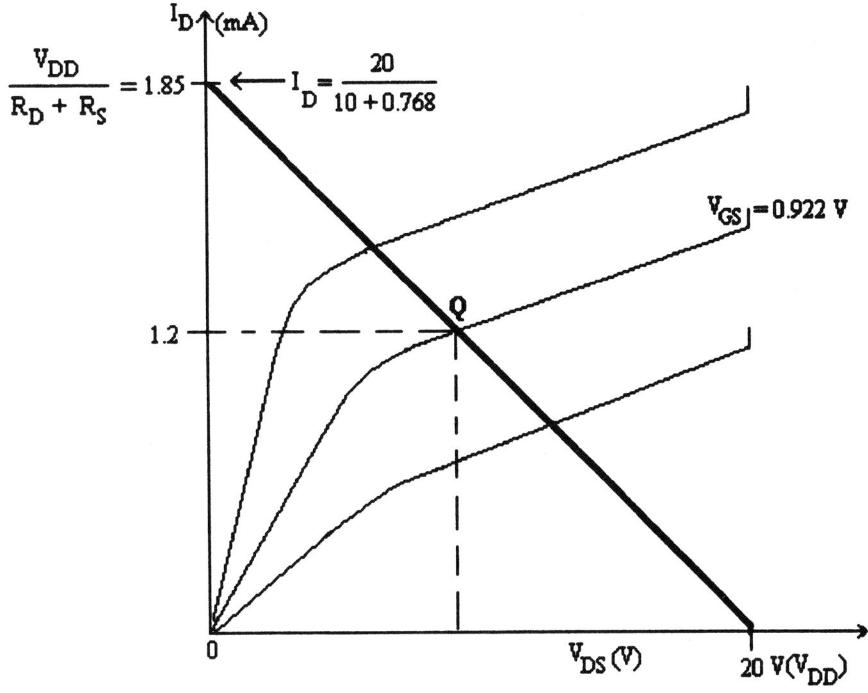

(a)

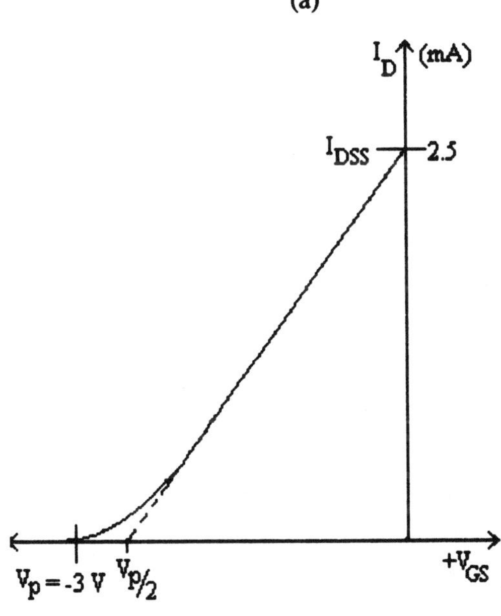

(b)

Fig. 5.13-2

(iv) *Voltage Gain: A_v*

Neglecting the capacitive reactances and the shunting effect of r_{DS} on R_D (as $r_{DS} >> R_D$), the equivalent circuit is shown in Fig. 5.13-3.

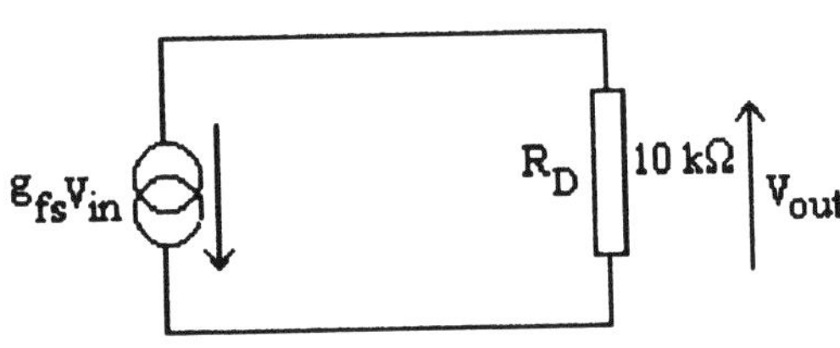

Fig. 5.13-3

$$A_v = V_{out}/V_{in}$$

$$= - g_{fs} V_{in} R_D/V_{in}$$

$$= - g_{fs} R_D$$

$$= 1.15 \times 10^{-3} \times 10 \times 10^3 \quad = \underline{\mathbf{11.5}}$$

Example 5. 14

The circuit diagram of a single-stage n-channel junction FET is shown in Fig. 5.14-0 and its associated output characteristics are shown in Fig. 5.14-1. Draw the d.c. load line for $V_{GS} = -1.6$ V. Draw also the a.c. load for the same value of V_{GS} and hence determine the voltage gain when the input signal is 0.4 V.

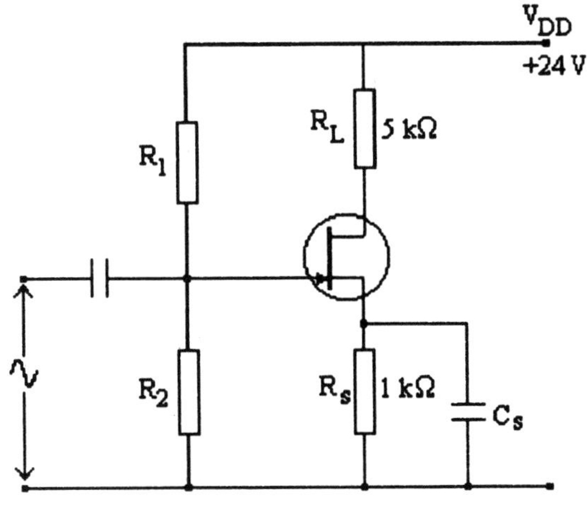

Fig. 5.14-0

Solution

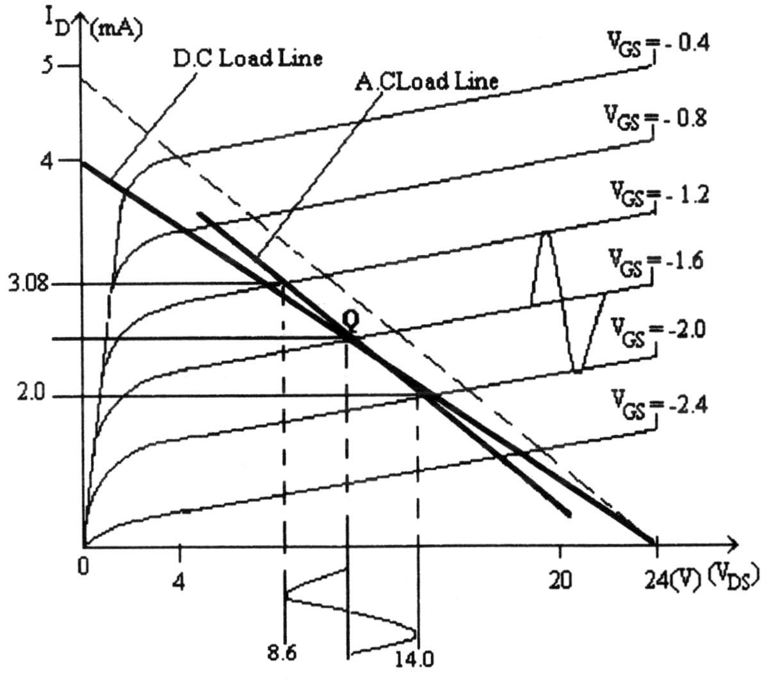

Fig. 5.14-1

(contd)

Refer to Fig. 5.14-1: *The D.C. Load Line*

The two points required to strike the d.c load line are at

$$I_D = {}^{V_{DD}}/(R_L + R_S)$$
$$= {}^{24}/(5 + 1) \times 10^{-3}$$
$$= \underline{4.0\ mA}$$

and $V_{DS} = \underline{24\ V}$

In summary, the two points for the d.c. load line are at

$I_D = 0$: $V_{DS} = \underline{24V}$

$I_D = \underline{4mA}$: $V_{DS} = 0$

The d.c. load line is drawn joining these two points.

The **Q** point (operating point) is on the characteristic $V_{GS} = -1.6$ V.

Refer to Fig. 5.14-1: *The A.C. Load Line*

The slope of the a.c load line $= {}^{-1}/R_L$
$= [{}^{1}/(5 \times 10^3)]$

The two points of the broken line are at:

$I_D = 0$: $V_{DS} = \underline{24\ V}$

and $I_D = {}^{V_{DS}}/R_L = {}^{24}/(5 \times 10^3)$
$= \underline{4.8\ mA}$: $V_{DS} = 0$

The a.c. load line is drawn parallel to the broken line intersecting the **Q** point.

From the a.c. load line, the voltage gain is
(using peak values)

$$A_V = \frac{\text{Output Voltage } (V_{DS})}{\text{Input Voltage } (-V_{GS})}$$

$$= \frac{(14.0 - 8.6)/2}{(-2.0 - (-1.2)/2}$$

$$= {}^{2.7}/_{-0.4} \qquad = -\underline{\mathbf{6.75}}$$

(using peak-to-peak values)

$$A_V = \frac{14.0 - 8.6}{-2.0 - (-1.2)} = 5.4/-0.8 \qquad = -\ \underline{\mathbf{6.75}} \text{ (no change)}$$

Example 5.15

(a) *Define the amplification factor (μ) of an FET and give the relationship to g_{fs} and r_{DS}.*

(b) *Figure 5.15-0 is a simple common-source amplifier. The parametric values of the FET are g_{fs} = 1.5 mS and r_{DS} = 80 kΩ, and the load resistor is 20 kΩ. Sketch the equivalent circuit of the amplifier and determine the voltage gain, assuming that the input resistance is extremely high and can be neglected.*

(c) *In the amplifier of Fig. 5.15-0, V_p = 2.5 V and I_{DSS} = 2.0 mA. It is desired to bias the circuit at I_D = 1.5 mA, the drain supply (V_{DD}) being 15 V. Determine the value of R_L to give a voltage gain of 10, assuming that $r_{DS} >> R_L$.*

(d) *Determine the value of R_S and hence, briefly explain how bias in the circuit of Fig. 5.15-0 is achieved.*

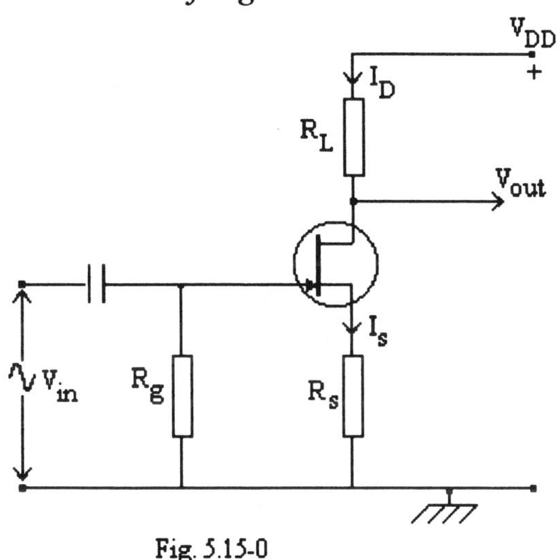

Fig. 5.15-0

Solution

(a) *Amplification Factor: μ*

The amplification factor of an FET, denoted by μ, is defined as the ratio of the small change in drain-source voltage (δV_{DS}) to the small change in gate-source voltage (δV_{GS}) which produces the same change in the drain current (δI_D).

(contd) *(a)*

$$\mu = \frac{\text{small change in } V_{DS}}{\text{small change in } V_{GS} \text{ producing the same change in } I_D} = \frac{\delta V_{DS}}{\delta V_{GS}}$$

The amplification factor, μ of an FET is related to the product of g_{fs} and r_{DS}. Thus

$$g_{fs} \times r_{DS} = \frac{\delta I_D}{\delta V_{GS}} \times \frac{\delta V_{DS}}{\delta I_D}$$

$$= \frac{\delta V_{DS}}{\delta V_{GS}} = \mu$$

$$\therefore \quad \mu = g_{fs}\, r_{DS} \qquad \text{(Eq. 5.15-0)}$$

(b)

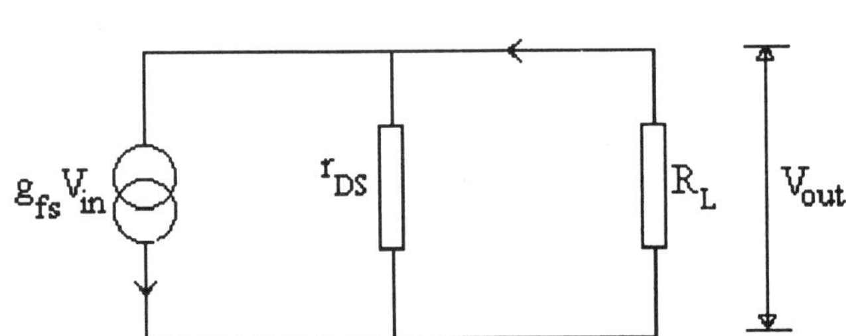

Fig. 5.15-1

$$V_{out} = -g_{fs}V_{in} \times \frac{r_{DS}R_L}{r_{DS} + R_L} \qquad \text{(i)}$$

$$\therefore \quad A_v = \frac{V_{out}}{V_{in}} = -g_{fs} \times \frac{r_{DS}R_L}{r_{DS} + R_L}$$

Substituting values

$$A_v = -1.5 \times 10^{-3} \times \frac{(80 \times 20) \times 10^6}{(80 + 20) \times 10^3}$$

$$= -1.5 \times 16 \qquad = -\underline{\mathbf{24}}$$

Note: Substituting $g_{fs}\, r_{DS}$ = μ in (i) above, we get

$$\mathbf{A_v = -\mu R_L/(r_{DS} + R_L)} \qquad \text{(Eq. 5.15-1)}$$

(contd)

(c) Since $r_{DS} >> R_L$, r_{DS} can be omitted from the calculation with no adverse effect in the result.

$$\text{Voltage gain,} \quad A_v = V_{out}/V_{in} = g_{fs} R_L$$

$$\text{However,} \quad g_{fs} = \frac{-2I_{DSS}}{V_p}\left(1 - \frac{V_{GS}}{V_p}\right) \text{ S} \qquad \text{(i)}$$

$$\text{but} \quad I_D = I_{DSS}(1 - V_{GS}/V_p)^2$$

$$\therefore \quad (1 - V_{GS}/V_p) = (I_D/I_{DSS})^{1/2} \quad .... \qquad \text{(ii)}$$

Substituting (ii) into (i), we get

$$g_{fs} = \frac{-2\,I_{DSS}}{V_p} \times \left[\frac{I_D}{I_{DSS}}\right]^{1/2}$$

$$= \frac{-2 \times 2 \times 10^{-3}}{-2.5} \times \left[\frac{1.5}{2.0}\right]^{1/2}$$

$$= 1.6 \times 0.866 \times 10^{-3} = \underline{1.39 \text{ mS}}$$

$$\text{Now} \quad A_v = g_{fs} R_L$$

$$\therefore \quad R_L = A_v/g_{fs} = 10/(1.39 \times 10^{-3}) = \underline{\mathbf{7.19\ k\Omega}}$$

(d) From (ii),

$$V_{GS} = -V_p\,[1 - (I_D/I_{DSS})^{1/2}] = -2.5\,[1 - (1.5/2.0)^{1/2}] = -2.5\,[1 - 0.866] = -\underline{0.335 \text{ V}}$$

$$\text{Source resistance,} \quad R_S = V_{GS}/I_D = 0.335/(1.5 \times 10^{-3}) = \underline{\mathbf{223\ \Omega}}$$

(contd)

(d)

The gate bias is derived from the voltage drop across R_S

$$V_{RS} = I_S \times R_S$$
$$= I_D \times R_S \quad (I_S = I_D \text{ since } I_G \text{ is negligible})$$

The d.c. potential across R_g is zero; that is, there is no voltage drop across R_g. The gate is effectively at ground potential.

The source, however, is 0.335 V *positive* with respect to ground, hence the gate is effectively *negative* with respect to the source, and as a consequence , bias is achieved.

CHAPTER 6

POWER AMPLIFIERS

Introduction

In small-signal amplification, the power output was not the major concern. In a large-signal amplifier, power output is the most significant factor. To obtain high power output, the active devices (transistors, FETs, etc.) are operated under Class B and Class C conditions. Power amplifiers are not without distortion problems. Class B push-pull amplifiers are known to cancel most distortion which improves efficiency. In Class C operation the efficiency is high when the amplifier is operated as a tuned amplifier. The tuned amplifier uses a resonance circuit to obtain high selectivity by selecting the fundamental frequency and rejecting others, with the result that the output waveform is near sinusoidal with minimal distortion. Cascading amplifiers to obtain higher power output are also very practical. In all amplifiers proper biasing of the active device's operating point within the manufacturer's specification is paramount for efficient operation.

Example 6.1

(a) *With the aid of a transistor characteristic, briefly explain the action of an amplifier operating in Class A, Class B and Class C modes.*

(b) *The characteristics of an n.p.n transistor are given in Table 6.1-0. The transistor is used in the simple common-emitter circuit shown in Fig. 6.1-0; the collector load resistance is 2.2 kΩ and supply voltage is 10 V. Assuming Class A operation, estimate using the load line*

- (i) *the total power dissipated in the circuit*
- (ii) *the quiescent collector power dissipation at a quiescent base current of 25 μA*
- (iii) *the current gain h_{fe} of the transistor at the operating point*

(contd)

Table 6.1-0

I_B (μA)	I_c (mA) for collector voltages of	
	2 V	10 V
10	0.9	1.7
20	1.8	2.8
30	2.8	4.2
40	3.9	5.5

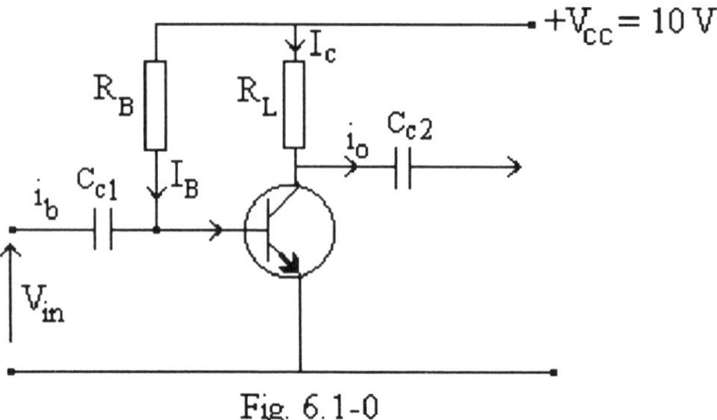

Fig. 6.1-0

Solution

(a)

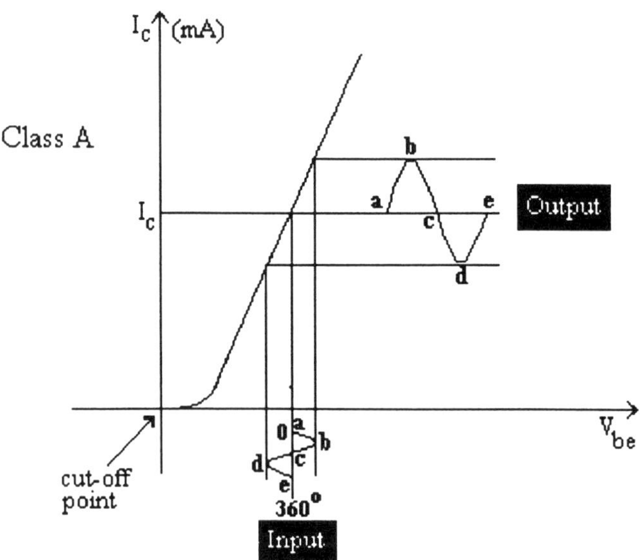

Fig. 6.1-1. Amplifier classification: Class A

(contd)

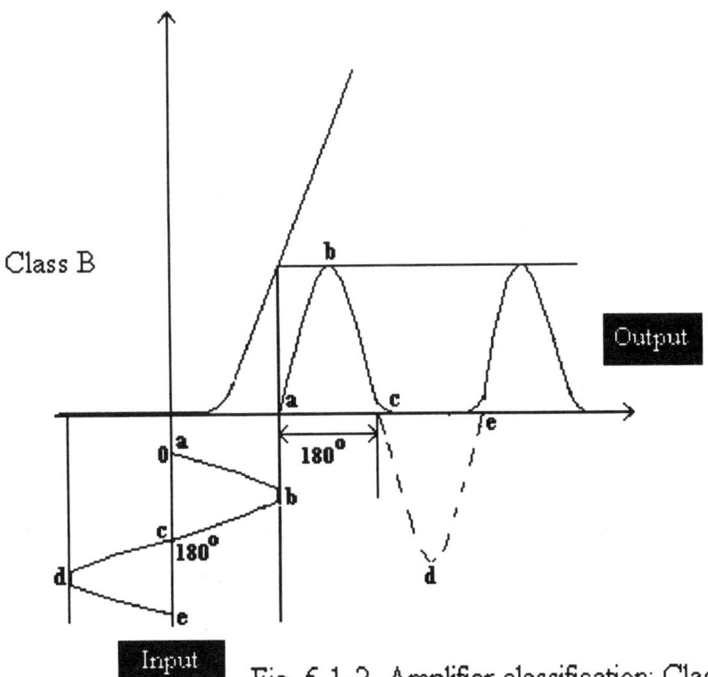

Fig. 6.1-2. Amplifier classification: Class B

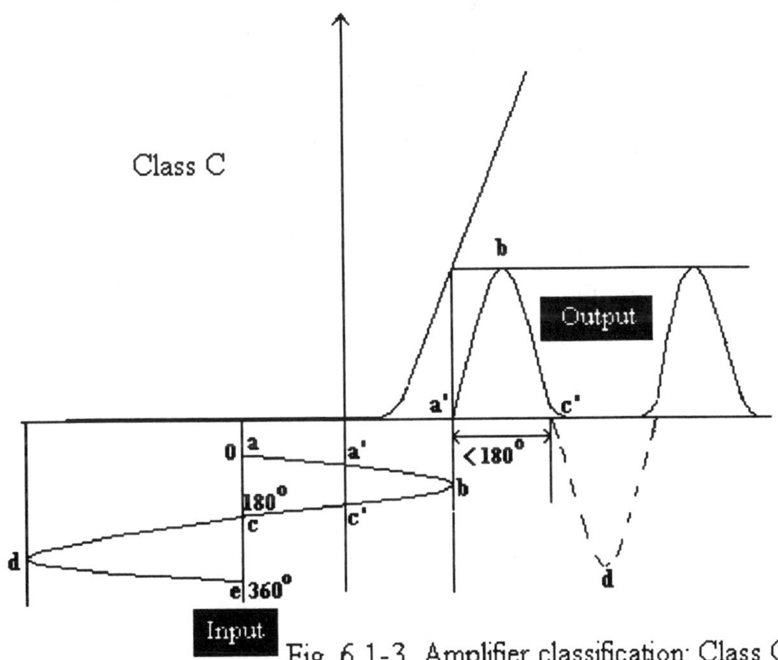

Fig. 6.1-3. Amplifier classification: Class C

(contd)

Class A Operation: Refer to Fig. 6.1-1

The operating point (Q point) is placed in the linear portion of the characteristic and biased such that output current flows during the whole period of the input signal; that is, the output signal flows for a complete cycle of 360°. Class A operation has very low distortion.

Class B Operation: Refer to Fig. 6.1-2

The amplifier is biased to cut-off. Therefore, the output signal flows only during the positive half-cycle of the input signal; that is, the input signal is 360° (biased at cut-off), and the output signal waveform is 180°.

In Class B operation the output waveform is very much distorted. It is useful when operated in a push-pull circuit because the second transistor of the push-pull circuit conducts during the negative half-cycle of the waveform and hence cancels the distortion, including the harmonics.

Class C operation: Refer to Fig. 6.1-3

In this mode of operation the amplifier is biased beyond cut-off and as a result the output current flows during only a small portion of the input signal waveform. That is, the input signal is 360° (biased beyond cut-off), and the output signal waveform is less than 180°.

In Class C operation the output is very much distorted. However, with a tuned load impedance the distortion is resolved; the tuned circuit selects the fundamental frequency and rejects all others, resulting in a sinusoidal output waveform.

(contd)

(b)

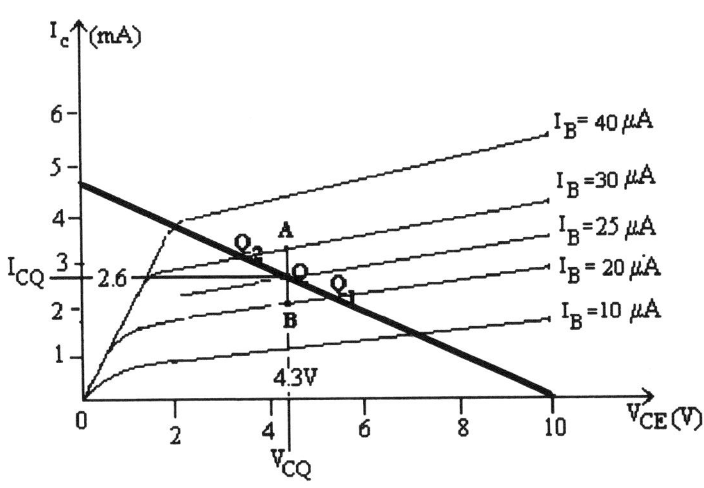

Fig. 6.1-4

<u>Refer to Fig. 6.1-4</u>

The d.c. load line is drawn connecting the two points, which are

$$V_{CE} = V_{CC} = 10V \quad \text{(Point 1)}$$

and

$$I_C = V_{CC}/R_L$$

$$= 10/2.2 \times 10^{-3} = 4.55 \text{ mA} \quad \text{(Point 2)}$$

(i)

$$V_{CQ} = 4.3 \text{ V}$$

$$I_{CQ} = 2.6 \text{ mA}$$

Power dissipated in the circuit = Average power drawn from supply

$$= V_{CC}(I_{CQ} + I_{BQ})$$

$$= 10\,(2.6 + 0.025) = \underline{\mathbf{26.25\ mW}}$$

(ii) The quiescent collector power dissipation is

$$= V_{CQ}\, I_{CQ}$$

$$= 4.3 \times 2.6 = \underline{\mathbf{11.18\ mW}}$$

(iii) The value of h_{fe} is calculated on the assumption that I_C changes between the points **A** and **B**; that is

$$h_{fe} = \left[\frac{\delta I_C}{\delta I_B}\right]_{V_C \text{ constant}}$$

$$h_{fe} = \frac{\text{change in } I_C}{\text{change in } I_B}$$

$$= \frac{(3.2 - 2.2) \times 10^{-3}}{(30 - 20) \times 10^{-6}} = \mathbf{100}$$

Example 6.2

(a) *Explain what is meant to match a load by transformer coupling.*

(b) *Calculate a suitable turns ratio for the transformer in the circuit of Fig. 6.2-0 if the effective collector load is to be 16 Ω and the load resistor R_L is 4 Ω. Assume an ideal transformer.*

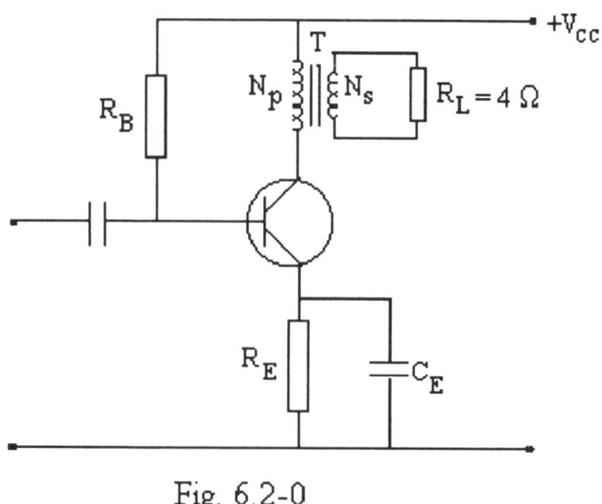

Fig. 6.2-0

(a) The loads used in audio amplifiers are usually loudspeakers or aerial coils with very low impedances. These output impedances are too small to be connected directly in the collector circuit to achieve appreciable power for most practical applications.

To obtain maximum power gain, the load must be matched to the transistor effective load resistance; that is, the effective load resistance must equal the amplifier output resistance.

The method of matching is achieved by the use of a step-down transformer (turns ratio) from the collector circuit to load as shown in Fig. 6.2-1. Maximum power transfer occurs when the load resistance is equal to the source resistance.

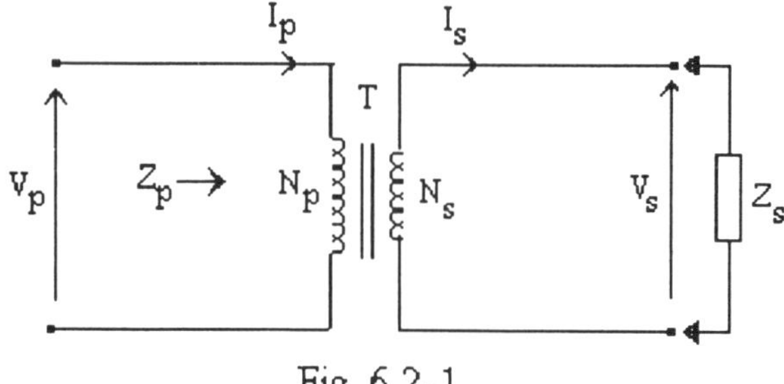

Fig. 6.2-1

(contd)

(a) Refer to Fig. 6.2-1

Legend

N_p = primary windings or turns

N_s = secondary windings or turns

I_p = current in primary windings

I_s = current in secondary windings

V_p = primary input voltage

V_s = secondary output voltage

The transformer consists of two coils, with N_p primary and N_s secondary windings or turns wound on a common ferrous core.
The change of flux in the core induces voltages in the windings proportional to the turns.
For an ideal transformer, the voltage ratio is

$$V_p/V_s = N_p/N_s \quad \text{..............} \quad \text{(Eq. 6.2-0)}$$

The current ratio is

$$I_p/I_s = V_s/V_p \quad \text{...............} \quad \text{(Eq. 6.2-1)}$$

The general transformation ratio is

$$I_p/I_s = V_s/V_p = N_s/N_p \quad \text{...} \quad \text{(Eq. 6.2-2)}$$

If an impedance Z_s is connected in the secondary, the impedance seen at the primary is

$$Z_p = V_p/I_p$$

$$V_p = (N_p/N_s)\, V_s \qquad \text{(Refer to Eq. 6.2-0)}$$

$$I_p = (N_s/N_p)\, I_s \qquad \text{(Refer to Eq.6.2-1)}$$

$$\therefore \quad Z_p = \frac{(N_p/N_s)V_s}{(N_s/N_p)I_s}$$

$$= (N_p/N_s)^2\, V_s/I_s$$

$$= (N_p/N_s)^2\, Z_s \qquad (Z_s = V_s/I_s)$$

(contd)

(b) For an ideal transformer

$$Z_p = (N_p/N_s)^2 Z_s \qquad \text{(Note: } Z_s = R_L\text{)}$$

Substituting values

$$16 = 4\,(N_p/N_s)^2$$

$$\therefore \quad (N_p/N_s)^2 = 16/4$$

$$= 4:1$$

$$\therefore \quad \text{Turns ratio, } N_p/N_s = \sqrt{(4:1)} = \underline{2:1}$$

Example 6.3

A silicon power transistor is employed in the design of the audio frequency amplifier shown in Fig. 6.3-0. The characteristics of the amplifier are shown in Fig. 6.3-1. The manufacturer's specification are

$I_{c(max)} = 1.0\ mA \qquad V_{CE(max)} = 50\ V$

$h_{FE} = 25 \qquad P_{D(max)} = 5\ W$ *(case at 30°C)*

Determine the component values in the circuit for maximum power to the load resistor $R_L = 800\ \Omega$. *Assume capacitive reactances are large enough to be neglected at the operating frequency.*

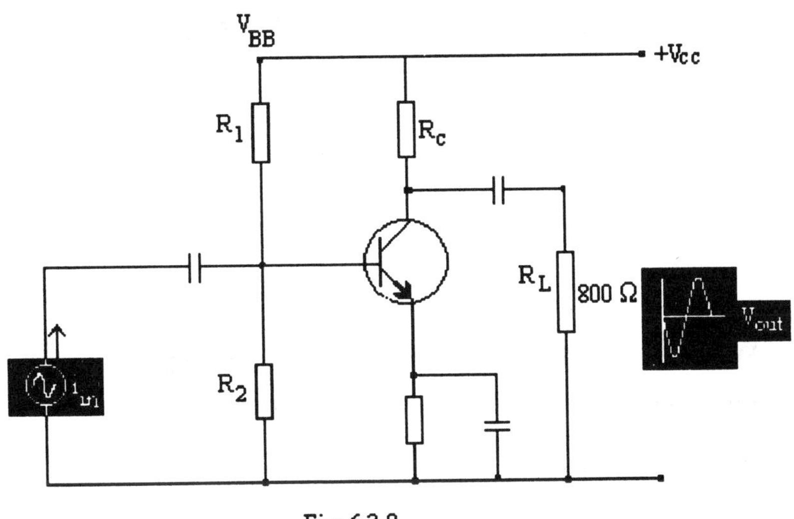

Fig. 6.3-0

For this design, superimpose on the transistor characteristics in Fig. 6.3-1

(i) the maximum power dissipation curve ($V_{CE}I_c = 5$ W)

(ii) the a.c load line tangent to the power dissipation curve (the hyperbola)

Devise a table for the power dissipation curve ($V_{CE}I_c = 5$ W) as follows:

V_{CE}	=	5	10	15	20	25	30	35	40	45	50 (volts)
I_c	=	1.0	0.5	0.33	0.25	0.2	0.16	0.14	0.125	0.11	0.1 (amp)

(contd)

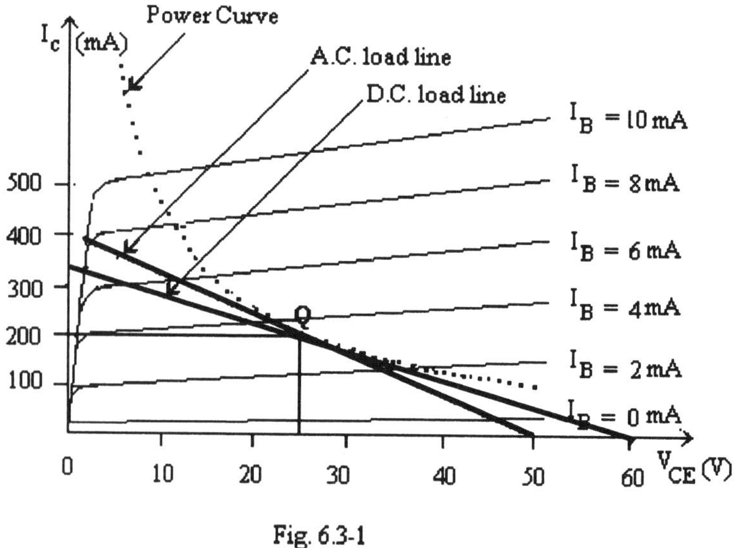

Fig. 6.3-1

Refer to Fig. 6.3-1

When the points are joined together, the hyperbola is obtained, representing the maximum power dissipation curve. It will be seen that the a.c. load line is tangential to the curve at V_{CE} = 25 V and I_c = 200 mA

For Class A operation, $V_{CE} = V_{cc}/2$

$= 50/2 \quad = \underline{25\ V}$

Power dissipation, $P_D = V_{CE}\, I_c$

$\therefore \quad I_c = P_D/V_{CE}$

$= 5/25 \quad = \underline{0.2\ A}$

The Q point is taken at V_{CE} = 25 V

and I_c = 0.2 A

$\therefore \quad R_{a.c.} = V_{CE}/I_c$

$= 25/0.2 \quad = \mathbf{\underline{125\ \Omega}}$

(contd)

To determine the collector resistance, R_c

$$^1/R_{a.c} = {}^1/R_c + {}^1/R_L$$

$$R_c = \frac{1}{1/R_{a.c} - 1/R_L}$$

$$= \frac{1}{1/125 - 1/800} \qquad = \mathbf{\underline{148\ \Omega}}$$

To design the bias circuit

Assume an emitter voltage drop $V_{RE} = 5V$

and $I_c = I_E$

$\therefore \quad R_E = {}^V{}_{RE}/I_c$

$= {}^5/0.2 \qquad = \mathbf{\underline{25\ \Omega}}$

To determine the required supply voltage V_{cc}

$$V_{cc} = V_{RE} + V_{CE} + I_cR_c$$

$$= 5 + 25 + (0.2 \times 148)$$

$$= 30 + 29.6 \qquad = 59.6$$

$$\approx \mathbf{\underline{60\ V}}$$

To stabilize changes in h_{FE}

Assume that $h_{FE}\, R_E = 10R_{B'}$

[where $R_{B'} = R_1R_2/(R_1 + R_2)$]

$\therefore \quad R_{B'} = h_{FE}\, R_E/10$

$= (25 \times 25)/10 \qquad = \mathbf{\underline{62.5\ \Omega}}$

To determine V_{BB}

Assume that $V_{BE} = 0.7$ V (A typical value for Si)

From the characteristic the Q point is at $I_B \approx 3$ mA

$\therefore \quad V_{BB} = I_BR_{B'} + V_{BE} + I_ER_E$

[Note: $I_E = (I_B + I_c) = 0.003 + 0.2$] $\quad = (0.003 \times 62.5) + 0.7 + (0.203)\ 25$

$= 0.1875 + 0.7 + 5.075 \approx \mathbf{\underline{6.0\ V}}$

(contd)

Refer to the equations on the derivation of R_1 and R_2

$$R_1 = R_{B'} ({}^{V}cc/V_{BB})$$

$$= 62.5 \times {}^{60}/6 \qquad = \underline{\mathbf{625\ \Omega}}$$

$$R_2 = ({}^{R_1}\, {}^{R_{B'}})/(R_1 - R_{B'})$$

$$= {}^{(625 \times 62.5)}/(625 - 62.5)$$

$$= {}^{39062.5}/562.5 \qquad = \underline{\mathbf{69.4\ \Omega}}$$

Derivation of equations for R_1 and R_2

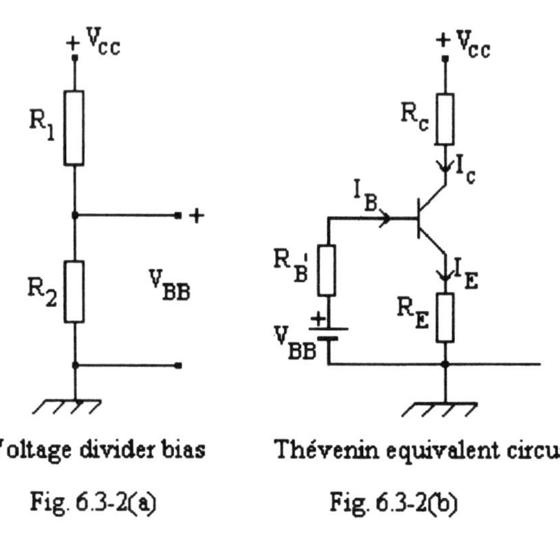

Voltage divider bias — Fig. 6.3-2(a)

Thévenin equivalent circuit — Fig. 6.3-2(b)

$$V_{BB} = \frac{R_2}{R_1 + R_2} V_{cc} \qquad \text{(i)}$$

$$R_{B'} = \frac{R_1 R_2}{R_1 + R_2} \qquad \text{(ii)}$$

$$\frac{V_{cc}}{V_{BB}} = \frac{R_1 + R_2}{R_2} \qquad \text{(iii)}$$

(contd) Derivation of equations for R_1 and R_2

Multiply both sides of (Eq.iii) by $R_1R_2/(R_1 + R_2)$ (that is, by $R_{B'}$).

$$\frac{R_1 + R_2}{R_2} \times \frac{R_1 R_2}{R_1 + R_2} = \frac{V_{cc}}{V_{BB}} \times \frac{R_1 R_2}{R_1 + R_2}$$

$$\therefore \quad R_1 = R_{B'} \frac{V_{cc}}{V_{BB}} \quad \text{......................} \quad (iv)$$

Substituting (iii) into (iv), we get

$$R_1 = R_{B'} \frac{R_1 + R_2}{R_2}$$

$$\therefore \quad R_1R_2 = R_{B'}R_1 + R_{B'}R_2$$

$$R_2 (R_1 - R_{B'}) = R_{B'}R_1$$

$$R_2 = \frac{R_{B'}.R_1}{R_1 - R_{B'}} \quad \text{.............} \quad (v)$$

Example 6.4

A single-ended transistor power amplifier takes a mean collector current of 1.0 A from a 12-Vd.c. supply and delivers an a.c. power of 3.6 W to a transformer-coupled loudspeaker. Calculate, neglecting all losses

(i) the collector efficiency

(ii) the collector power dissipation

The transistor, now used in a Class A audio frequency power amplifier, takes a collector bias current of 0.25 A. When a sinusoidal input signal is applied to the amplifier the collector voltage varies from –2 V to –18 V and the collector current from –0.3 A to –0.2 A. Calculate

(iii) the d.c. power taken from the supply

(iv) the a.c. power output

Solution

(i) The d.c input power, $P_{IN} = V_{CQ} \times I_{CQ}$

$= V_{CE} \times I_{CQ}$

$= 12 \times 1.0 \qquad = \underline{12\ W}$

The collector efficiency, $\eta = \frac{\text{output power}}{\text{input power}}$

$= {}^{3.6}/_{12} \qquad = \underline{\mathbf{0.3\ (30\%)}}$

(ii) Collector power dissipation, $P_d = P_{IN} - P_{OUT(a.c.)}$

$= 12 - 3.6 \qquad = \underline{\mathbf{8.4\ W}}$

[Note: Power is lost in the primary windings of the output transformer, the emitter resistor R_E and the collector. However, it is assumed that the losses in the transformer and R_E are very small in comparison to the collector.]

(iii) The d.c. power taken from the supply is

$P_{IN} = 12 \times 0.25 \qquad = \underline{\mathbf{4\ W}}$

(iv) The a.c. power output, $P_{OUT(a.c.)} = \frac{I_{max} - I_{min}}{2\sqrt{2}} \times \frac{V_{max} - V_{min}}{2\sqrt{2}}$

$= \frac{(0.3 - 0.2) \times (18 - 2)}{8} \qquad = \underline{\mathbf{0.2\ W}}$

$\underline{\mathbf{(200\ mW)}}$

Example 6.5

A push-pull audio amplifier working under Class A conditions employs two transistors, each of which has h_{fe} = 49 and h_{oe} = 100 x 10^{-6} S. The output transformer has a turns ratio of 5:1 and feeds a load resistance of 4 Ω. The collector supply is –25V and the direct current taken from the supply is 0.1 A. Calculate the efficiency of the amplifier when a sinusoidal input signal of 1.0 mA (rms) is applied to the base of each transistor.

Solution

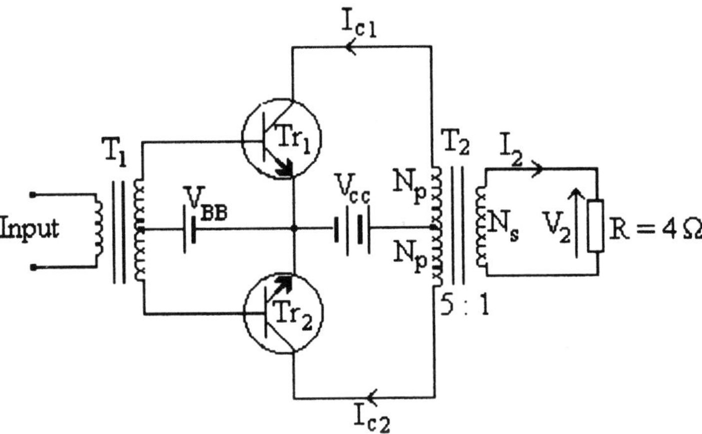

Fig. 6.5-0. Basic push-pull amplifier

Refer to Fig. 6.5-0

The circuit shows a simple output stage using center-tap input and output transformers.

The a.c. load resistance reflected into each half of the primary (N_p) of the output transformer (T_2) is

$$R_{a.c.} = (N_p/N_s)^2 R$$
$$= n^2 R \text{ (where } n^2 = \text{transformation ratio)}$$

The effective collector-to collector load resistance is

$$R_{a.c.} = n^2 R$$
$$= 5^2 \times 4 \quad = \underline{100\ \Omega}$$

The load on each transistor $= 100/2 \quad = \underline{50\ \Omega}$

(contd)

The current gain (A_i) of each transistor is

$$A_i = h_{fe}/(1 + h_{oe}R_L)$$

$$= 49/[1 + (100 \times 10^{-6})50] \approx \underline{49}$$

Current gain, $A_i = I_c/I_b$

$\therefore \quad I_c = A_i I_b$

The load current (I_L) of each transistor is

$$I_L = A_i I_b$$ [Note: load current = collector current]

$$= 49 \times 1.0 \times 10^{-3} = \underline{0.049\ A}$$

The power output from each transistor is

$$P_{out} = I_L^2 R_{a.c.}$$

$$= 0.049^2 \times 50 = \underline{0.12\ W}$$

Total power output is

$$P_{out(T)} = 2 \times 0.12 = \underline{0.24\ W}$$

The input power (d.c.) taken from the supply is

$$P_{IN} = 25 \times 0.1 = \underline{2.5\ W}$$

The collector efficiency,

$$\eta = P_{out(T)}/P_{IN}$$

$$= (0.24/2.5) \times 100 = \underline{\mathbf{9.6\%}}$$

[Note on Fig. 6.5-0: The expensive and bulky transformers can be eliminated by using *npn* and *pnp* transistors in the complementary symmetry configuration.]

CHAPTER 7

NEGATIVE FEEDBACK

Introduction

One of the most important features associated with negative feedback is the controlling action it imparts to closed-loop amplifiers and control systems. Generally, a fraction of the output is fed back in antiphase to the input and as a result corrective action is enforced.

Overview of Basic Theory

In a negative feedback amplifier a fraction of the output signal is fed back in antiphase with the input signal. The different methods employed to derive the fraction of the feedback signal lead to the classification of the four types of negative feedback amplifiers.

1. Series Voltage Feedback
   or
   Voltage-Voltage Feedback

2. Series Current Feedback
   or
   Voltage-Current Feedback

3. Shunt Voltage Feedback
   or
   Current-Voltage Feedback

4. Shunt Current Feedback
   or
   Current-Current Feedback

Basically, when the feedback signal is proportional to the output voltage of the amplifier, voltage feedback is said to be applied.
Current feedback is said to be applied when the feedback signal is proportional to the output current.

(contd)

1. *Series Voltage Feedback*

A fraction of the output voltage is fed back in series with the input signal voltage. The feedback voltage is arranged to be in antiphase with the input signal.

Refer to Fig. 7.0-0

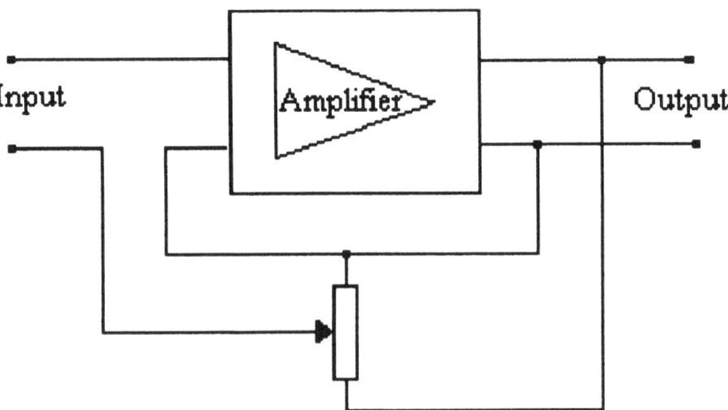

Fig. 7.0-0. Series voltage feedback

2. *Series Current Feedback*

The feedback voltage is proportional to the output current and is applied in series opposition with the input signal.

Refer to Fig. 7.0-1

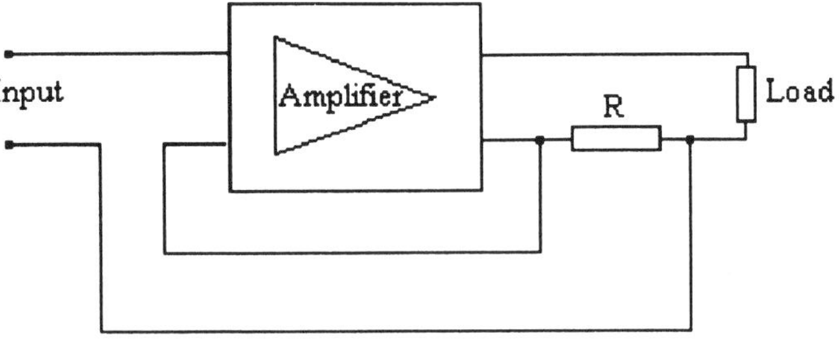

Fig. 7.0-1. Series current feedback

(contd)

3. *Shunt Voltage Feedback*

A current proportional to the output voltage is fed back in antiphase with the input signal current.

Refer to Fig. 7.0-2

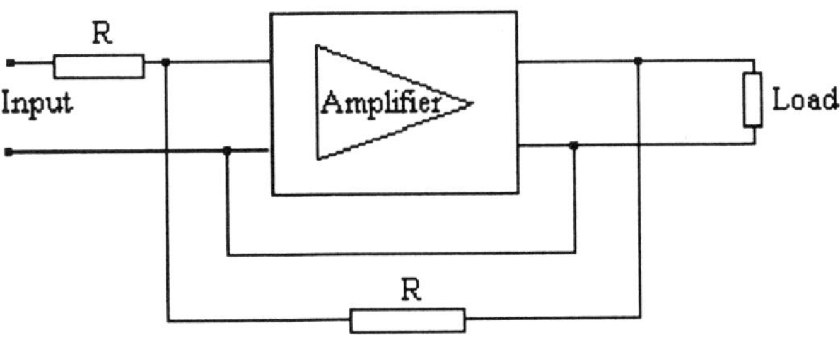

Fig. 7.0-2. Shunt voltage feedback

4. *Shunt Current Feedback*

A fraction of the output current is fed back in antiphase with the input signal current.

Refer to Fig. 7.0-3

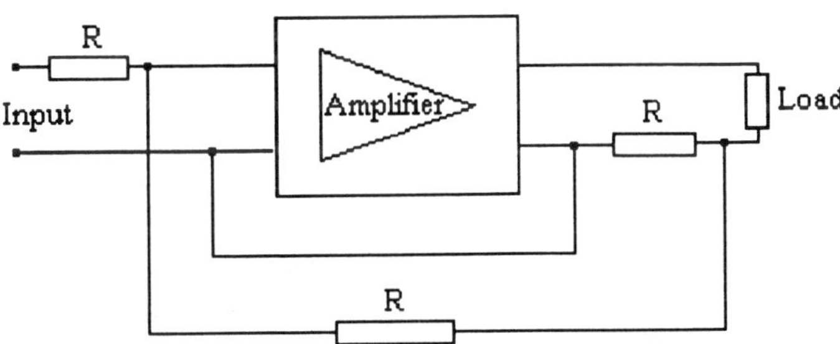

Fig. 7.0-3. Shunt current feedback

Example 7.1

(a) *If the gain of an amplifier stage without feedback is represented by A_v, derive an expression for the gain when a fraction of the output voltage is fed back in antiphase to the input.*

(b) *An amplifier has a gain of 100 without feedback. Calculate its voltage gain if 3/100 of the output voltage is fed back to the input in opposition to the input signal.*

(c) *If due to ageing the gain without feedback falls to 80, calculate the the percentage reduction in gain*

(i) *without feedback*

(ii) *with feedback*

Comment on the significance of the results in (i) and (ii).

Solution

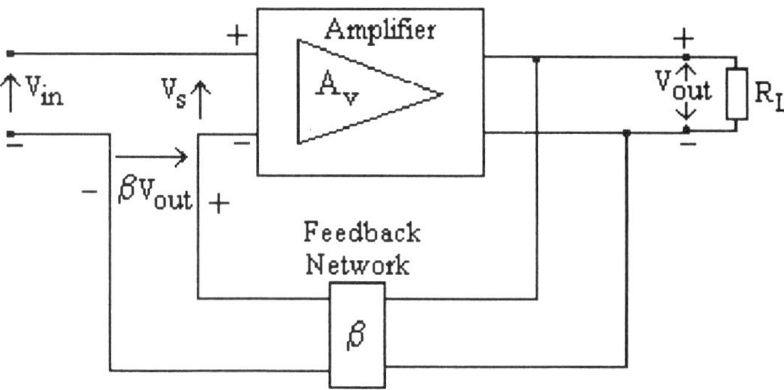

Fig. 7.1-0. Series voltage feedback

(a) Refer to Fig. 7.1-0

The output voltage of the amplifier	= Vout
Voltage gain, A_V of the amplifier without feedback	= Vout/V_{in}
The fraction of the voltage fed back	= β
Voltage gain $A_{V'}$ of the amplifier with feedback	= V'out/V_{in}
Voltage fedback in anti-phase to the input	= $-\beta V_{out}$

The expression for the gain with negative feedback will be considered from two norms; however, the end result will be the same.

(contd) *(a)*

(Case 1)	(Case 2)
$V_s = V_{in} - \beta V_{out}$	$V_s = V_{in} + \beta V_{out}$
$V_{out} = A_V V_s$	$= A_V V_s$
$= A_V (V_{in} - \beta V_{out})$	$= A_V(V_{in} + \beta V_{out})$
$= A_V V_{in} - \beta A_V V_{out}$	$= A_V V_{in} + \beta A_V V_{out}$
$V_{out} + \beta A_V V_{out} = A_V V_{in}$	$V_{out} - \beta A_V V_{out} = A_V V_{in}$
$V_{out}(1 + \beta A_V) = A_V V_{in}$	$V_{out}(1 - \beta A_V) = A_V V_{in}$
$V_{out} = \dfrac{A_V V_{in}}{1 + \beta A_V}$	$V_{out} = \dfrac{A_V V_{in}}{1 - \beta A_V}$
In case 1, the feedback fraction βA_V was initially treated as negative.	In case 2, the feedback fraction $(-\beta A_V)$ will have to be introduced in the expression.

Let the voltage gain with *negative feedback* be denoted by $\mathbf{A_V'}$. Then

$$A_V' = \frac{V_{out}}{V_{in}} = \frac{A_V}{1 - \beta A_V}$$

Since the feedback is negative (antiphase to the input), βA_V is *negative*

$$\therefore \quad A_V' = \frac{A_V}{1 - (-\beta A_V)} = \frac{A_V}{1 + \beta A_V} \quad \text{.....................} \quad \text{Eq. 7.1-0)}$$

(b)

$$A_V' = \frac{A_V}{1 + \beta A_V} = \frac{100}{1 - (-3/100) \times 100}$$

$$= \frac{100}{1 + (3/100) \times 100} = \frac{100}{4}$$

$$= \mathbf{25}$$

(c) (i) Percentage reduction in gain without feedback is

$$= \frac{100 - 80}{100} \times 100\% = \underline{\mathbf{20\%}}$$

(ii) New gain with feedback, A_V' $= 80/[1 + (3/100 \times 80)$

$= 80/3.4 = \underline{\mathbf{23.53}}$

∴ Reduction in gain with feedback

$= 25 - 23.53 = \underline{\mathbf{1.47}}$

Expressed as a percentage $= [(25 - 23.53)/25] \times 100$

$= \underline{\mathbf{5.89\%}}$

Negative feedback stabilizes the amplifier against ageing of components at the expense of a small reduction in gain.

Example 7.2

(a) *Sketch the circuit diagram of an emitter-follower amplifier and derive an expression for the voltage gain with negative feedback.*

(b) *An emitter-follower amplifier employs a transistor whose h-parameters are $h_{ie} = 1000\ \Omega$, $h_{fe} = 50$, $h_{oe} = 80 \times 10^{-6}$ S and $h_{re} = 0.1 \times 10^{-3}$. Calculate its voltage gain when the emitter resistor is 2.5 kΩ.*

Solution

(a)

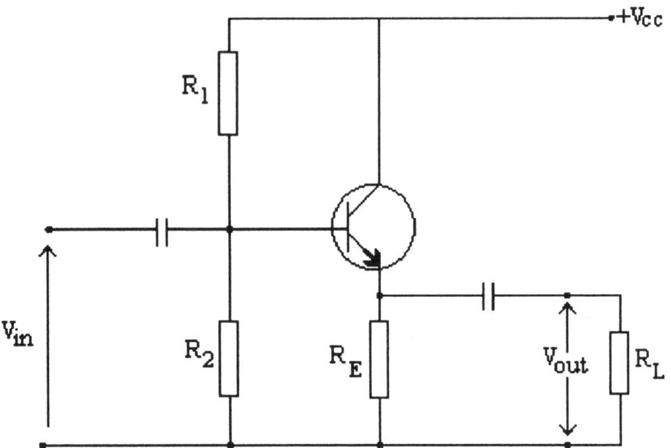

Fig. 7.2-0. Emitter-follower (common collector)

$$\text{Voltage gain,} \quad A_v = \frac{A_i R_L}{R_{in}} \qquad \text{(i)}$$

$$R_{in} = \frac{V_{be}}{I_b} = h_{ie}$$

$$\text{Current gain,} \quad A_i = \frac{I_c}{I_b} = \frac{-h_{fe}}{1 + h_{oe} R_L}$$

Substituting for A_i and R_{in} in (i), we get

$$A_v = \frac{-h_{fe}}{1 + h_{oe} R_L} \times \frac{R_L}{h_{ie}}$$

(contd) *(a)*

The relationships connecting the common-emitter *h*-parameters to the emitter-follower parameters are

$$h_{ic} = h_{ie} \qquad h_{fc} = -(h_{fe} + 1)$$

$$h_{oc} = h_{oe} \qquad h_{rc} = 1 - h_{re}$$

The load R_L is assumed to have a negligible shunting effect on R_E.

Hence $$A_v = \frac{-h_{fc}}{1 + h_{oc} R_E} \times \frac{R_E}{h_{ie}}$$

In this circuit (emitter-follower) the whole output voltage is fed back in series opposition to the input signal, so that

$$V_{be} = V_{in} - V_{out}$$

and $\beta = 1$ (that is, 100% negative feedback)

$$\therefore \quad A'_v = \frac{A_v}{1 + \beta A_v}$$

$$= \frac{\dfrac{(-h_{fc} R_E)}{h_{ic}(1 + h_{oc} R_E)}}{1 + \dfrac{(-h_{fc} R_E)}{h_{ic}(1 + h_{oc} R_E)}}$$

$$= \frac{\dfrac{(-h_{fc} R_E)}{h_{ic}(1 + h_{oc} R_E)}}{\dfrac{h_{ic}(1 + h_{oc} R_E) + (-h_{fc} R_E)}{h_{ic}(1 + h_{oc} R_E)}}$$

$$= \frac{(-h_{fc} R_E)}{h_{ic}(1 + h_{oc} R_E)} \times \frac{h_{ic}(1 + h_{oc} R_E)}{h_{ic}(1 + h_{oc} R_E) + (-h_{fc} R_E)}$$

$$= \frac{(-h_{fc} R_E)}{h_{ic} + h_{oc} h_{ic} R_E + (-h_{fc} R_E)}$$

$$= \frac{-h_{fc} R_E}{h_{ic} + R_E(-h_{fc} + h_{oc} h_{ic})} \qquad \text{(Eq. 7.2-0)}$$

(contd)

(b)

$$h_{ic} = h_{ie} = 1000\ \Omega$$

$$h_{fc} = -(h_{fe} + 1) = -51$$

$$h_{oc} = h_{oe} = 80 \times 10^{-6}\ \text{S}$$

$$h_{rc} = (1 - h_{re}) = 1 - 0.1 \times 10^{-3} \approx 1$$

$$A'_v = \frac{-h_{fc}R_E}{h_{ic} + R_E(-h_{fc} + h_{oc}h_{ic})}$$

$$= \frac{51 \times 2.5 \times 10^3}{1000 + 2.5 \times 10^3[51 + (80 \times 10^{-6} \times 1000)]}$$

$$= \frac{51 \times 2.5 \times 10^3}{1000 + 127.5 \times 10^3 + 200}$$

$$= \frac{127.5 \times 10^3}{128.7 \times 10^3} = \underline{\mathbf{0.9907}}$$

The emitter-follower has a voltage gain of less than unity.

Example 7.3

(a) *In a series current feedback amplifier derive an expression for the fraction β of the output fed back to the input circuit.*

(b) *The amplifier shown in Fig. 7.3-1 has $R_L = 5000\ \Omega$, $R_E = 1000\ \Omega$, $h_{ie} = 800\ \Omega$, $h_{re} = 0$, $h_{oe} = 100 \times 10^{-6}$ S and $h_{fe} = 100$. Assuming that the shunting effects of R_1 and R_2 are negligible, calculate the voltage gain of the amplifier.*

Solution

(a)

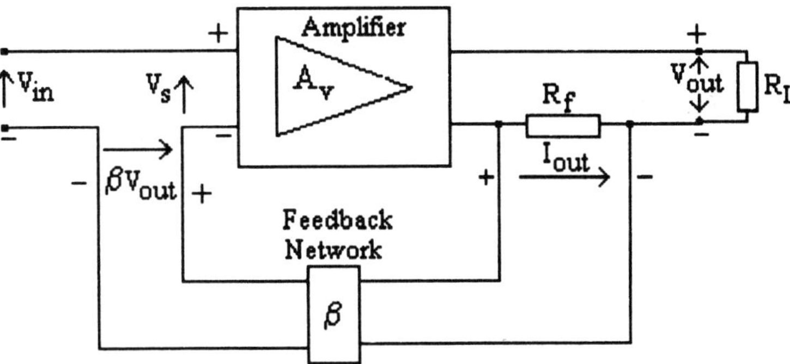

Fig. 7.3-0. Series current feedback

The output current I_{out} flows in the feedback resistor R_f and the load resistor R_L.

$$\text{Output voltage, } V_{out} = I_{out} R_L$$

$$\text{Fraction of } V_{out} \text{ fed back} = I_{out} R_f$$

$$= -\beta V_{out}$$

$$\beta V_{out}/V_{out} = (I_{out} R_f)/(I_{out} R_L)$$

$$\therefore \quad \beta = R_f/R_L \qquad \text{................(Eq. 7.3-0)}$$

(contd)

(b)

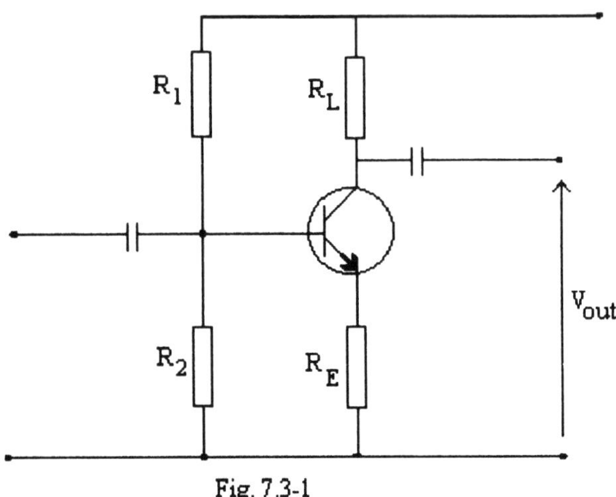

Fig. 7.3-1

Neglecting the shunting effects of R_1 and R_2, the voltage gain is

$$A_v = A_i \frac{R_L}{R_{in}}$$

$$= \frac{h_{fe}}{1 + h_{oe}R_L} \times \frac{R_L}{h_{ie}}$$

Gain with feedback, $$A'_v = \frac{A_v}{1 + \beta A_v} \qquad \left[\beta = \frac{R_E}{R_L}\right]$$

$$= \frac{\dfrac{h_{fe}R_L}{(1 + h_{oe}R_L)\,h_{ie}}}{1 + \dfrac{h_{fe}R_L}{(1 + h_{oe}R_L)h_{ie}} \times \dfrac{R_E}{R_L}}$$

$$= \frac{\dfrac{h_{fe}R_L}{(1 + h_{oe}R_L)h_{ie}}}{\dfrac{(1 + h_{oe}R_L)h_{ie} + h_{fe}R_E}{(1 + h_{oe}R_L)h_{ie}}}$$

$$= \frac{h_{fe}R_L}{(1 + h_{oe}R_L)h_{ie} + h_{fe}R_E} \qquad \text{(Eq. 7.3-1)}$$

If it is assumed that $h_{fe}R_E >> (1 + h_{oe}R_L)h_{ie}$, then (Eq. 7.3-1) reduces to

$$\mathbf{A'_v} = \mathbf{R_L/R_E}$$

$$= {}^1/\beta \qquad \text{(Eq. 7.3-2)}$$

(contd) *(b)*

Using (Eq. 7.3-1) and substituting values, we get

$$A'_V = \frac{100 \times 5 \times 10^3}{(1 + 100 \times 10^{-6} \times 5 \times 10^3) 800 + 100 \times 10^3}$$

$$= \frac{500 \times 10^3}{(800 + 400) + (100 \times 10^3)} = \underline{\mathbf{4.94}}$$

Using (Eq. 7.3-2) and substituting values, we get

$$A'_V = \frac{5 \times 10^3}{1 \times 10^3} = \underline{\mathbf{5}}$$

Example 7.4

(a) *Derive an expression for the feedback factor β in a shunt voltage (current-voltage) feedback amplifier.*

(b) *A simple shunt voltage feedback amplifier is shown in Fig. 7.4-0. Calculate the current gain if $R_L = 2.5\ k\Omega$, $R_f = 56\ k\Omega$, $h_{ie} = 800\ \Omega$, $h_{oe} = 50 \times 10^{-6}$ S, $h_{fe} = 100$ and h_{re} is negligible.*

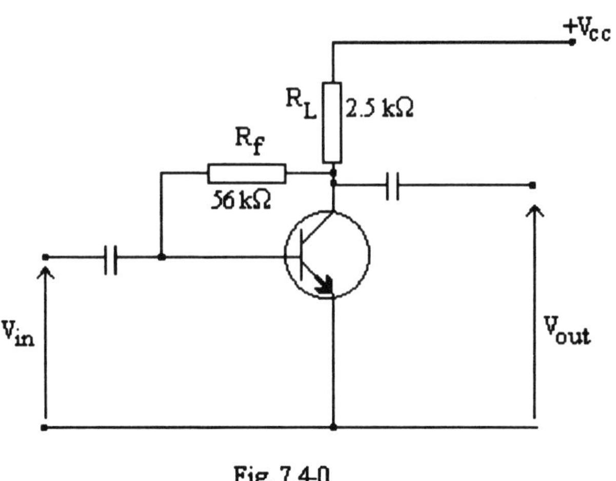

Fig. 7.4-0

Solution

(a)

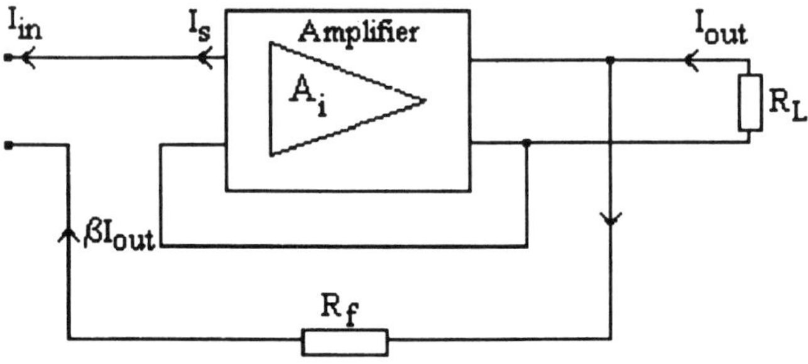

Fig. 7.4-1. Shunt voltage feedback

$$\text{Output current, } I_{out} = V_{out}/R_L$$

$$\beta I_{out}/I_{out} = (V_{out}/R_f)/(V_{out}/R_L)$$

$$\therefore \quad \beta = \mathbf{R_L/R_f} \qquad \text{......... (Eq. 7.4-0)}$$

(contd)

(b)

The effective collector load, $R'_L = \dfrac{R_f R_L}{R_f + R_L}$

$= \dfrac{56 \times 2.5}{56 + 2.5} = 2.393\ k\Omega$

Current gain before feedback, $A_i = \dfrac{h_{fe}}{1 + h_{oe}R'_L}$

$= \dfrac{100}{1 + (50 \times 10^{-6} \times 2.393 \times 10^3)}$

$= \dfrac{100}{1.12}$ = **89.29**

The feedback factor, $\beta = R_L/R_f$ (Refer to Eq. 7.4-0)

$= 2.5/56$

Current gain with feedback, $A'_i = A_i/(1 + \beta A_i)$

$= 89.29/[1 + (2.5/56) \times 89.29]$

$= 89.29/4.986$ = **17.91**

Example 7.5

If the current gain of an amplifier stage without feedback is represented by A_i, derive an expression for the current gain when a fraction of the output current is fed back in antiphase with the input signal current.

Solution

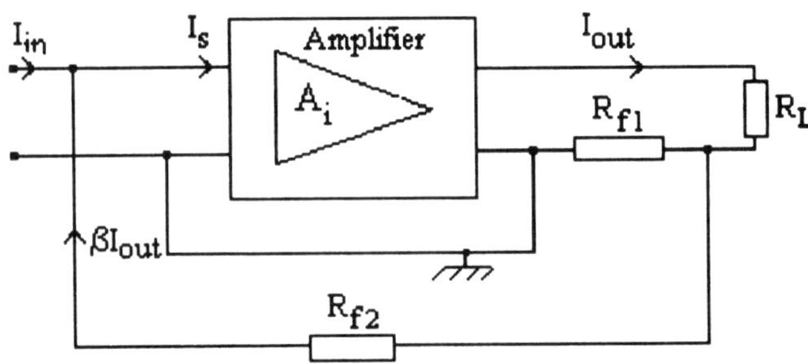

Fig. 7.5-0. Shunt current feedback (current-current feedback)

The input impedance Z_{in} of the amplifier is assumed to be very low. The current source I_{in} is assumed to have a very high impedance. Assuming that $R_{f2} >> Z_{in}$, the fraction of current (βI_{out}) fed back is

$$\beta I_{out} = \frac{I_{out} R_{f1}}{R_{f1} + R_{f2}}$$

$$\beta = \frac{R_{f1}}{R_{f1} + R_{f2}}$$

The feedback current βI_{out} is in opposition to the input current I_{in} so that

$$I_s = I_{in} - \beta I_{out}$$

$$I_{out} = A_i I_s$$

$$= A_i (I_{in} - \beta I_{out})$$

$$I_{out} (1 + \beta A_i) = A_i I_{in}$$

$\therefore$ Gain with feedback, $\mathbf{A'_i} = I_{out}/I_{in}$

$$= \mathbf{A_i/(1 + \beta A_i)} \quad \text{.........} \quad \text{(Eq. 7.5-0)}$$

Example 7.6

Show that an amplifier with negative voltage feedback in series with the input voltage increases the input impedance (Z_{in}) and decreases the output impedance (Z_o) of the amplifier.

Solution

[Note: In general, feedback can affect both the input and output impedance levels of an amplifier (which is usually multistage). The way in which feedback is introduced into the input circuit influences the output and input impedances. **Generally, voltage feedback reduces the output impedance while current feedback increases it.]**

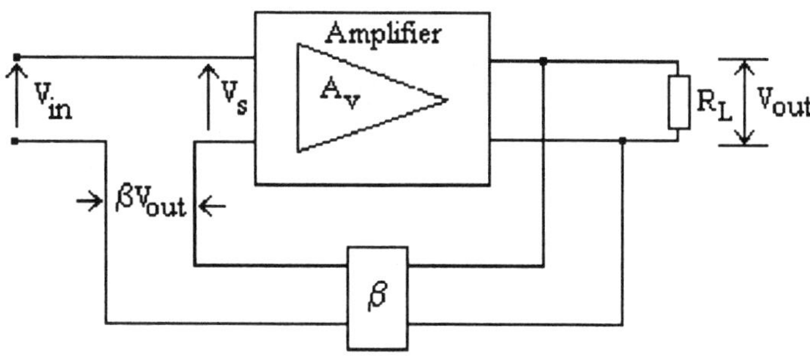

Fig. 7.6-0

Series Negative Voltage Feedback

<u>*Input Impedance Z'_{in}* - Refer to Fig. 7.6-0</u>

With no feedback, $Z_{in} = V_{in}/I_{in}$

$$I_{in} = V_{in}/Z_{in} \quad \text{...... (Eq. 7.6-0)}$$

In series negative feedback the input voltage required for a given input current must be increased. Consequently, the input impedance is increased. Let the net input voltage be V'_{in}. Then

$$I_{in} = (V'_{in} - \beta V_{out})/Z_{in}$$

$$\therefore \quad V'_{in} = I_{in} Z_{in} + \beta V_{out}$$

(contd) - Z'_{in}

Input impedance with negative feedback is

$$Z'_{in} = V'_{in}/I_{in}$$

$$= \frac{I_{in}Z_{in} + \beta V_{out}}{I_{in}}$$

$$= Z_{in} + \frac{\beta V_{out}}{I_{in}}$$

Substituting for I_{in} from Eq. 7.6-0, we get

$$Z'_{in} = Z_{in} + \frac{\beta V_{out}}{V_{in}/Z_{in}}$$

$$= Z_{in} + \beta V_{out} \cdot \frac{Z_{in}}{V_{in}}$$

$$= Z_{in}(1 + \frac{\beta V_{out}}{V_{in}})$$

$$= Z_{in}(1 + \frac{\beta A_v V_{in}}{V_{in}})$$

$$\therefore \quad \mathbf{Z'_{in} = Z_{in}(1 + \beta A_v)} \quad \text{............} \quad \text{(Eq. 7.6-1)}$$

Therefore, negative feedback in series with the input signal voltage increases the input impedance by a factor of $(1 + \beta A_v)$; that is

$$\mathbf{Z'_{in} = Z_{in}(1 + \beta A_v)} \quad \text{(Refer to Eq. 7.6-1)}$$

Output Impedance, Z'_o - Refer to Fig. 7.6-0

Looking into the amplifier from its output terminals, we may regard it as a voltage generator having an e.m.f **E** with its own internal impedance $\mathbf{Z_o}$ as shown in Figs. 7.6-1(a) and 7.6-1(b).

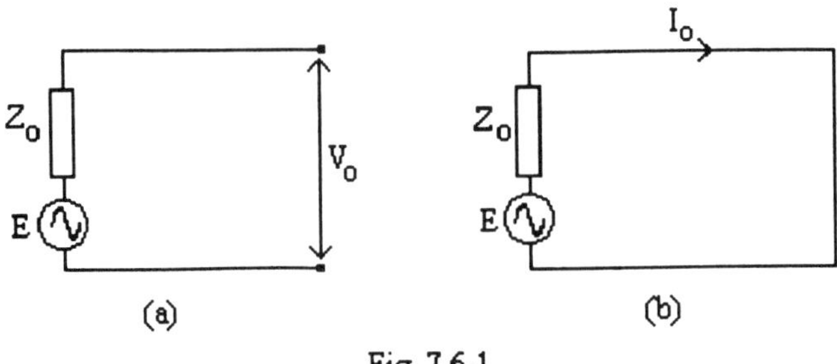

Fig. 7.6-1

(contd) - Z'_o : Refer to Figs. 7.6-1(a) and 7.6-1(b)

Open-circuit output voltage, $V_o = E$

Short-circuit output current, $I_o = E/Z_o$

Output impedance, $Z_o = E/I_o$

The open-circuit and short-circuit conditions of this amplifier are shown in Figs. 7.6-2(a) and 7.6-2(b).

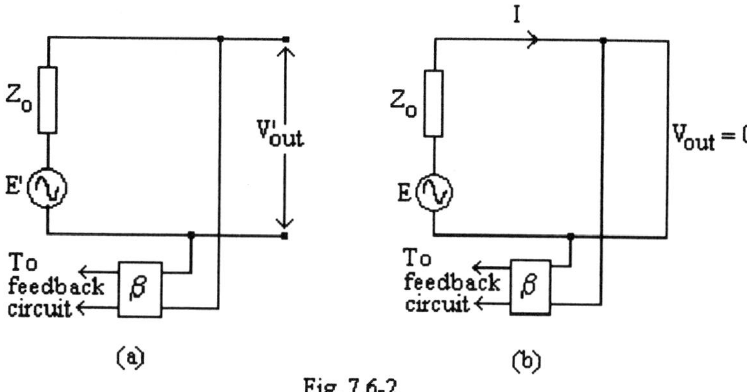

Fig. 7.6-2

On open circuit, $E' = V'_{out}$

[Refer to Example 7.1(a) Case 1 and Eq. 7.1-0]

$$= A_v V_{in}/(1 + \beta A_v) \qquad \text{..(Eq. 7.6-2)}$$

On short circuit, $V_{out} = 0$. Hence there will be no feedback, so that

$$E' = E = A_v V_{in}$$

and

$$I = E/Z_o$$

$$= A_v V_{in}/ Z_o \quad \ldots\ldots \qquad \text{(Eq.7.6-3)}$$

Therefore, the output impedance with negative feedback (Z'_o) is

[Divide Eq. 7.6-2 by Eq. 7.6-3]

$$\mathbf{Z'_o} = E'/I$$

$$= \frac{\dfrac{A_v V_{in}}{1 + \beta A_v}}{\dfrac{A_v V_{in}}{Z_o}}$$

$$= \frac{\mathbf{Z_o}}{\mathbf{1 + \beta A_v}} \qquad \ldots\ldots\ldots\ldots\ldots \qquad \text{(Eq. 7.6-4)}$$

(contd) - Z'_o

Hence, series voltage feedback in opposition to the input signal decreases the output impedance by a factor of $[^1/(1 + \beta Av)]$; that is

$$\mathbf{Z'_0 = Z_0 \times {}^1/(1 + \beta A_v)} \quad \text{(Refer Eq. 7.6-4)}$$

Alternative Method - Thévenin's Equivalent Circuit - Z'_o:

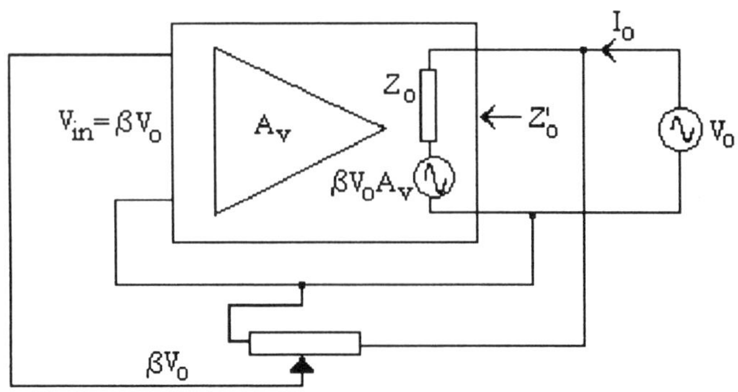

Fig. 7.6-3

Refer to Fig. 7.6-3

A constant-voltage generator is connected to the output terminals and the input terminals are short-circuited.

A fraction β of the output voltage V_o is fed back to the input.

Z_o is the output impedance without feedback.
Z'_o is the output impedance with feedback.

The current, $$I_o = \frac{V_o + \beta V_o A_v}{Z_o}$$

$$Z'_o = \frac{V_o}{I_o} = \frac{V_o}{\dfrac{V_o + \beta V_o A_v}{Z_o}}$$

$$= \frac{V_o Z_o}{V_o (1 + \beta A_v)}$$

$$\mathbf{Z'_o = \frac{Z_o}{1 + \beta A_v}} \quad \text{............. (Eq. 7.6-5)}$$

Example 7.7

Show that negative current feedback in series with the input signal increases both the input and output impedances (Z_{in} and Z_o) of an amplifier.

Solution

Series Current Negative Feedback - Z'_{in}

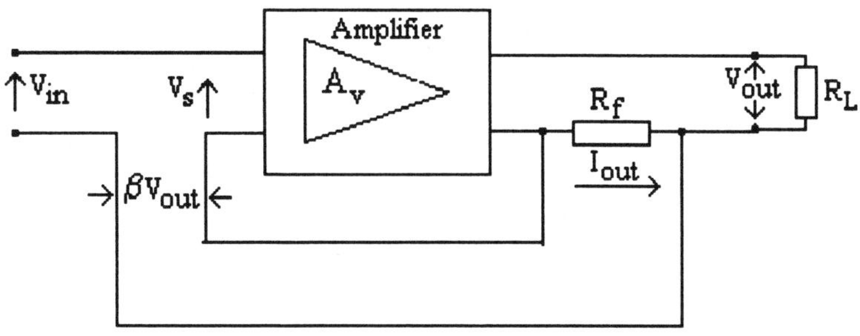

Fig. 7.7-0

$$V_{out} = I_{out} R_L$$

Feedback voltage $= I_{out} R_f$

$$\beta = R_f/R_L$$

With no feedback, $Z_{in} = V_{in}/I_{in}$

$\therefore \quad I_{in} = V_{in}/Z_{in}$ .... (Eq.7.7-0)

With negative feedback

$$I_{in} = (V'_{in} - \beta V_{out})/Z_{in}$$

$$V'_{in} = I_{in} Z_{in} + \beta V_{out}$$

$$Z'_{in} = V'_{in}/I_{in}$$

$$= (I_{in} Z_{in} + \beta V_{out})/I_{in}$$

$$= Z_{in} + \beta V_{out}/I_{in}$$

Substituting for I_{in} from Eq. 7.7-0, we get

$$\mathbf{Z'_{in}} = Z_{in} + \beta V_{out}/ \frac{V_{in}}{Z_{in}}$$

$$= \mathbf{Z_{in}(1 + \beta A_v)}$$ .... (Eq.7.7-1)

(contd) - Z'_{in}

Equation 7.7-1 for the input impedance is in the same form as for series voltage feedback.

Hence, it can be concluded that series negative feedback always increases the input impedance of an amplifier.

Series Current Negative Feedback - Z'_o

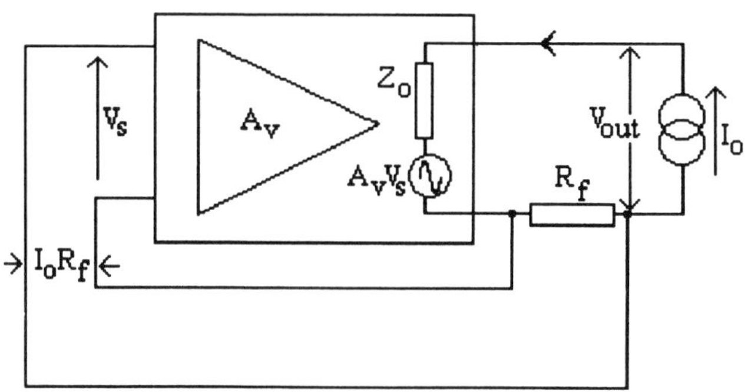

Fig. 7.7-1

For the output impedance, a constant-current generator, I_o, is connected across the output as shown in Fig. 7.7-1.

Voltage fed back to the input = I_oR_f = V_s

Voltage gain of amplifier = A_v

Refer to Fig. 7.7-1

$$V_{out} = I_o(Z_o + R_f) + A_vV_s$$

$$= I_o(Z_o + R_f) + I_oR_fA_v$$

$$= I_oZ_o + I_oR_f + I_oR_fA_v$$

∴ With feedback, $\mathbf{Z'_o} = \mathbf{V_{out}/I_o}$

$$= \mathbf{Z_o + (1 + A_v)\,R_f} \quad ..... \quad \text{(Eq. 7.7-2)}$$

Hence, in series current negative feedback the output impedance is higher with feedback than without feedback.

(contd) *Alternative Method - Z'_o*

Refer to Fig. 7.7-2

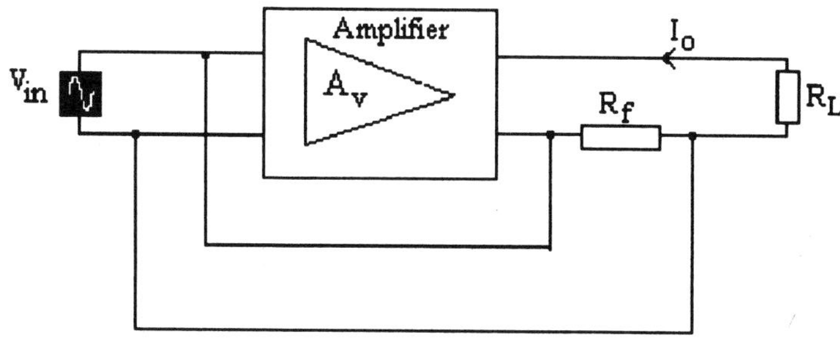

Fig. 7.7-2

Voltage fed back $= I_oR_f$

Feedback fraction, $\beta = \dfrac{I_oR_f}{I_o(R_f + R_L)}$

$$= \frac{R_f}{R_f + R_L}$$

$R_f << R_L$ so that

$$\beta = R_f/R_L$$

The open-circuit and short-circuit conditions for this amplifier are shown in Figs. 7.7-3(a) and 7.7-3(b).

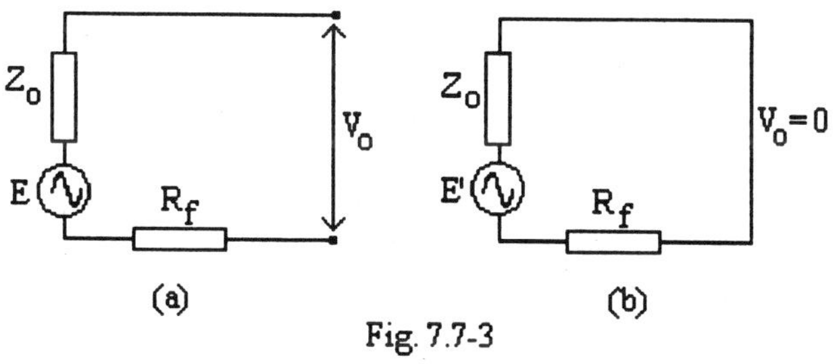

Fig. 7.7-3

On open-circuit [Fig. 7.7-3(a)], there will be no feedback voltage.

$\therefore$ Open-circuit voltage, $V_o = E = A_vV_{in}$ (Eq. 7.7-3)

(contd) - *Alternative Method* - Z'_o

On short-circuit [Fig.7.7-3(b)], the short-circuit current is

$$I_o = E'/(Z_o + R_f)$$

Assuming $R_f << Z_o$,

$$I_o = E'/Z_o$$

From Eq. 7.6-2 $\quad E' = A_v V_{in}/(1 + \beta A_v)$

$$\therefore \quad I_o = A_v V_{in}/[(1 + \beta A_v)Z_o] \quad \text{...(Eq. 7.7-4)}$$

With feedback,

$$\mathbf{Z'_o} = \frac{E'}{I_o}$$

$$= \frac{A_v V_{in}}{\dfrac{A_v V_{in}}{(1+\beta A_v)Z_o}}$$

$$= \mathbf{Z_o(1 + \beta A_v)} \text{................} \quad \text{(Eq. 7.7-5)}$$

Equations 7.7-5 and 7.7-2 will yield the same result for the output impedance Z'_o

See Example 7.8.

Example 7.8

The transistor amplifier shown in Fig .7.8-0 employs series current negative feedback. $R_f = 100\ \Omega$, $R_L = 2500\ \Omega$, $h_{ie} = 1500\ \Omega$, $h_{fe} = 80$, *and* $h_{oe} = 10^{-4}$ *S.*

Determine approximately the input and output impedances of the amplifier.

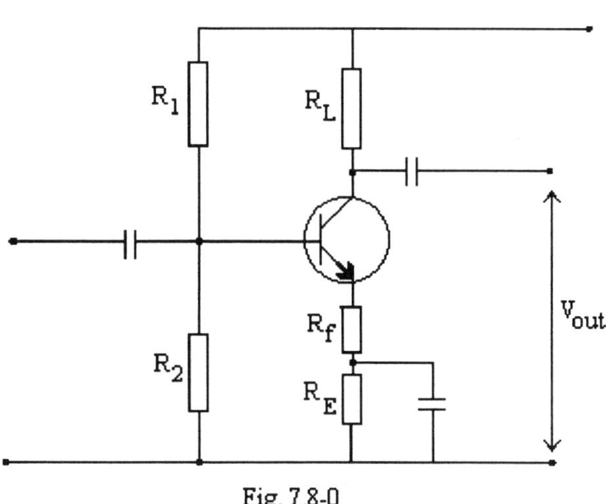

Fig. 7.8-0

Solution

Input impedance with feedback , $Z'_{in} = Z_{in}(1 + \beta A_v)$

where $Z_{in} = R_{in} = h_{ie}$ (neglecting the effects of R_1 and R_2)

$$\beta = R_f/R_L$$

$$A_v = A_i\, R_L/R_{in}$$

$$= A_i\, R_L/h_{ie}$$

$$A_i = h_{fe}/(1 + h_{oe}R_L)$$

$$\therefore \quad Z'_{in} = h_{ie}\left(1 + \frac{R_f}{R_L} \times \frac{h_{fe}}{1 + h_{oe}R_L} \times \frac{R_L}{h_{ie}}\right)$$

$$= h_{ie} + \frac{R_f h_{fe}}{1 + h_{oe}R_L}$$

$$= 1500 + \frac{100 \times 80}{1 + 10^{-4} \times 2.5 \times 10^3}$$

$$= 1500 + \frac{8000}{1.25} \quad = \underline{\mathbf{7900\ \Omega}}$$

(contd)

Output impedance with feedback is

$$Z'_o = Z_o + (1 + A_v)\, R_f \qquad \text{(Refer to Eq. 7.7-2)}$$

OR

$$Z'_o = Z_o\,(1 + \beta A_v) \qquad \text{(Refer to Eq.7.7-5)}$$

where $Z_o = {}^1/h_{oe}$

Either equation will yield the same result.

$Z'_o = Z_o + (1 + A_v)\, R_f$ [Eq. 7.7-2]

$$A_v = A_i \frac{R_L}{R_{in}} \approx A_i \frac{R_L}{h_{ie}}$$

$$A_i = \frac{h_{fe}}{1 + h_{oe}R_L}$$

$$\therefore Z'_o = \frac{1}{h_{oe}} + \left(1 + \frac{h_{fe}}{1 + h_{oe}R_L} \times \frac{R_L}{h_{ie}}\right) R_f$$

$$\approx \frac{1}{h_{oe}} + \frac{R_f h_{fe}}{h_{oe} h_{ie}}$$

$Z'_o = Z_o(1 + \beta A_v)$ [Eq. 7.7-5]

$$A_v = A_i \frac{R_L}{R_{in}} \approx A_i \frac{R_L}{h_{ie}}$$

$$= \frac{h_{fe}}{1 + h_{oe}R_L} \times \frac{R_L}{h_{ie}}$$

$$\beta = \frac{R_f}{R_L}$$

$$\therefore Z'_o = \frac{1}{h_{oe}}\left(1 + \frac{R_f}{R_L} \times \frac{h_{fe}}{1 + h_{oe}R_L} \times \frac{R_L}{h_{ie}}\right)$$

$$\approx \frac{1}{h_{oe}} + \frac{R_f h_{fe}}{h_{oe} h_{ie}}$$

$$\therefore Z'o = \frac{\mathbf{1}}{\mathbf{h_{oe}}} + \frac{\mathbf{R_f h_{fe}}}{\mathbf{h_{oe}\, h_{ie}}}$$

$$= 10^4 + \frac{10^4 \times 100 \times 80}{1500} \qquad = \underline{\mathbf{63.33\ k\Omega}}$$

Example 7.9

A common-emitter amplifier uses a transistor which has an h_{ie} = 1000 Ω The overall gain without feedback is 10000. If 5% negative series voltage feedback is applied, determine the input impedance Z'_{in} if the load resistance is 75 Ω
If the output impedance without feedback Z_o is 15 kΩ, determine the output impedance Z'_o when feedback is applied.

Solution

Without feedback (neglecting the effect of d.c. bias resistors), the input impedance is

$$Z_{in} = R_{in} \approx h_{ie} = \underline{1000\ \Omega}$$

Overall voltage gain,

$$A_v = A_i R_L/R_{in}$$

$$= 10000 \times {}^{75}/1000 = \underline{750}$$

With feedback, input impedance is

$$Z'_{in} = Z_{in}(1 + \beta A_v) \quad \text{(Refer to Eq. 7.6-1)}$$

$$= Z_{in}(1 + 0.05 \times 750)$$

$$= 1000 \times 38.5 = \underline{\mathbf{38.5\ k\Omega}}$$

With feedback, output impedance is

$$Z'_o = Z_o/(1 + \beta A_v)$$

$$= {}^{15000}/[1 + (0.05 \times 750)]$$

$$= {}^{15000}/38.5 = \underline{\mathbf{389.6\ \Omega}}$$

Example 7.10

The two-stage transistor amplifier shown in Fig. 7.10-0 has R_{E1} chosen to give 2% negative feedback in the first stage.

The values of R_f and R_{E2} are chosen to give 1% negative feedback over the two stages.

If h_{ie} for Tr1 is 1000 Ω, determine the new value of input impedance.

The voltage gain with feedback to the first stage is 150 and that of the second stage is 200.

Briefly comment on the feedback path.

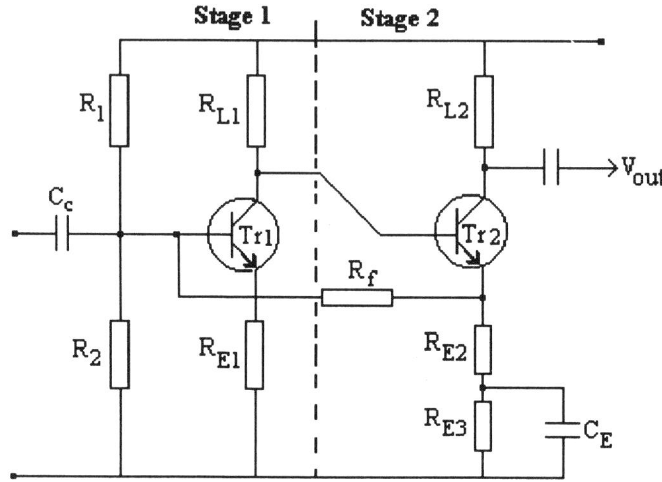

Fig. 7.10-0. Two-stage amplifier employing negative feedback (compound feedback)

Solution

Stage 1

The emitter resistor R_{E1} gives negative series current feedback to the first stage. Hence the input impedance with feedback is increased from $R_{in} \approx h_{ie}$.

$$\therefore \quad Z'_{in1} = h_{ie}(1 + \beta_1 A_{V1})$$

The gain of the stage falls to

$$A'_{V1} = A_{V1}/(1 + \beta A_{V1})$$

$$= 150/[1 + (0.02 \times 150)]$$

$$= 150/4 \qquad = \underline{37.5}$$

(contd)

The overall gain of both stages 1 and 2 with **feedback in stage 1** only is

$$A'_V = A'_{V1} \times A_{V2}$$

$$= 37.5 \times 200 \qquad = \underline{7500}$$

The new overall input impedance is reduced by the negative feedback in stage 2 from Z'_{in1} to

$$Z_{in\,f} = Z'_{in1}/(1 + \beta_2 A'_V)$$

$$= h_{ie}(1 + \beta_1 Av_1)/(1 + \beta_2 A'_V)$$

$$= \frac{1000(1 + 0.02 \times 150)}{1 + (0.01 \times 7500)}$$

$$= \frac{1000(1 + 3)}{1 + 75}$$

$$= 4000/76 \qquad = \underline{\mathbf{52.6\ \Omega}}$$

It is of interest to note that for shunt negative feedback the input current can be divided between the amplifier and the feedback path with the consequence that the input impedance is reduced.

We may conclude from this example that by suitable choice of component values the input and output impedance levels of multistage amplifiers employing compound negative feedback can be made to meet design criteria.

Example 7.11

An amplifier has an open loop-gain of 270 which is found to fall by 10% due to changes in supply voltage.

If the gain is to be stabilized so that it falls by 1%, calculate the amount of negative feedback required.

If the original upper 3 dB frequency is 50 kHz, calculate its value when this amount of negative feedback is applied.

Solution

The gain of the amplifier with negative feedback is

$$A'_V = A_V/(1 + \beta A_V) \qquad \text{(Eq. 7.11-0)}$$

When A_V falls by 10%, A'_V must fall to only 1%.

That is, when $A_V = 0.9A_V$, $A'_V = 0.99A'_V$.

Substituting in Fig. 7.11-0, we get

$$0.99A'_V = 0.9A_V/(1 + \beta 0.9A_V) \qquad \text{(Eq. 7.11-1)}$$

Dividing Eq. 7.11-1 by Eq. 7.11-0, we get

$$\frac{0.99A'_V}{A'_V} = \frac{0.9A_V/(1+\beta 0.9A_V)}{A_V/(1+\beta A_V)}$$

$$\frac{0.99A'_V}{A'_V} = \frac{0.9A_V}{1+\beta 0.9A_V} \times \frac{1+\beta A_V}{A_V}$$

$$0.99 = \frac{0.9(1+\beta A_V)}{1+\beta 0.9A_V}$$

$$0.99(1 + \beta 0.9A_V) = 0.9(1 + \beta A_V)$$

$$0.99 + 0.99 \times 0.9\beta A_V = 0.9 + 0.9\beta A_V$$

$$0.99 - 0.9 = 0.9\beta A_V - 0.99 \times 0.9\beta A_V$$

$$0.09 = 0.9(1 - 0.99)\beta A_V$$

$$\therefore \quad \beta A_V = 0.09/[0.9(1 - 0.99)]$$

$$= 0.1/(1 - 0.99)$$

$$= 0.1/0.01 \qquad = \underline{10}$$

(contd)

The loop gain A_v is given as 270.

$$\therefore \qquad \beta = {}^{10}/_{270} = {}^{1}/_{27} \qquad \text{(i)}$$

At high frequencies the amplifier gain is

$$A_v = A_o/(1 + {}^{jf}/_{f_2}) \qquad \text{(ii)}$$

where A_o = midband gain

f_2 = upper 3 dB frequency

With negative feedback, $A'_v = A_v/(1 + \beta A_v)$ (iii)

Substituting for A_v from (ii) into (iii), we get

$$A'_v = \frac{A_o/(1 + jf/f_2)}{1 + \beta A_o/(1 + jf/f_2)}$$

$$= \frac{A_o/(1 + jf/f_2)}{\dfrac{1 + jf/f_2 + \beta A_o}{1 + jf/f_2}}$$

$$= \frac{A_o}{1 + jf/f_2} \times \frac{1 + jf/f_2}{1 + jf/f_2 + \beta A_o}$$

$$= \frac{A_o}{jf/f_2 + 1 + \beta A_o}$$

Dividing top and bottom by $(1 + \beta A_o)$, we get

$$A'_v = \frac{A_o/(1 + \beta A_o)}{(jf/f_2 + 1 + \beta A_o)/(1 + \beta A_o)}$$

$$= \frac{A_o/(1 + \beta A_o)}{jf/f_2(1 + \beta A_o) + 1}$$

$$= \frac{A_o/(1 + \beta A_o)}{1 + j[f/f_2(1 + \beta A_o)]}$$

$$= \frac{A'_o}{1 + j[f/f_2(1 + \beta A_o)]} \qquad \text{(Eq. 7.11-2)}$$

(contd)

The new upper 3 dB frequency is defined as the frequency at which the midband frequency gain has fallen to $1/\sqrt{2}$ of its value.

New upper 3 dB frequency $= A'_o/\sqrt{2}$

Let this new upper 3 dB frequency be f'_2. Then from Eq. 7.11-2

$$f'_2/f_2(1 + \beta A_o) = 1$$

$$f'_2 = f_2(1 + \beta A_o)$$

$$= 50\left(1 + \frac{1}{27} \times 270\right) \quad \text{(refer to (i), } \beta = 1/27\text{)}$$

$$= 50(1 + 10) \qquad = \underline{\mathbf{550\ kHz}}$$

Example 7.12

An extract from the specifications of an integrated circuit (I.C) amplifier is as follows

Nominal gain = *86*

95% sample range = *80 – 90 dB*

Z_{in} = *100 kΩ*

Upper 3 dB cut-off frequency is = *16 Hz*

This amplifier is to be used with voltage negative feedback in series with the input to give a nominal gain of 1000. For the feedback amplifier, determine

(a) *the gain spread of the system*

(b) *the input impedance of the system*

(c) *the bandwidth of the system*

Solution

(a) With **no feedback**, nominal gain is 86 dB

$$\therefore \quad 20 \log A_v = 86$$

$$A_v = \underline{19952.6}$$

With **feedback**, nominal gain is

$$A'_v = 1000$$

$$= A_v/(1 + \beta A_v)$$

That is $1000 = 19952.6/(1 + \beta 19952.6)$

$$1 + \beta 19952.6 = 19952.6/1000 = \underline{19.953} \quad \text{(Eq. 7.12-0)}$$

$$\beta = (19.953 - 1)/19952.6$$

$$= 18.953/19952.6 = \underline{9.5 \text{x} 10^{-4}}$$

Maximum gain $\mathbf{A_{v1}}$ **without feedback** is 80 dB.

$$\therefore \quad 20 \log A_{v1} = 80$$

$$A_{v1} = \underline{10000}$$

(contd) *(a)*

Maximum gain $\mathbf{A'_{V1}}$ **with feedback** is

$$A'_{V1} = A_{V1}/(1 + \beta A_{V1})$$

$$= {}^{10000}/[1 + (9.5 \times 10^{-4} \times 10000)]$$

$$= {}^{10000}/(1 + 9.5) \quad = \underline{\mathbf{952.38}}$$

$$= \underline{\mathbf{59.58\ dB}}$$

Maximum gain $\mathbf{A_{V2}}$ **without feedback** is 90dB.

$$\therefore \quad 20 \log A_{V2} = 90$$

$$A_{V2} = \underline{31622.8}$$

Maximum gain $\mathbf{A'_{V2}}$ **with feedback** is

$$A'_{V2} = A_{V2}/(1 + \beta A_{V2})$$

$$= {}^{31622.8}/[1 + (9.5 \times 10^{-4} \times 31622.8)]$$

$$= {}^{31622.8}/(1 + 30.04) \quad = \underline{\mathbf{1018.78}}$$

$$= \underline{\mathbf{60.1\ 6dB}}$$

∴ **<u>Gain spread of the system is from 59.58 to 60.16</u>**

(b) Input impedance with negative series voltage feedback is

$$Z'_{in} = Z_{in}(1 + \beta A_V) \qquad \text{(Refer to Eq. 7.6-1)}$$

$$(1 + \beta A_V) = 19.953 \qquad \text{(Refer to Eq. 7.12-0)}$$

$$\therefore \quad Z'_{in} = 100(19.953) \quad = \underline{\mathbf{1995.3\ k\Omega}}$$

$$\approx \underline{2\ M\Omega}$$

(c) The upper 3 dB frequency is the frequency at which the midband frequency gain has fallen to $^1/\sqrt{2}$ of its value. The bandwidth is the frequency range between the lower and upper 3 dB points. Ignoring the lower 3 dB cut-off frequency, the bandwidth is equal to the upper 3dB cut-off frequency

Without feedback, the upper 3 dB cut-off frequency is 16 Hz.

With negative feedback, the upper 3 dB cut-off frequency will be

$$f'_2 = f_2(1 + \beta A_V) \qquad \text{(Refer to Eq. 7.11-2)}$$

$$= 16 \times 19.953 \quad = \underline{\mathbf{319\ Hz}}$$

Example 7.13

An amplifier has an open-loop gain of 1000 $\angle 70^o$ and the feedback factor is –0.02 $\angle 20^o$. Determine the gain of the amplifier with negative feedback. What limiting value and phase of the feedback fraction β would be required to render the amplifier unstable.

Solution

Substituting values in the general expression for gain of an amplifier with negative feedback, we get

$$A'_v = \frac{A_v}{1 + \beta A_v}$$

$$= \frac{1000\angle 70^\circ}{1 + (-0.02\angle 20^\circ \times 1000\angle 70^\circ)}$$

$$= \frac{1000\angle 70^\circ}{1 - 20\angle 90^\circ}$$

$$= \frac{1000\angle 70^\circ}{1 - j20}$$

$$= \frac{1000\angle 70^\circ}{\sqrt{1^2 + 20^2}\angle \tan^{-1}(-20/1)}$$

$$= \frac{1000\angle 70^\circ}{20.02\angle -87.13^\circ}$$

$$= 49.95\angle 70^\circ + (-87.13^\circ)$$

$$= \mathbf{49.95\angle -17.13^o}$$

For the amplifier to be unstable, the feedback factor β must be positive (i.e., positive feedback), which will add to the signal voltage.

For positive feedback, $A_v' = A_v/1 - (+\beta A_v)$

$$\mathbf{A_v' = A_v/(1 - \beta A_v)}$$

(contd) - *Instability in amplifier*

If $\beta A_v = 1$, the gain with feedback becomes infinitely large.

That is

$$A'_v = A_v/(1-1)$$

$$= A_v/0 \quad = \infty$$

With the gain at infinity, the amplifier becomes unstable and behaves as an oscillator.

In practice, $\beta A_v = 1$ implies that the loop gain is unity and that zero phase shift occurs around the loop.

$$\therefore \quad \beta A_v = 1\angle 0^o$$

$$\beta = 1\angle 0^o/(1000\angle 70^o)$$

$$= \mathbf{1 \times 10^{-3} \angle{-70^o}}$$

Example 7.14

An amplifier with an output resistance of 1.2 kΩ has an overall gain of 10000∠180° when connected to a load of 1.2 kΩ.
The voltage gain is to be reduced to 100 by the simultaneous application of both current and voltage feedback in series with the input so that the output impedance remains unchanged. Calculate the value of the current feedback resistance to be connected in series with the 1.2 kΩ load and the percentage voltage feedback.

Solution

Gain with negative feedback is

$$A'_v = A_v/(1 + \beta A_v)$$

Substituting the given values, we get

[Note: $10000\angle 180^{\circ} = j^2 10000 = -10000$]

$$\therefore \quad 100 = -10000/(1 - \beta 10000)$$

$$1 - 10000\beta = -10000/100$$

$$= -100$$

$$\therefore \quad \beta = 101/10000 = \underline{\mathbf{0.0101}}$$

$$= \underline{\mathbf{1.01 \times 10^{-2}}}$$

Let the current feedback factor = β_i.
Let the voltage feedback factor = β_v.
The output resistance change will be in the ratio of $(1+ \beta_i)/(1 + \beta_v)$.
However, the output resistance must remain unchanged with the application of feedback.

$$\therefore \quad \beta_i = \beta_v = \beta/2$$

$$= 0.0101/2 = \underline{\mathbf{0.00505}}$$

$$= \underline{\mathbf{5.05 \times 10^{-3}}}$$

$$\beta_i = R_f/1.2\ \text{k}\Omega$$

$$\therefore \quad R_f = 1.2 \times 10^3 \times 5.05 \times 10^{-3}$$

$$= 1.2 \times 5.05 = \underline{\mathbf{6.06\ \Omega}}$$

CHAPTER 8

OSCILLATORS, TRANSIENTS AND SYNCHRONIZING

Introduction

Oscillation is obtained and sustained in an amplifier when the feedback signal or gain is large enough to make the feedback factor unity. Oscillators are the generating sources (referred to as the carrier frequency) for radio transmitting stations and because of their characteristics these transmitting stations operate simultaneously with minimal interference.

Synchronizing a free-running oscillator is to lock it into an external signal source.

The design of video amplifiers, in particular for television, demands the understanding of response to transients as well as the steady state signal.

Example 8.1

(a) *An amplifier having a voltage gain A_v without feedback has a fraction β of its output voltage fed back in series with its input. Derive a general expression for the amplifier gain when the feedback loop is closed, with the requirement to maintain oscillations with positive feedback.*

(b) *Draw the circuit diagram of a tuned-collector oscillator and explain in brief its principle of operation.*

Solution

(a) <u>Refer to Fig. 7.1-0</u>

Since the feedback is positive, βV_{out} will add to the signal voltage.

$$\therefore \quad V_s = V_{in} + \beta V_{out}$$

$$V_{out} = A_v V_s \qquad (A_v = \text{gain without feedback})$$

$$= A_v(V_{in} + \beta V_{out})$$

$$= A_v V_{in} + \beta A_v V_{out}$$

(contd)

$$V_{out} - \beta A_v V_{out} = A_v V_{in}$$

$$V_{out}(1 - \beta A_v) = A_v V_{in}$$

$$V_{out} = A_v V_{in}/(1 - \beta A_v)$$

$$A'_v = V_{out}/V_{in} = A_v/(1 - \beta A_v)$$

Therefore, the expression for the gain of an amplifier with positive feedback is

$$\mathbf{A'_v = A_v/(1 - \beta A_v)}$$

Since the feedback is positive, then βA_v is positive. Thus

$$A'_v = A_v/[1 - (+\beta A_v)]$$

In this condition A'_v is greater than A_v. If $\beta A_v = 1$, then the gain with feedback becomes infinitely large. Thus

$$A'_v = A_v/(1 - 1)$$

$$= A_v/0 \qquad = \infty$$

When $\beta A_v = 1$, the loop gain is unity and the phase shift around the loop is zero. The feedback voltage is therefore identical to the signal voltage. Hence, the input source can be removed and replaced by the feedback voltage; the circuit will continue to give a signal output, for the amplifier has become an oscillator.

In this condition the signal source is the noise in the circuit selectively amplified to give the output. In general, both β and A_v may be complex, which implies conditions of magnitude and phase.

In magnitude, $\beta A_v = 1$.

In phase, $\beta A_v = 0$ or any multiple of 2π.

In practice, for sinusoidal oscillators the open loop is designed to be frequency dependent so that oscillation only occurs when $\beta A_v = 1$.

(contd)

(b)

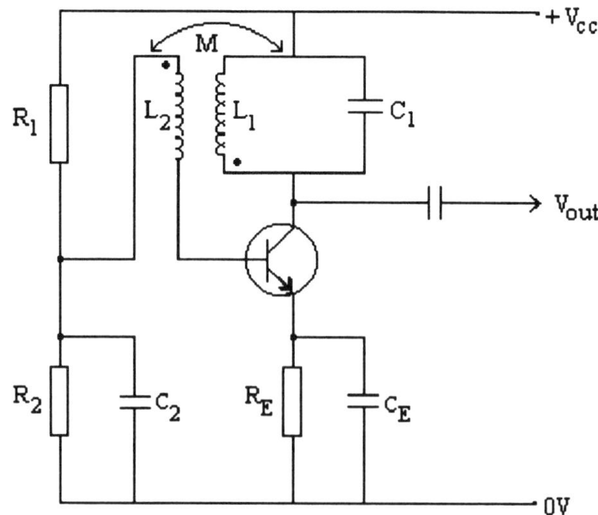

Fig. 8.1-0. Tuned-collector oscillator
(dots represent points of similar polarity)

Refer to Fig. 8.1-0

At the instant of switching on, noise or small voltage fluctuation in the input circuit is amplified and appears at the collector of the transistor. The current flowing in L_1 induces an e.m.f in L_2, which is applied to the base of the transistor.

In this oscillator the transistor induces a phase shift of 180° and a further 180° is introduced by the transformer secondary, giving the total phase shift of 360° (i.e., a loop phase shift of zero) required for oscillation. The current gain h_{fe} must be high enough to make the loop gain $\beta A_v = 1$ to sustain oscillation.

The frequency of oscillation is given by the expression

$$f_{osc} \approx {}^1/[2\pi\sqrt{(L_1C_1)}]$$

Example 8.2

The circuit diagram of a Hartley oscillator is shown in Fig. 8.2-0.

(a) *Briefly explain its principle of operation.*

(b) *Show how the circuit can be modified to form a Colpitts oscillator.*

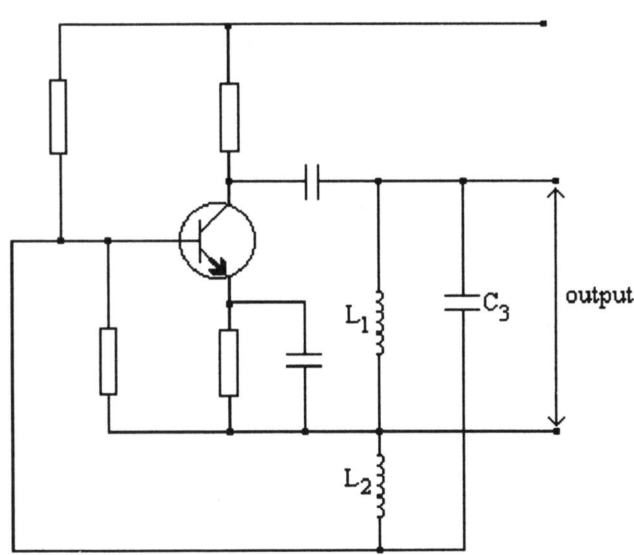

Fig. 8.2-0. Hartley oscillator

(a) The tuned circuit consists of L_1, L_2 and C_3. L_1 and L_2 are in series, with C_3 in parallel. The transistor introduces a 180° phase shift, and to obtain a loop phase shift of 360°, there is a 180° phase shift between the feedback voltage and the output voltage (feedback voltage V_{L2} and output voltage V_{L1} tend to180^o). The transistor must have a high current gain to ensure a loop gain of unity.

(Note: Mutual inductance, M - two coils are said to be mutually coupled when a change in magnetic flux produced by one coil causes an e.m.f to be induced in the other.)

If there is no mutual coupling (M) between coils L_1 and L_2, then the frequency of oscillation is given by the equation

$$f_{osc} = {}^1/\{2\pi\sqrt{[C(L_1 + L_2)]}\} \text{.....} \quad \text{(Eq. 8.2-0)}$$

and the gain condition by the expression

$$h_{fe} >> {}^{L_1}/L_2 \quad \text{(Eq. 8.2-1)}$$

(contd)

If $M \neq 0$, then Eqs. 8.2-0 and 8.2-1 are modified to read

$$f_{osc} = 1/\{2\pi\sqrt{[C(L_1 + L_2 + 2M)]}\} \quad \text{(Eq. 8.2-2)}$$

and $$h_{fe} = (L_1 + M)/(L_2 + M)$$

(b) The tuned circuit can be modified to form a Colpitts oscillator as shown in Fig. 8.2-1. Capacitor C acts as the base-decoupling capacitor for the signal.

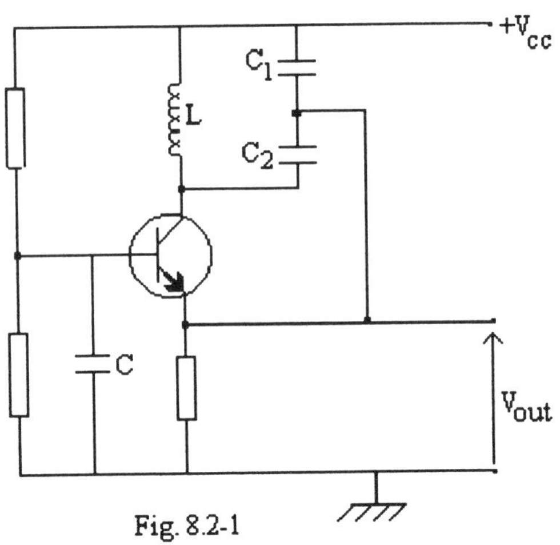

Fig. 8.2-1

For the Colpitts oscillator, the frequency of oscillation is given by the equation

$$f_{osc} = \frac{1}{2\pi\left[\left(L\,\frac{C_1 C_2}{C_1 + C_2}\right)\right]^{1/2}}$$

and the condition to maintain oscillation by the expression

$$h_{fe} >> C_2/C_1$$

Note: The frequency of oscillation is given by the equation

$$1/j\omega C_2 + j\omega L + 1/j\omega C_1 = 0$$

from which $$f_{osc} = 1/\{2\pi\sqrt{[L(C_1C_2)/(C_1 + C_2)]}\}$$

The gain condition to maintain oscillation is given by the equation

$$j\omega L + (1 + h_{fe})/j\omega C_2 = 0$$

i.e., $$1 + h_{fe} = \omega^2 LC_2 = C_2/(C_1 + 1)$$

Hence $$h_{fe} >> C_2/C_1$$

Example 8.3

The block diagram of a common-emitter connected transistor amplifier with feedback is shown in Fig. 8.3-0. Show that the circuit will oscillate when the amplifier input is short-circuited if

$$(h_{11} + h_{ie})(h_{22} + h_{oe}) - (h_{12} + h_{re})(h_{21} + h_{fe}) = 0$$

where h_{11}, h_{12}, h_{22} and h_{21} are the network h-parameters and h_{ie}, h_{re}, h_{oe} and h_{fe} are the transistor h-parameters.

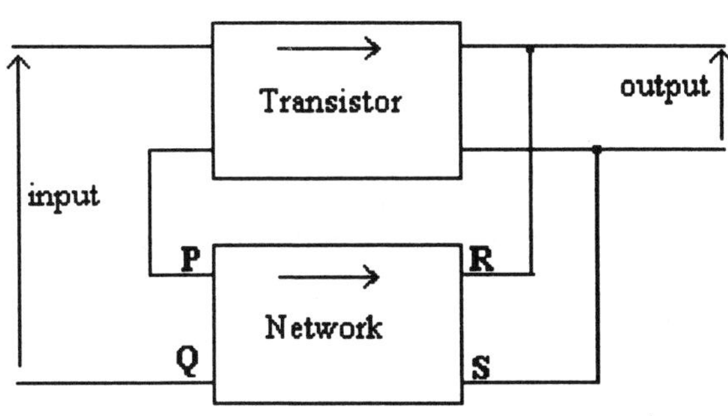

Fig. 8.3-0

For the network shown in Fig.8.3-1, determine the value of h_{fe} required to maintain oscillations and the resulting frequency of oscillation if

$$h_{11} >> h_{ie}, \quad h_{22} >> h_{oe} \text{ and } h_{12} >> h_{re}.$$

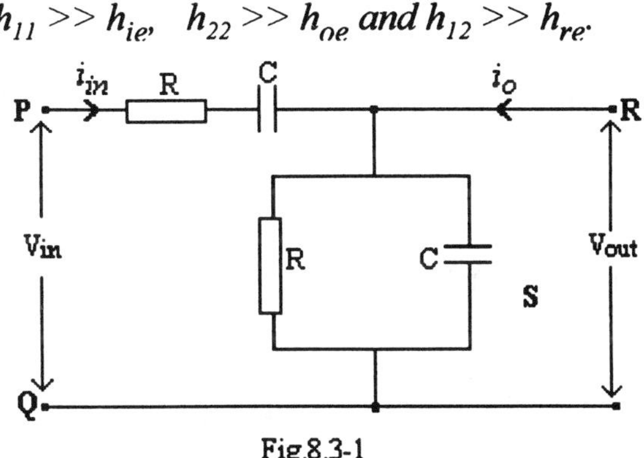

Fig.8.3-1

Solution

Theory - Network *h*-parameters

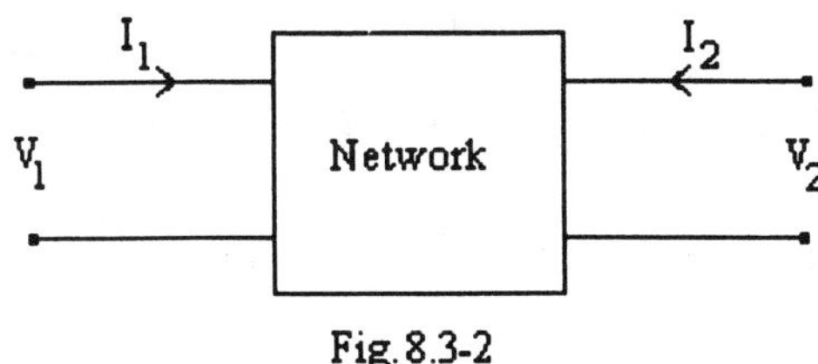

Fig. 8.3-2

(contd)

Refer to Fig. 8.3-2

$$V_1 = h_{11}I_1 + h_{12}V_2$$

$$I_2 = h_{21}I_1 + h_{22}V_2$$

where

$h_{11} = V_1/I_1$ = input resistance with S.C output

$h_{21} = I_2/I_1$ = forward current gain with S.C output

$h_{12} = V_1/V_2$ = reverse voltage ratio with input O.C

$h_{22} = I_2/V_2$ = output admittance with O.C input

[Note: O.C = open circuit

S.C = short circuit]

The *h*-parameter equivalent circuit for the network and transistor of Fig. 8.3-0 is now shown in Fig. 8.3-3.

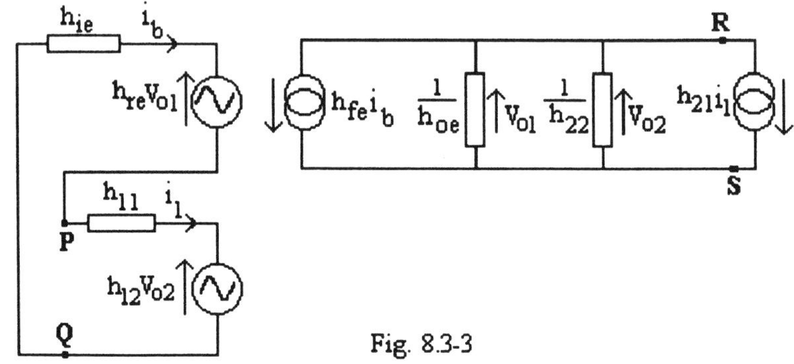

Fig. 8.3-3

Analyzing the output of Fig. 8.3-3 we see that

$$i_1 = i_b$$

and

$$V_{O1} = V_{O2}$$

$$= \frac{i_1(h_{fe} + h_{21})}{h_{oe} + h_{22}} \qquad \text{(Eq. 8.3-0)}$$

Considering the input circuit, we get

$$V_{O1}(h_{re} + h_{12}) = -i_1(h_{ie} + h_{11})$$

Substituting for V_{O1} from Eq. 8.3-0, we get

$$-i_1(h_{re} + h_{12})\frac{(h_{fe} + h_{21})}{(h_{oe} + h_{22})} = -i_1(h_{ie} + h_{11})$$

(contd)

For the circuit to oscillate, we are given that

$$(h_{11} + h_{ie})(h_{22} + h_{oe}) - (h_{12} + h_{re})(h_{21} + h_{fe}) = 0 \qquad \text{(Eq. 8.3-1)}$$

The procedure is to determine the *h*-parameters of the network of Fig. 8.3-1, first with S.C output and then with O.C input.

With S.C output, we have

$$h_{11} = V_{in}/i_{in} = R + {}^{1}/j\omega C \qquad \text{(i)}$$

$$h_{21} = i_o/i_{in} = -1 \qquad \text{(ii)}$$

[Note: The input current also flows in the S.C but opposite to i_o.]

With O.C input, we have

$$h_{12} = V_{in}/V_o = 1 \qquad \text{(iii)}$$

$$h_{22} = i_o/V_o = \frac{1 + j\omega CR}{R} \qquad \text{(iv)}$$

Conditions given to maintain oscillations are

$$h_{11} >> h_{ie}$$

$$h_{22} >> h_{oe}$$

$$h_{12} >> h_{re}$$

Now to determine the value of h_{fe}, we rewrite Eq. 8.3-1 using the conditions given:

$$(h_{11})(h_{22}) - (h_{12})(h_{21} + h_{fe}) = 0$$

That is

$$h_{11}h_{22} - h_{12}(h_{21} + h_{fe}) = 0 \qquad \text{(Eq. 8.3-2)}$$

Substituting in Eq. 8.3-2 the values for h_{11}, h_{21}, h_{12} and h_{22} obtained in (i), (ii), (iii), and (iv), we get

$$(R + {}^{1}/j\omega C)\left(\frac{1 + j\omega CR}{R}\right) - 1\,(-1 + h_{fe}) = 0$$

$$(R + 1/j\omega C)\left(\frac{1 + j\omega CR}{R}\right) + 1 - h_{fe} = 0$$

(contd)

$$\frac{(j\omega CR + 1)}{j\omega C}\,\frac{(1 + j\omega CR)}{R} + 1 - h_{fe} = 0$$

$$\frac{1 + j^2\omega^2C^2R^2 + j2\omega CR + 1 - h_{fe}}{j\omega CR} = 0$$

$$1 + j^2\omega^2C^2R^2 + j2\omega CR + j\omega CR - h_{fe}j\omega CR = 0$$

$$1 - \omega^2C^2R^2 + j2\omega CR + j\omega CR - h_{fe}j\omega CR = 0$$

Considering real and imaginary quantities, we have

Imaginary Quantities

$$j2\omega CR + j\omega CR - h_{fe}j\omega CR = 0$$

$$\therefore \quad - h_{fe}j\omega CR = -j2\omega CR - j\omega CR$$

$$h_{fe}j\omega CR = 3j\omega CR$$

$$\mathbf{h_{fe} = 3}$$

This is the minimum value of the transistor current gain required for oscillation.

Real Quantities

$$\omega^2C^2R^2 = 1$$

$$4\pi^2f^2C^2R^2 = 1$$

$$\therefore \quad \mathbf{f_{osc}} = 1/\sqrt{(4\pi^2C^2R^2)}$$

$$= \mathbf{1/(2\pi CR)}$$

This is the frequency of oscillation.

Example 8.4

An astable multivibrator uses two npn transistors in the common-emitter connection. Sketch a circuit diagram of this multivibrator and with the aid of collector and base waveforms, explain its operation.

Solution

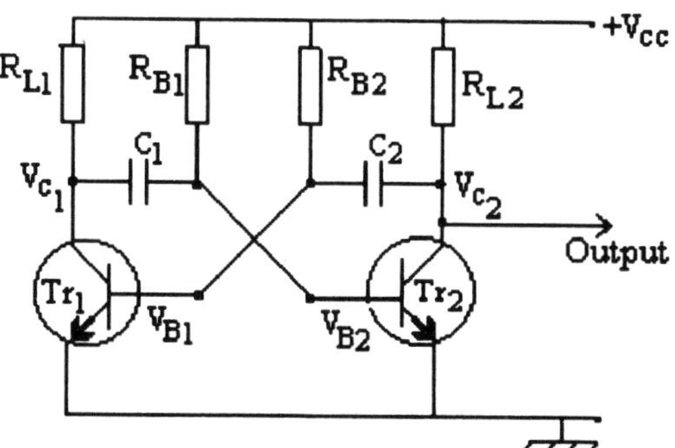

Fig. 8.4-0. Astable multivibrator

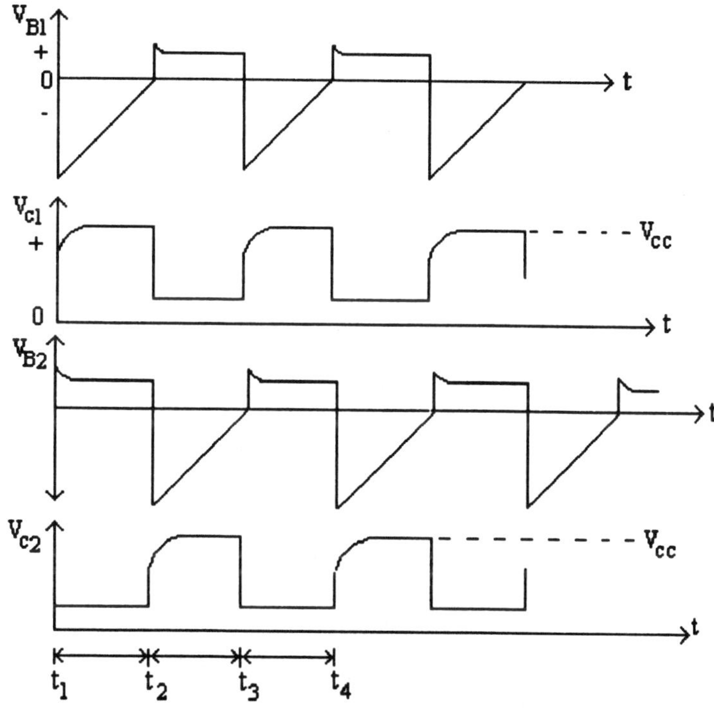

Fig. 8.4-1. Waveforms of an astable multivibrator

The operation of the astable multivibrator can best be explained from the analysis of the waveforms in Fig. 8.4-1.

(contd)

At the instant the d.c. supply is switched on, both transistors will carry equal rising currents. However, some random variation will cause a slight increase or decrease of current in one transistor with respect to the other transistor.

The waveform commences with Tr_1 cut-off at t_1.

1. At time t_1 Tr1 is cut off, V_{B1} is negative with respect to ground and its collector potential is rising rapidly to V_{CC}. That is

   V_{B1} is reverse biased
   Tr_1 is cut off
   V_{c1} is approximately V_{CC}

2. At the same time, the base potential of Tr_2 is sufficiently positive and forward biased to drive Tr_2 to conduct and thus reduces its collector potential to approximately 0 V. That is

   V_{B2} is forward biased
   Tr_2 is conducting
   V_{c2} is approximately 0 V

3. Tr_2 is now in a stable state but Tr_1 is not because its base potential is going positive as C_2 discharges through R_{B2}, giving a forward bias current flow in the base of Tr_1.

4. At time t_2 the base potential of Tr_1 is sufficiently positive to forward bias Tr_1 to conduct so that its collector potential changes from V_{CC} to approximately 0 V. The voltage across a capacitor cannot change instantaneously; the negative charge in C_1 is transferred to the base of Tr_2, thus cutting off Tr_2. This in turn causes V_{c2} to raise to V_{CC}. That is

V_{B1} is forward biased	V_{B2} is reverse biased
Tr_1 is conducting	Tr_2 is cut off
V_{c1} is approximately 0 V	V_{c2} approximately V_{CC}

5. During the period t_2-t_3 the current is governed by C_1 discharging through R_{B1} until the base potential of transistor Tr_2 is sufficiently positive to drive Tr_2 to conduct. The regenerative action this time results in Tr_2 conducting and Tr_1 cut off.

   The whole cycle repeats itself continuously.

(contd)

6. During the period t_1-t_2 relatively static conditions prevail:

V_{C2} is approximately 0 V

V_{C1} equals + V_{CC}

V_{B2} is approximately 0 V (slightly positive)

7. However, the voltage across R_{B2} is now $2V_{CC}$ and the current through R_{B2} charges C_2 from $-V_{CC}$ to $+V_{CC}$. The time constant (T) of this exponential charging current is

$$T = C_2R_2$$

During the interval t_1-t_2 Tr_1 is cut off, and during the interval t_2-t_3 Tr_2 is cut off.

The equation of this interval of relaxation is

$$V_{B1} = V_{CC} - 2V_{CC}e^{-t/C_2R_{B2}} \quad \text{(Eq. 8.4-0)}$$

$$V_{B2} = V_{CC} - 2V_{CC}e^{-t/C_1R_{B1}} \quad \text{(Eq. 8.4-1)}$$

Thus during the interval t_1-t_2 (relaxation period)

$$0 = V_{CC} - 2V_{CC}e^{-T_1/C_2R_{B2}} \quad \text{(Eq. 8.4-2)}$$

Equating this equation, we get

$$\exp(T_1/C_2R_{B2}) = 2$$

$$T_1 = C_2R_{B2} \log_e 2 \quad \text{(Eq. 8.4-3)}$$

$$= 0.69\ C_2R_{B2}$$

Similarly

$$T_2 = C_1R_{B1} \log_e 2 \quad \text{(Eq. 8.4-4)}$$

$$= 0.69\ C_1R_{B1}$$

The periodic time of the waveform and

pulse repetition frequency (PRF) are

$$PRF = {}^1/(T_1 + T_2)$$

$$= {}^1/[0.69(C_1R_{B1} + C_2R_{B2})]$$

For a symmetrical multivibrator

$$C_1R_{B1} = C_2R_{B2} = CR$$

$$\therefore \quad PRF = {}^1/(2\ CR \log_e 2)$$

$$= {}^1/(1.38\ CR)$$

Example 8.5

Explain with the aid of time-related waveforms the operation of the astable multivibrator shown in Fig. 8.5-0. Determine graphically the frequency of oscillation.

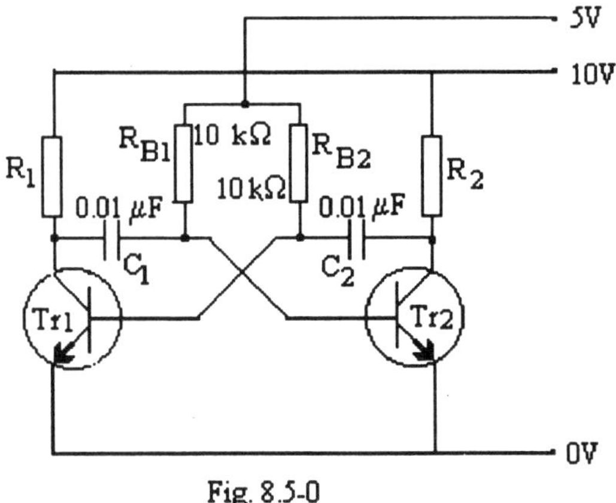

Fig. 8.5-0

Solution

The operation of the astable multivibrator has been described in Example 8.4. However, in this example the base resistors are fed from a separate source of supply.

To determine the frequency of oscillation consider the circuit conditions when Tr_2 has just been cut off as shown in Fig. 8.5-1. The capacitor C_1 will discharge from –10 V to +5 V. As a result, Tr_2 will switch on again when its base potential reaches approximately zero. The discharge curve is shown in Fig. 8.5-2.

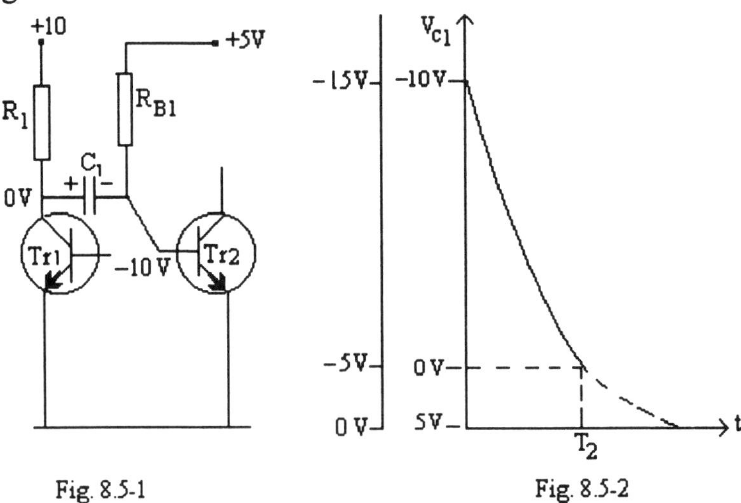

Fig. 8.5-1

Fig. 8.5-2

(contd)

The analysis of Fig. 8.5-2 is simplified by using a false zero.

The time T_2 for which Tr_2 is cut off is given by the expression

$$V_B = V_{cc}\, e^{(-T_2/C_1R_{B1})}$$

where $V_{cc} = -15V$

and $V_B = -5V$.

That is

$$-5 = -15e^{(-T_2/C_1R_{B1})}$$

$$3 = e^{T_2/C_1R_{B1}}$$

$$T_2 = C_1R_{B1} \log_e 3$$

$$= 0.01 \times 10^{-6} \times 10 \times 10^3 \log_e 3$$

$$= 1.0986 \times 10^{-4} \approx \underline{0.11\ mS}$$

Since the circuit uses components of equal values, i.e.

$$C_1 = C_2 \quad \text{and} \quad R_{B1} = R_{B2}$$

the periodic time for one complete cycle will be 2 x 0.11 mS;

that is

$$T = 0.22\ mS$$

$$\therefore \quad f = {}^1/T$$

$$= {}^1/(0.22 \times 10^{-3}) = \underline{\mathbf{4.54\ kHz}}$$

This example illustrates that the multivibrator frequency can be adjusted by varying the supply voltage to the base resistors.

Example 8.6

For the simple circuit shown in Fig. 8.6-0, sketch the graphs of the voltages across the resistor R and the capacitor C and the charging current at any instant of time after switch S is closed. Hence derive the equations for

(i) the charging current

(ii) the voltage across capacitor C with switch S closed

(iii) the voltage across C at any instant of time after S is opened

From the values obtained in (i), (ii) and (iii), explain the meaning of the term 'time constant', denoted by T.

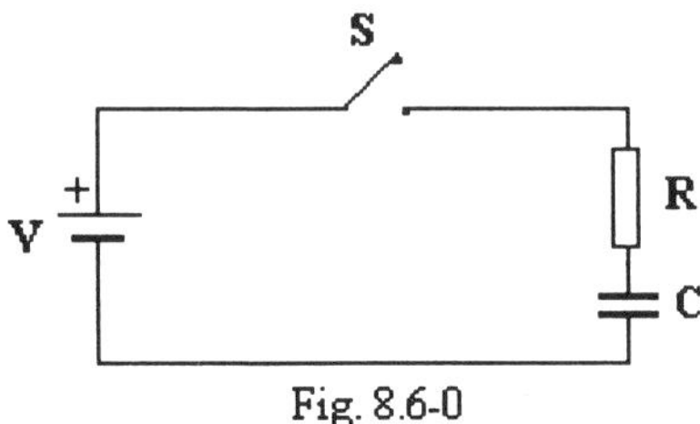

Fig. 8.6-0

Solution

With Switch S Closed

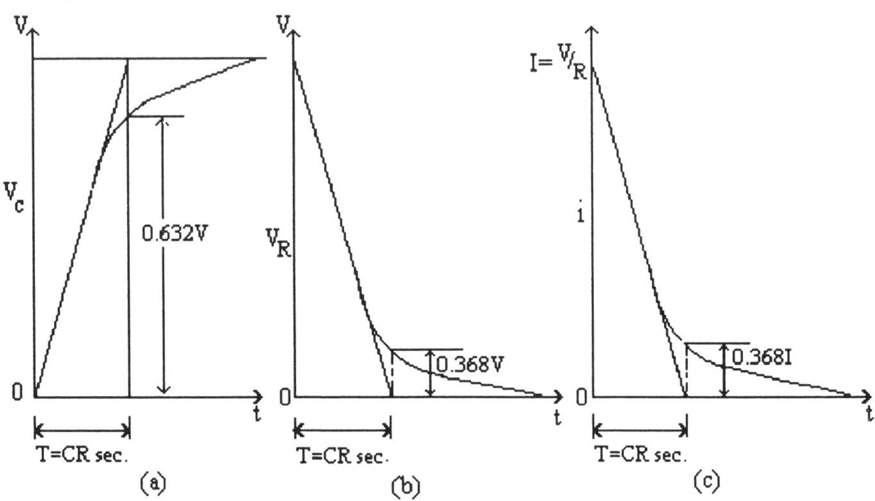

Fig. 8.6-1

(contd)

Refer to Fig. 8.6-1

The graphs represent the following:

(a) graph of voltage across C at any instant of time

(b) graph of voltage across R at any instant of time

(c) charging current in the circuit at any instant of time

(i) *Charging Current*

Total voltage V at any instant = voltage across R + voltage across C

That is $V = iR + {}^{1}/_{C}\int i\, dt$

$$^{dV}/_{dt} = R^{di}/_{dt} + {}^{i}/_{C}$$

Since the supply voltage V is constant

$$^{dV}/_{dt} = 0$$

$$\therefore \quad 0 = R^{di}/_{dt} + {}^{i}/_{C}$$

$$-R^{di}/_{dt} = {}^{i}/_{C}$$

$$^{di}/_{dt} = -{}^{i}/_{CR}$$

$$^{dt}/_{di} = -{}^{CR}/_{i}$$

$$dt = -CR\, ^{di}/_{i}$$

$$\int dt = \int -CR\, ^{di}/_{i}$$

$$\therefore \quad t = -CR \log_e i + C_1 \; \ldots\ldots \qquad \text{(Eq. 8.6-0)}$$

(C_1 is the constant of integration)

When $t = 0$

$$i = {}^{V}/_{R}$$

Substituting for t and i in Eq. 8.6-0, we get

$$0 = -CR \log_e {}^{V}/_{R} + C_1$$

$$C_1 = CR \log_e {}^{V}/_{R}$$

Substituting for C_1 in Eq. 8.6-0, we get

$$t = -CR \log_e i + CR \log_e {}^{V}/_{R}$$

$$-t = CR \log_e i - CR \log_e {}^{V}/_{R}$$

$$= CR (\log_e i - \log_e I)$$

$$\therefore \quad ^{-t}/_{CR} = \log_e i - \log_e I$$

$$^{-t}/_{CR} = \log_e {}^{i}/_{I}$$

$$e^{-t/CR} = {}^{i}/_{I}$$

$$\therefore \quad \mathbf{i = Ie^{-t/CR}} \qquad \text{(Eq. 8.6-1)}$$

where R is in ohms, C is in farads and t = CR sec.

(contd)

(ii) Voltage across C with switch S closed

$$\text{Voltage across } R = V - v_c \quad (v_c = \text{voltage across C})$$

$$\therefore \quad i = \frac{V - v_c}{R}$$

$$\text{But} \quad i = C\,{}^{dv}/_{dt}$$

$$\therefore \quad \frac{V - v_c}{R} = C\,{}^{dv}/_{dt}$$

$$({}^{1}/C)dt/dv = {}^{R}/(V - v_c)$$

$$({}^{1}/CR)\,dt = {}^{dv}/(V - v_c)$$

$$\int({}^{1}/CR)\,dt = \int{}^{dv}/(V - v_c)$$

$${}^{t}/CR = -\log_e V - v_c + C_1 \qquad \text{(Eq. 8.6-2)}$$

Again in this equation C_1 is the constant of integration.

$$\text{When} \quad t = 0, \quad v_c = 0$$

Substituting for t and v_c in Eq. 8.6-2, we get

$$0 = -\log_e V + C_1$$

$$C_1 = \log_e V$$

Substitute for C_1 in Eq. 8.6-2, we get

$${}^{t}/CR = -\log_e (V - v_c) + \log_e V$$

$$-{}^{t}/CR = \log_e (V - v_c) - \log_e V$$

$$\therefore \quad Ve^{-t/CR} = V - v_c$$

$$v_c = V - Ve^{-t/CR}$$

$$\therefore \quad \mathbf{v_c = V(1 - e^{-t/CR})} \qquad \text{(Eq. 8.6-3)}$$

where V is the voltage across the capacitor C when t = 0.

R is the resistor in ohms.

C is the capacitor in farads.

(contd)

(iii) *Voltage across C at any instant of time after switch S is opened*

When $t = 0$

$v_c = V_s$ (the supply voltage)

But $V = 0$ (the final voltage)

Substituting for t, v_c and V in Eq. 8.6-2, we get

$$^t/CR = -\log_e (V - v_c) + C_1 \quad \text{(Refer to Eq. 8.6-2)}$$

$$\therefore \quad 0 = -\log_e (0 - V_s) + C_1$$

$$-C_1 = -\log_e (-V_s)$$

$$C_1 = \log_e (-V_s)$$

$$\therefore \quad ^t/CR = -\log_e (V - v_c) + \log_e (-V_s)$$

$$-^t/CR = \log_e (V - v_c) - \log_e (-V_s)$$

$$= \log_e (-v_c) - \log_e(-V_s) \quad (V = 0)$$

$$= \log_e (^{-v}c/-V_s)$$

$$= \log_e (^{v}c/V_s)$$

$$\mathbf{v_c = V_s e^{-t/CR}} \qquad \text{(Eq. 8.6-4)}$$

Time Constant

Equation 8.6-1 gives the current

$$i = Ie^{-t/CR}$$

and $t = CR$ sec

Substituting for CR in the above equation (Eq. 8.6-1), we get

$$i = Ie^{-t/t}$$

$$= Ie^{-1}$$

$$= \mathbf{0.368I}$$

*The time constant, denoted by **T** is the time taken for the current to fall to 0.368 (36.8%) of its original value. Refer to Fig. 8.6-1(c).*

(contd)

Refer to Eq. 8.6-3

$$v_c = V(1 - e^{-t/CR})$$
$$= V(1 - e^{-CR/CR}) \quad (t = CR)$$
$$= V(1 - 0.368)$$
$$= \mathbf{0.632V}$$

That is, the time constant T is the time taken for the voltage to rise to to 0.632, or 63.2%, of its final value. Refer to Fig. 8.6-1(a).

Refer to Eq. 8.6-4

$$v_c = V_s e^{-t/CR}$$
$$= V_s e^{-1} \quad (t = CR \text{ sec})$$
$$= \mathbf{0.368V_s}$$

This is the case of the decay of voltage across the capacitor C and hence defines the time constant T as the time taken for the voltage to fall to 0.368, or 36.8%, of its original value.

Example 8.7

(a) *The circuit of Fig. 8.7-0 is a simple L R circuit connected to a d.c. supply through a single pole switch, S. Explain with the aid of sketches the circuit behavior when switch S is closed.*

(b) *Derive expressions for*

(i) *rise of current in the circuit*

(ii) *decay of current in the circuit*

(c) *The components of an L R circuit are shown in Fig. 8.7-1.*

Calculate *(i)* *the value of the current after 0.5 sec*

(ii) *the time constant of the circuit*

(iii) *the time taken for the current to reach 1.0 A*

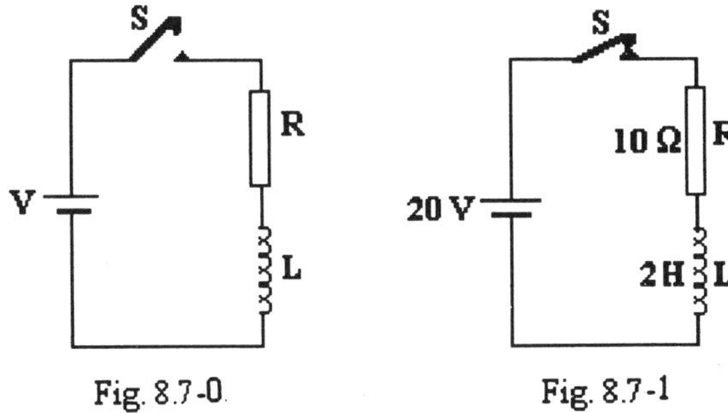

Fig. 8.7-0. Fig. 8.7-1

Solution

Consider Fig. 8.7-0

At the instant of closing switch S the current does not immediately rise to its final value, $I = {}^{V}/R$.

Let t = time after closing switch S

and i = instantaneous current

Now $V = iR + L\,{}^{di}/dt$

$\therefore \quad V - L\,{}^{di}/dt = iR$

On closing switch 'S', $i = 0$

$\therefore \quad iR = 0$

and $V = L\,{}^{di}/dt$ or ${}^{di}/dt = {}^{V}/L$

This is the initial rate of change of current.

When the current I has reached its final steady value

${}^{di}/dt = 0$ and $V = IR$

Final value of $I = {}^{V}/R$

(contd)

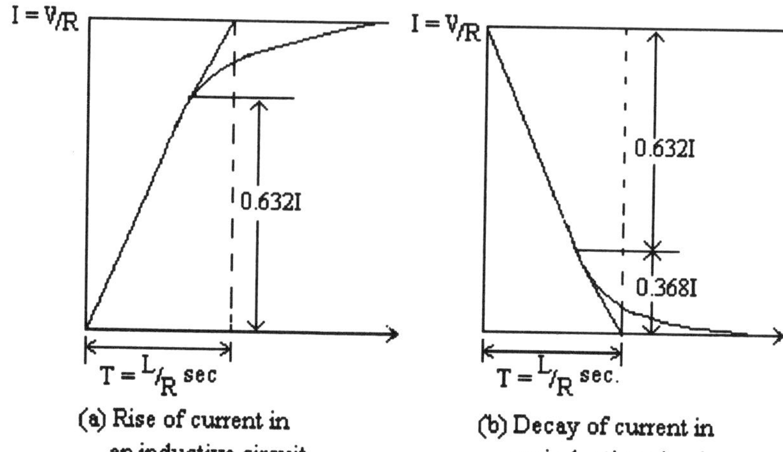

(a) Rise of current in an inductive circuit

(b) Decay of current in an inductive circuit

Fig. 8.7-2

(b) (i) Rise of current in an inductive circuit.

Refer to Fig. 8.7-2(a)

$$V = iR + L\,{}^{di}/{}_{dt}$$

Dividing by R: $${}^{V}/{}_{R} = i + ({}^{L}/{}_{R})\,{}^{di}/{}_{dt}$$

$$\therefore \quad I = i + ({}^{L}/{}_{R})\,{}^{di}/{}_{dt}$$

Inverting: $$({}^{R}/{}_{L}){}^{dt}/{}_{di} = {}^{1}/{}_{(I-i)}$$

Integrating: $$\int ({}^{R}/{}_{L}){}^{dt}/{}_{di} = \int {}^{1}/{}_{(I-i)}$$

Let $I - i = Z$; then $${}^{di}/{}_{dZ} = -1$$

$$\therefore \quad \int ({}^{R}/{}_{L}){}^{dt}/{}_{di} = \int {}^{1}/{}_{Z}$$

$$\int ({}^{R}/{}_{L})dt = \int {}^{di}/{}_{Z}$$

$$\int ({}^{R}/{}_{L})dt = \int ({}^{1}/{}_{Z})({}^{di}/{}_{dZ})dZ$$

$${}^{Rt}/{}_{L} = \log_e Z \ x \ (-1) + C$$

$${}^{Rt}/{}_{L} = -\log_e(I - i) + C \quad \text{............ (Eq. 8.7-0)}$$

When $t = 0$, $i = 0$

$$\therefore \quad 0 = -\log_e I + C$$

and $$C = \log_e I$$

$$\therefore \quad {}^{Rt}/{}_{L} = -\log_e(I - i) + \log_e I$$

$${}^{-Rt}/{}_{L} = \log_e [{}^{(I - i)}/{}_{I}]$$

$$\therefore \quad e^{-Rt/L} = {}^{(I - i)}/{}_{I}$$

$$Ie^{-Rt/L} = I - i$$

$$i = I - Ie^{-Rt/L}$$

$$\mathbf{i = I\,(1 - e^{-Rt/L})} \qquad \text{(Eq. 8.7-1)}$$

(contd) <u>Refer to Eq. 8.7-1</u>

The time constant **T** for an inductive circuit is **L/R sec**; it is the time taken for the current to rise to 0.632, or 63.2%, of its final value. See Fig. 8.7-2(a).

$$\mathbf{i} = I(1 - e^{-R/L \times L/R})$$

$$= I(1 - e^{-1}) \qquad (t = T = L/R)$$

$$= \mathbf{0.632I}$$

(b) (ii) *Decay of current in an inductive circuit*

<u>Refer to Fig. 8.7-2(b), the decay curve</u>

From Eq. 8.7-0 determine the value of C.

That is $\quad {}^{Rt}/_{L} = -\log_e(I - i) + C$

When $t = 0, \quad i = {}^{V}/_{R}$

Final value of $\quad I = 0$

Substituting these values in the above expression (Eq. 8.7-0), we get

$$0 = -\log_e(0 - {}^{V}/_{R}) + C$$

from which $\quad C = \log_e(-{}^{V}/_{R})$

$$\therefore \quad {}^{Rt}/_{L} = -\log_e(I - i) + \log_e(-{}^{V}/_{R})$$

But final value of $I = 0$.

$$\therefore \quad {}^{Rt}/_{L} = -\log_e(-i) + \log_e(-{}^{V}/_{R})$$

$$-Rt/L = \log_e {}^{i}/({}^{V}/_{R})$$

$$\therefore \quad e^{-Rt/L} = {}^{i}/({}^{V}/_{R})$$

$$\therefore \quad \mathbf{i} = \mathbf{Ie^{-Rt/L}} \qquad ({}^{V}/_{R} = I, \text{ the final value})$$

In the case of the decay of current, the time constant, T of an inductive circuit is L/R sec and is the time taken for the current to fall to 0.368, or 36.8%, of its initial value.

$$i = Ie^{-Rt/L}$$

$$= Ie^{-1} \qquad (t = T = L/R)$$

$$\mathbf{i} = \mathbf{0.368I}$$

(contd)

(c) (i)

$$i = I(1 - e^{-Rt/L})$$

$$I = {}^{V}/_{R} = {}^{20}/_{10} = \underline{2\ A}$$

$$\therefore \quad i = 2\,[1 - e^{-(10 \times 0.5)/2}]$$

$$= 2\,[1 - e^{-2.5}]$$

$$= 2\,[1 - 0.0821] = \underline{\mathbf{1.836\ A}}$$

(ii) Time constant, T

$$T = {}^{L}/_{R} \text{ sec}$$

$$= {}^{2}/_{10} = \underline{\mathbf{0.2\ sec}}$$

(iii)

$$i = I(1 - e^{-Rt/L})$$

$$= 2(1 - e^{-5t})$$

Given that i = 1 A

$$1 = 2 - 2e^{-5t}$$

$$2e^{-5t} = 2 - 1$$

$$e^{-5t} = {}^{1}/_{2}$$

$$^{1}/_{e^{5t}} = {}^{1}/_{2}$$

$$e^{5t} = 2$$

$$5t = \log_e 2$$

$$= 0.6931$$

$$\therefore \quad t = {}^{0.6931}/_{5} = \underline{\mathbf{0.1386\ sec}}$$

Example 8.8

A 5 MHz LC oscillator has an inductor whose inductance increases by 60 parts in 10^6 per ^oC rise in temperature and a capacitor whose capacitance falls by 40 parts in 10^6 per ^oC temperature rise. Determine the temperature which will produce a frequency change of 250Hz. How can better stability be achieved in an LC oscillator circuit.

Solution

Let us assume that the original frequency of oscillation is

$$f_{osc} = 1/[(2\pi\sqrt{(LC)}]\ldots\ldots \quad \text{(Eq. 8.8-0)}$$

We are given that with increase in temperature in degrees Centigrade (^oC)

inductance L will increase by $= L + \delta L$ and

capacitance C will decrease by $= C - \delta C$.

Therefore the new frequency of oscillation will be

$$f'_{osc} = 1/\{2\pi\sqrt{[(L + \delta L)(C - \delta C)]}\}$$

Let δL be represented by $= xL$

And δC be represented by $= yc$.

$$\therefore \quad f'_{osc} = 1/\{2\pi\sqrt{[(L + xL)(C - yC)]}\}$$

$$= 1/\{2\pi\sqrt{[L(1 + x)\, C(1 - y)]}\}$$

$$= 1/\{2\pi\sqrt{[LC(1 + x)(1 - y)]}\}$$

$$= 1/\{2\pi\sqrt{[LC(1 + x - y - xy)]}\}$$

xy is very small and can be neglected.

$$\therefore \quad f'_{osc} = 1/\{2\pi\sqrt{[LC(1 + x - y)]}\} \quad \text{(Eq. 8.8-1)}$$

Dividing Eq. 8.8-1 by Eq .8.8-0, we get

$$\frac{f'_{osc}}{f_{osc}} = \frac{\dfrac{1}{2\pi\sqrt{[LC(1 + x - y)]}}}{\dfrac{1}{2\pi\sqrt{(LC)}}}$$

$$= \frac{1}{\sqrt{(1 + x - y)}}$$

$$\approx 1 - \frac{x - y}{2} \quad \text{(by binomial expansion)}$$

$$f'_{osc} \approx f_{osc}(1 - \frac{x - y}{2}) \ldots\ldots \quad \text{(Eq. 8.8-2)}$$

(contd)

From information given and in this particular exercise

$$x = 60 \times 10^{-6} \text{ per } 1\ {}^{o}C$$

and $$y = 40 \times 10^{-6} \text{ per } 1\ {}^{o}C$$

$$f_{osc} = 5 \times 10^{6}\ Hz$$

and frequency change $= f'_{osc} - f_{osc} = 250Hz$

$$\therefore \quad 250 = -5 \times 10^{6}\ [(60 - 40)/_{2}] \times 10^{-6} \times t\ {}^{o}C$$

$$= -50t\ {}^{o}C$$

$$\therefore \quad t\ {}^{o}C = -\,{}^{250}/_{50} \qquad = \mathbf{-5\ {}^{o}C}$$

This implies a temperature decrease of 5 ^{o}C will give a frequency change of 250 Hz. In other words, the frequency will change by 250 Hz when the temperature shifts by –5 ^{o}C.

For better stability of frequency it is recommended that a crystal oscillator be employed in place of the LC oscillator.

Example 8.9

(a) *A symmetrical square wave of frequency 1.0 kHz and amplitude ±10 V is applied to the input terminals of the circuits shown in Figs. 8.9-0 and 8.9-1. For each condition, sketch on the same time scale two cycles of the input and output waveforms. State which circuit may be used as*

(i) an integrator

(ii) a differentiator

(b) *Determine the ratio of the capacitive reactance (X_c) to the resistance (R) in each circuit assuming a sinusoidal waveform of 1.0 kHz. Comment on your results.*

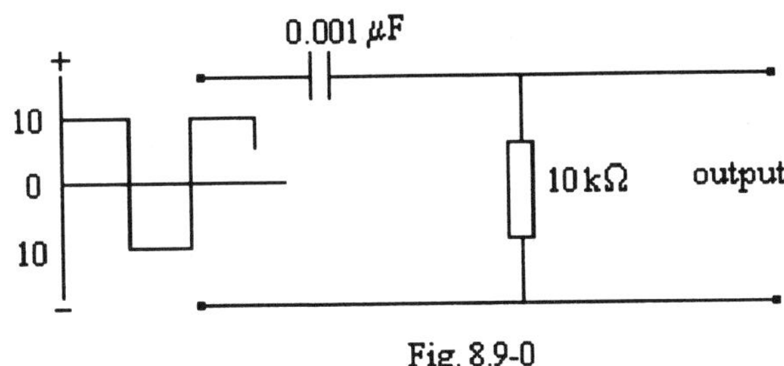

Fig. 8.9-0

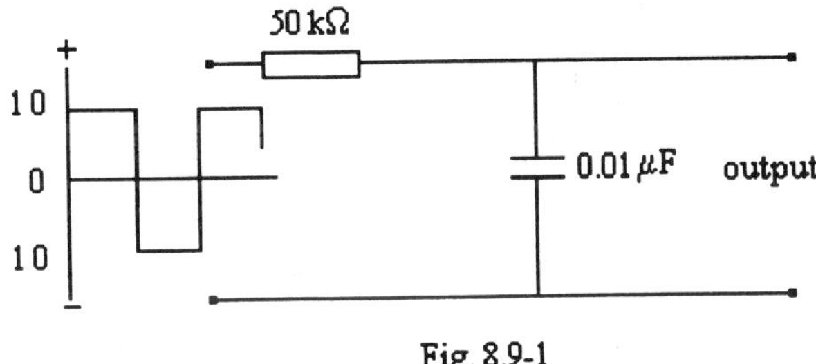

Fig. 8.9-1

Solution

(a) <u>Refer to Fig. 8.9-0</u>

Time constant, T = CR sec

= $0.001 \times 10^{-6} \times 10 \times 10^{3}$ = <u>0.01 mS</u>

<u>Refer to Fig. 8.9-1</u>

Time constant, T = CR sec

= $0.01 \times 10^{-6} \times 50 \times 10^{3}$ = <u>0.5 mS</u>

(contd) *(a)*

Periodic time of input signal $= {}^{1}/f$

$= {}^{1}/1000 \quad = \underline{1\ mS}$

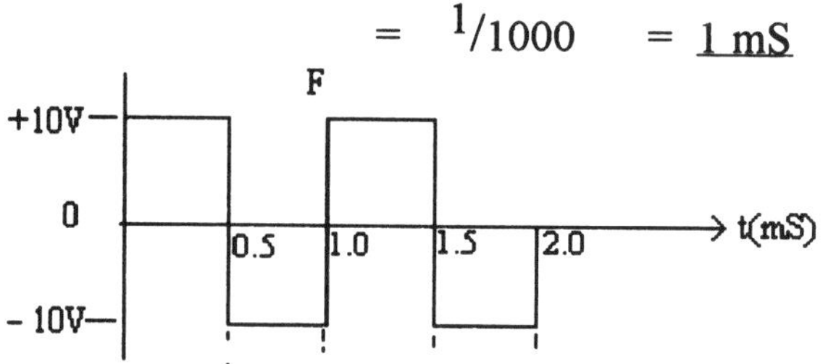

Fig. 8.9-2. Input waveform for Figs. 8.9-0 and 8.9-1

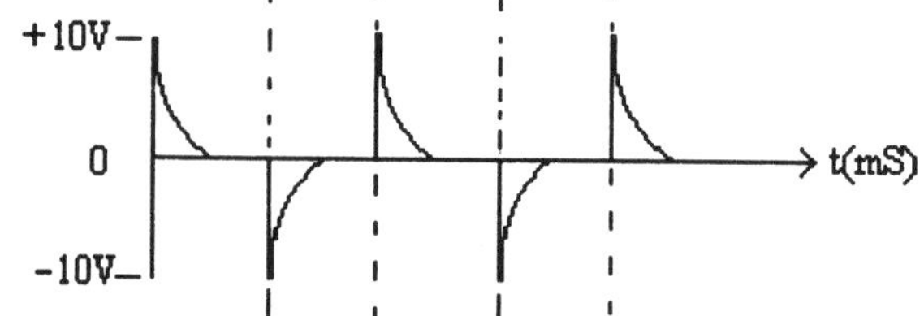

Fig. 8.9-3. Output waveform of circuit in Fig. 8.9-0

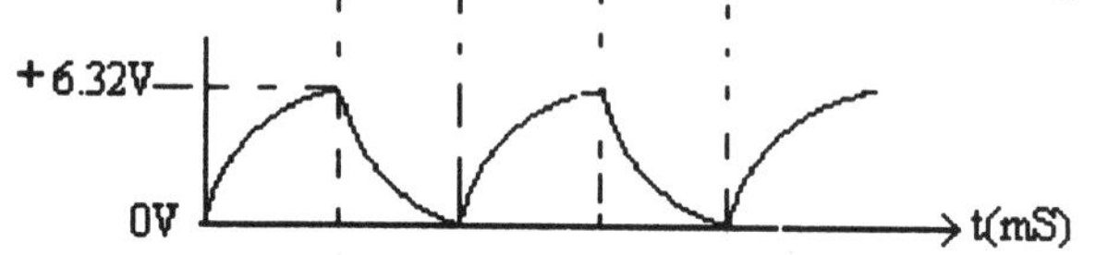

Fig. 8.9-4. Output waveform of circuit in Fig. 8.9-1

(i) The circuit of Fig. 8.9-0 produces an output voltage proportional to the slope of the input signal voltage. Therefore, the circuit is called a **differentiator.**

The time constant of the circuit is very small (0.01 mS or 10 μS) compared with the periodic time of the input signal (1.0 mS). The output waveform is nearly ideal except for the width of the spikes. Ideally, the spikes would have zero width.

(ii) The circuit of Fig. 8.9-1 is an **integrator** and it should produce an output proportional to the area under the input signal waveform, that is, a sawtooth waveform from the square wave input. Unfortunately, this is not the ideal output waveform due to the fact that the time constant of the circuit (0.5 mS) corresponds to the time of the input signal.

For a better sawtooth waveform from the square wave input, the time constant of the **integrator** circuit should be approximately five (5) times this value.

(contd)

(b) Refer to Fig. 8.9-0 : *The Differentiator Circuit*

The reactance, X_c $= {}^{1}/(2\pi fC)$

$= {}^{10^6}/(2 \times 3.14 \times 1000 \times 0.001)$

$\approx$ 159 kΩ

$\therefore$ Ratio of X_c:R $=$ 159:10 $\approx$ **16:1**

Refer to Fig. 8.9-1 : *The Integrator Circuit*

The reactance, X_c $= {}^{1}/(2\pi fC)$

$= {}^{10^6}/(2 \times 3.14 \times 1000 \times 0.01)$

$\approx$ 15.9 kΩ

$\therefore$ Ratio of X_c:R $=$ 15.9:50 $\approx$ 1:3

The reactance of the capacitor when a sinusoidal signal is applied to the *differentiator circuit* (Fig. 8.9-0) is approximately sixteen (16) times the value of R.
That is

$X_c >> R$

Ratio 16 : 1

Similarly, the reactance of the capacitor in the *integrator circuit* (Fig.8.9-1) is about one-third (${}^{1}/3$) the value of R.
That is

$X_c << R$

Ratio 1 : 3

Example 8.10

Sketch the circuit diagram of a bistable multivibrator suitable for counting in binary code from a single-pulse source.

Briefly describe with the aid of waveform diagrams its principle of operation.

Draw a block diagram to show how a number of stages may be connected to count in decades.

Solution

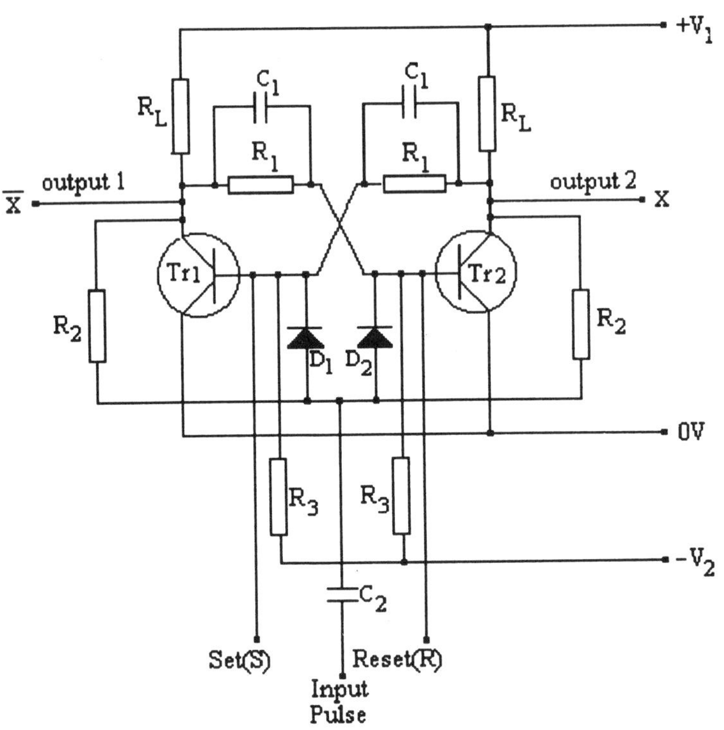

Fig. 8.10-0. Bistable multivibrator suitable for counting in binary code

In a bistable multivibrator Tr_2 conducts once for every two (2) input pulses and therefore the number of output pulses available from its output is half (½) the number of trigger pulses.

Since the bistable circuit has two stable states, it is customary to number them **0** and **1. The conducting transistor is in the 0 state.**

(contd)

Consider Fig. 8.10-0 which shows a typical circuit diagram of a practical multivibrator suitable for counting in binary code from a single input pulse source.

Fig. 8.10-1 shows the triggering pulses and output waveforms. The circuit is so designed that when one transistor is ON, the other is OFF.

1. Consider the circuit with Tr_1 initially conducting (ON) and Tr_2 cut off (OFF) by the combined action of the potential divider R_1 and R_3 between Tr_1 collector and the $-V_2$ supply rail.
   The circuit is in the first of its two stable states. Tr_1 collector voltage is approximately equal to zero and Tr_2 collector voltage is equal to V_1 [**Tr_1 is ON, Tr_2 is OFF**].

2. In this condition, if Tr_1 represents binary 0 (output bar X) and Tr_2 represents binary 1 (output X), the circuit can be used to store a binary digit or count in binary coded decimal.
   Since Tr_1 is saturated (ON), its collector-base junction is forward biased. Thus diode D_1 is reverse biased through R_2. Similarly, when Tr_2 is being cut-off (OFF), its collector-base junction is reverse biased. Thus D_2 is forward biased through R_2.

3. When the *first input pulse is applied,* it is differentiated by the action of C_2 and the input resistance of the transistor (C_2 forms a differentiated circuit with the input resistance of the transistors). Since the pulse is positive, it is steered by D_2 to Tr_2, thus driving Tr_2 to conduct (switch ON). As a result, Tr_2 is now ON and Tr_1 is cut off (OFF). The circuit is now in the second of its two stable states; Tr_1 is OFF and Tr_2 is ON. Each time an input pulse is applied the circuit changes state.

4. In practice, two further signal inputs are made available, the Set input (S) and the Reset input (R). When a positive pulse is applied to the S input, Tr_1 will conduct and Tr_2 will cut off, thereby setting the output to **1**.
   Similarly, a positive pulse at the R input always resets the output to **0.**

(contd)

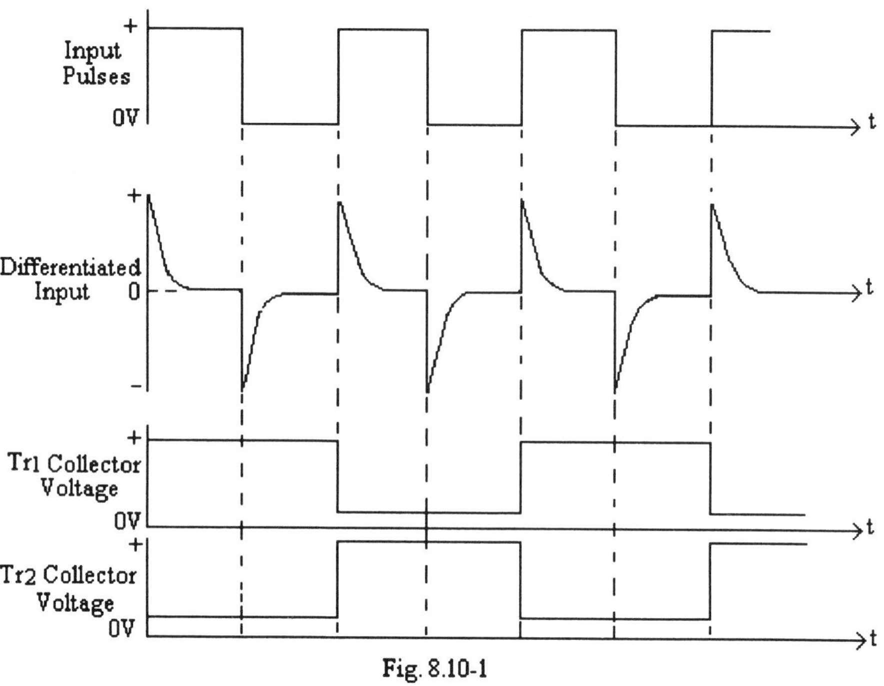

Fig. 8.10-1

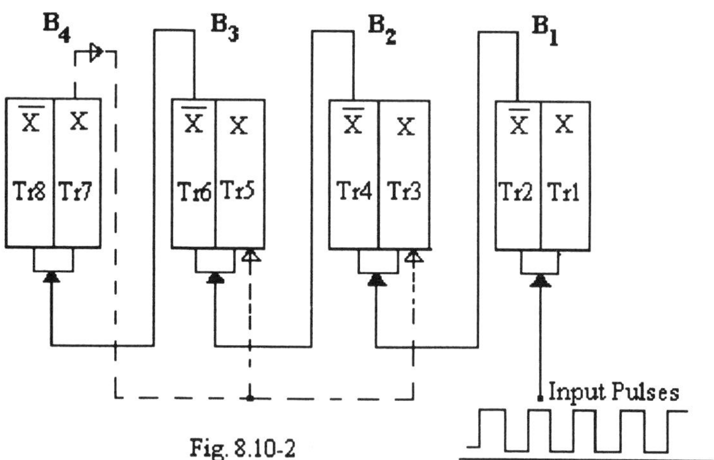

Fig. 8.10-2

If four bistable pairs (usually called flip-flops) are cascaded, they will count down by a power of 2, or in the binary system. In general, **n** bistables will count down by a factor of $2^{\mathbf{n}}$. However, feedback may be applied to reset the circuit to a previous condition after arrival of the nth pulse. In essence, **r** pulses are subtracted so as to provide a scaling factor equal $2^{\mathbf{n}} - \mathbf{r}$.

(contd)

The circuit in Fig. 8.10-2 will scale by the decimal factor 10, or count by decades. The four bistables provide a decimal count of 10, where the countdown ratio would normally be 16 (2^4). The circuit, by virtue of feedback, makes **r** = 6 and $2^{\mathbf{n}} - 6 = 10$; that is, $2^4 - 6 = 10$.

To determine the status of the count, a small lamp (L) is connected in series with the collectors of Tr_1, Tr_3, Tr_5 and Tr_7. If this lamp is not glowing, it indicates that the other transistor of a pair is conducting. Zero (0) state is associated with the conducting state .

In brief, the operation may be described as follows:

1. All bistables are reset to zero; Tr_2, Tr_4, Tr_6 and Tr_8 will be ON, and all lamps will be out, indicating zero count.

2. The 1st input, pulse, being positive, is differentiated and steered to switch ON Tr_1 (of B_1), and the indicator lamp (L_1) comes on. Tr_2 (of B_1) is then turned OFF and the monitor or counter reads 0001, i.e., decimal 1.

3. The second pair Tr_3 and Tr_4 (of B_2) is not affected because with Tr_2 turned OFF a negative pulse is generated at its output, which will be blocked by the diode, having been reverse biased.

4. The 2nd input pulse is steered to Tr_2 and turns it ON. As a result, Tr_1 is switched OFF and the indicator lamp (L_1) is out. At the same time, Tr_2 turns on, and a positive pulse is transmitted and steered to turn ON Tr_3 and turn OFF Tr_4 (of B_2). The indicator lamp L_2 is then on, and the counter or monitor reads 0010, i.e., decimal 2.

5. The 3rd input pulse transfers conduction back to Tr_1, lighting L_1. The 4th input pulse returns conduction to Tr_2 and a positive pulse to the second pair, causing Tr_5 to conduct and lighting lamp L_4.

6. By using feedback from Tr_7 on the 8th pulse, two counts are added to the second bistable and four counts to the third, thus advancing the count by 6. That is, on the 8th pulse the counter will read 1000 (decimal 8), but because of feedback the counter will read on the 8th pulse 1110 (decimal 14). After two more pulses, or a count of 10, all stages are returned to the initial condition. The counter or monitor action is tabulated in Tables 8.10-0 and 8.10-1.

(contd)

Table 8.10-0

	Decimal Value	Binary Coded Decimal				
		B_4 (8)	B_3 (4)	B_2 (2)	B_1 (1)	Output
	0	0	0	0	0	
	1	0	0	0	1	
	2	0	0	1	0	
	3	0	0	1	1	
	4	0	1	0	0	
	5	0	1	0	1	
	6	0	1	1	0	
	7	0	1	1	1	
Before Feedback	8	1	0	0	0	
After Feedback	*8 (14)	1	1	1	0	
	9 (15)	1	1	1	1	
	10 (16)	0	0	0	0	Output

Table 8.10-1

	Pulses	B_4		B_3		B_2		B_1	
		Tr7 (L4)	Tr8	Tr5 (L3)	Tr6	Tr3 (L2)	Tr4	Tr1 (L1)	Tr2
All bistable Reset to zero	0	Off	On	Off	On	Off	On	Off	On
	1	Off	On	Off	On	Off	On	On	Off
	2	Off	On	Off	On	On	Off	Off	On
	3	Off	On	Off	On	On	Off	On	Off
	4	Off	On	On	Off	Off	On	Off	On
	5	Off	On	On	Off	Off	On	On	Off
	6	Off	On	On	Off	On	Off	Off	On
	7	Off	On	On	Off	On	Off	On	Off
Before Feedback	8	On	Off	Off	On	Off	On	Off	On
After Feedback	8*	On	Off	On	Off	On	Off	Off	On
	9	On	Off	On	Off	On	Off	On	Off
	10	Off	On	Off	On	Off	On	Off	On → Output

The conditions for the successive stages after each pulse are shown in Table 8.10-1.

L_1, L_2, L_3 and L_4 represent lamps connected in series with the collectors of transistors Tr_1, Tr_3, Tr_5 and Tr_7.

Example 8.11

*A waveform at the transistor base of an astable multivibrator is shown in Fig. 8.11-0(a); the waveform has a designed slight unbalance. A series of triggering pulses (**sync pulses**); Fig. 8.11-0(b) is added to the waveform. The resulting waveform is shown in Fig.8.11-0(c) and represents a new frequency synchronized by the triggering **sync pulse**. Explain how synchronization is achieved.*

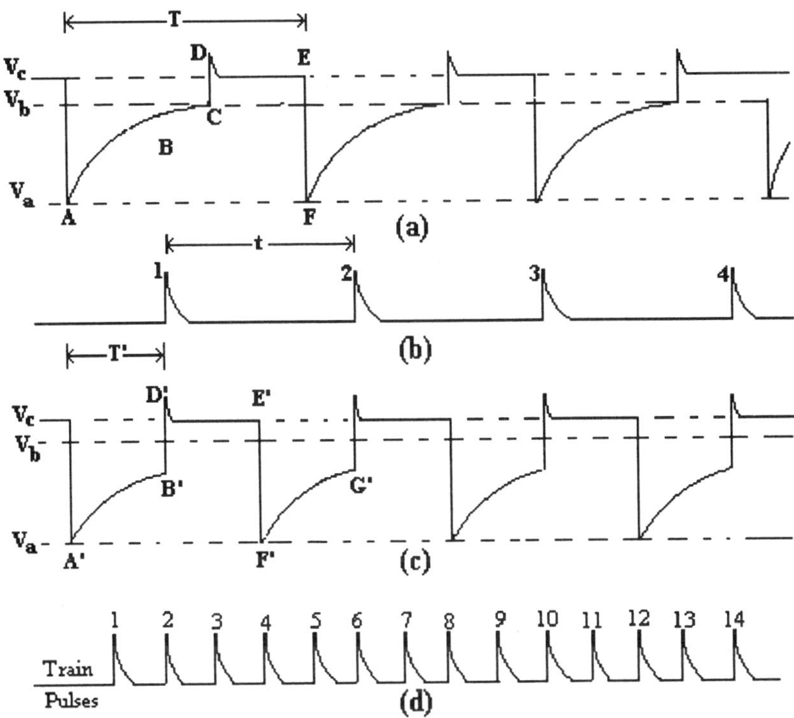

Fig. 8.11-0. Basic Synchronization

Solution

To ***synchronize or sync*** a free-running oscillator means to lock it into an external signal frequency. If the external frequency changes slightly, a well-designed synchronized circuit changes its frequency accordingly. Whenever this happens, the synchronized circuit will operate at a ***forced frequency*** instead of at a free-running frequency.

(contd)

Refer to the Waveform in Fig. 8.11-0(a)

It can be seen that the waveform changes exponentially from point A at V_a volts to point B at V_b volts. V_b is the voltage at which the multivibrator changes state. Consequently when the voltage reaches V_b, the waveform abruptly switches from C to D and remains saturated until E, when the circuit switches back from E to F.

Consider when the triggering pulses in Fig. 8.11-0(b) are added to the waveform in Fig. 8.11-0(a). While the waveform is changing from A to C, a triggering pulse appears at B (pulse 1) and forces the waveform up to D without waiting to discharge to C.

Pulse 1 has triggered the circuit and the waveform has now taken the shape shown in Fig. 8.11-0(c). This stage remains ON from D' to E'.

The injected synchronizing pulse has changed the time T (A to D) to the new time T' (A' to D') without disturbing the other stages of the circuit. Hence, the time D to E is identical to the time D' to E'.

The circuit inherently switches from E' to F'. A discharge occurs from F' to G'. At G' ***sync pulse*** 2 is injected and the circuit is forced to switch state. Each pulse triggers this stage from one state to the other, that is, from a nonconducting state to a conducting state.

It should be noted that the waveform of Fig.8.11-0(a) has been purposely drawn unbalanced so as to graphically dramatize how *a balanced waveform results only when* ***sync pulses*** *are applied.*

One of the requirements for synchronization is that the synchronizing voltage must be sufficiently large in amplitude to **sync**. Secondly, the time interval from pulse 1 to pulse 2 must be shorter than the time interval from A to E, i.e., ***t must be shorter than T.*** In brief, the ***sync frequency*** must be greater than the free-running frequency of the multivibrator.

Very often it is desirable to **sync** a low-frequency oscillator to a high-frequency signal. In this example, we assume the multivibrator frequency is 1000 Hz and the high frequency signal is 4000 Hz. This is a ratio of 4:1. Therefore, the multivibrator frequency is synchronized every fourth cycle by the high-frequency signal. The 4000 Hz signal is converted into sharp pulses as shown in Fig.8.11-0(d). The multivibrator is triggered on pulses 2, 6, 10, 14, etc.

Example 8.12

(a) *It is desirable to sync a free-running low-frequency oscillator to a high-frequency signal source. If the frequency of the source train pulses is 2.4 kHz and the free-running oscillator frequency is 480 Hz, determine the frequency when the oscillator is locked-in to pulses 2, 6, 10, and 14.*

(b) *Comment on possible waveform distortion in the process of synchronization and briefly describe a practical application of a sync control circuit.*

Solution

(a) Multivibrator frequency = 480 Hz (low frequency)
Frequency of train pulses = 2400 Hz (high frequency)
Multivibrator locked-in pulses are 2, 6, 10 and 14

∴ The multivibrator is triggered every 4 cycles; that is

$$6 - 2 = 4$$
$$10 - 6 = 4$$
$$14 - 10 = 4$$

∴ Locked frequency of oscillation $= {}^{2400}/4 = \underline{\textbf{600 Hz}}$.

Multivibrator locked-in to train pulses

Multivibrator frequency = 480 Hz
Frequency of train pulses = 2400 Hz
Multivibrator locked-in to train pulses 1, 4, 7, 10, 13

∴ Multivibrator is triggered every 3 cycles; that is,

$$4 - 1 = 3$$
$$7 - 4 = 3$$
$$10 - 7 = 3$$
$$13 - 10 = 3$$

∴ Sync frequency of multivibrator $= 2400/3 = \underline{\textbf{800 Hz}}$

(contd)

(b) It is undesirable to have too wide a difference between the free-running and lock-in frequency, which may result in distortion. In addition, the sync voltage should be just enough, for values above this may also result in waveform distortion. The following general rule should be adhered to:

Use enough sync voltage to obtain good synchronization.

Consider the sync control circuit shown in Fig.8.12- 0 (this is a practical sync control circuit found in an oscilloscope).

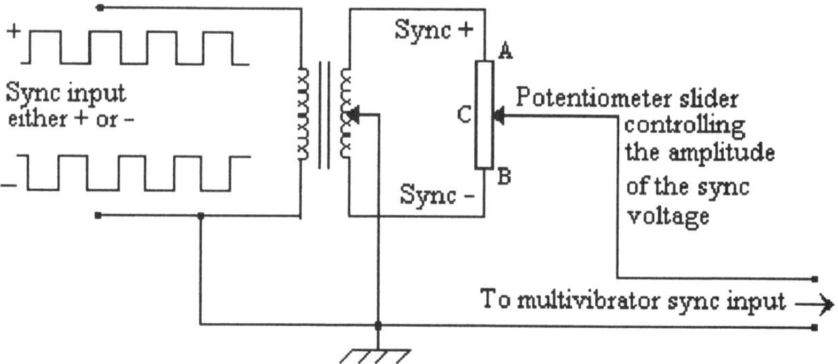

Fig. 8.12-0. Syn control circuit

When the slider of the potentiometer is set at the center C, the sync voltage to the multivibrator is zero. Hence it continues as a free-running oscillator. When the sync input signal pulses are positive, an upward shift of the potentiometer slider toward A produces the positive output pulses required for synchronization.

If, however, the incoming sync pulses are negative, a downward shift of the potentiometer slider toward B produces the positive output pulses required to sync the multivibrator.

If both positive and negative sync input pulses are available, either polarity can be selected for synchronization.

CHAPTER 9

THE THYRISTOR
(Silicon - Controlled Rectifier - SCR)

Introduction

The thyristor is a four-layer silicon semiconductor junction device which can be triggered into the conducting state and conducts unidirectionally until the potential across it is reduced to almost zero. The thyristor is also known by the name silicon-controlled rectifier (SCR). An important characteristic of the thyristor is its very fast switch-on and switch-off time. SCRs and triacs are all members of the thyristor family. They act as high-speed power switches.

In general, a semiconductor junction diode is an automatic unsophisticated switch; it will switch on whenever it is forward biased.
The transistor can also be used as a switch; collector current flows when there is adequate base current to switch it on. Transistors are widely used as switches in digital circuits but are limited by their inability to handle high power, in contrast to the thyristor, which is capable of controling a large amount of power with only a minimum amount of controlling energy.

The thyristor can operate at potentials up to several hundred volts and can handle currents up to ten of hundreds of amperes. As a result, thyristors are commonly used in sophisticated application to rectification (a.c. to d.c.), inversion (d.c. to a.c.), relaying, timing, ignition, speed control, a.c. and d.c. power control systems, heaters and alarms at power levels ranging from a few milliwatts to hundreds of kilowatts.

The following examples will help explain the basic characteristics of this outstandingly useful device, with a selection of basic industrial circuits for demonstration and educational endeavours.

Example 9.1

Using appropriate diagrams, describe the operation of a thyristor. Sketch the anode characteristics of the device and indicate zero and forward gate current.

Solution

The thyristor is a four-layer *pnpn* semiconductor device, represented in Fig. 9.1-0(a); the symbol shown in (b) resembles that of a normal rectifier with the exception of an additional terminal known as the **gate**.

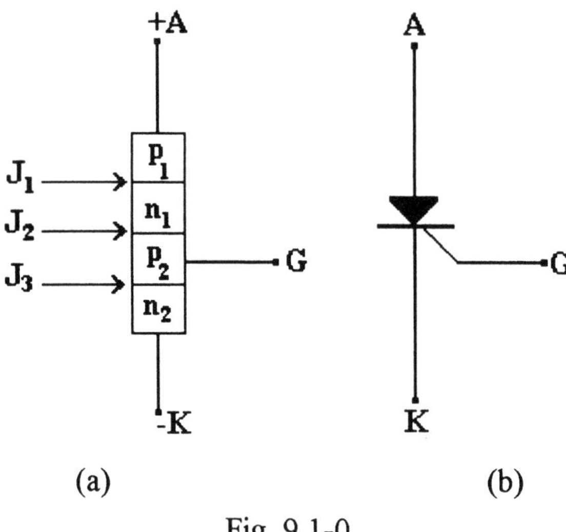

Fig. 9.1-0

When forward biased (A+ with respect to K), junction J_2 is reverse biased and the thyristor is in its **blocking state**, with only leakage current flowing. If the anode-cathode (A-K) voltage is increased sufficiently high, avalanche breakdown of J_2 occurs and this value of anode-cathode voltage is known as the breakdown voltage V_{BO}.

However, breakdown of junction J_2 is achieved at lower anode-cathode voltages by injecting a small signal current in the gate terminal.

Once the thyristor is in the conducting state, the gate current can be removed, since there is no further control by the gate on the thyristor and hence the anode current remains high.

(contd)

To return the thyristor to its blocking state is to reduce the anode current below a critical value known as the **holding current**, which is usually achieved by reducing the anode-cathode voltage to zero or making it negative.

When the thyristor is reverse biased, junctions J_1 and J_3 are reverse biased and only a small leakage current flows until the reverse voltage is increased sufficiently high to cause avalanche breakdown of J_1 and J_3.

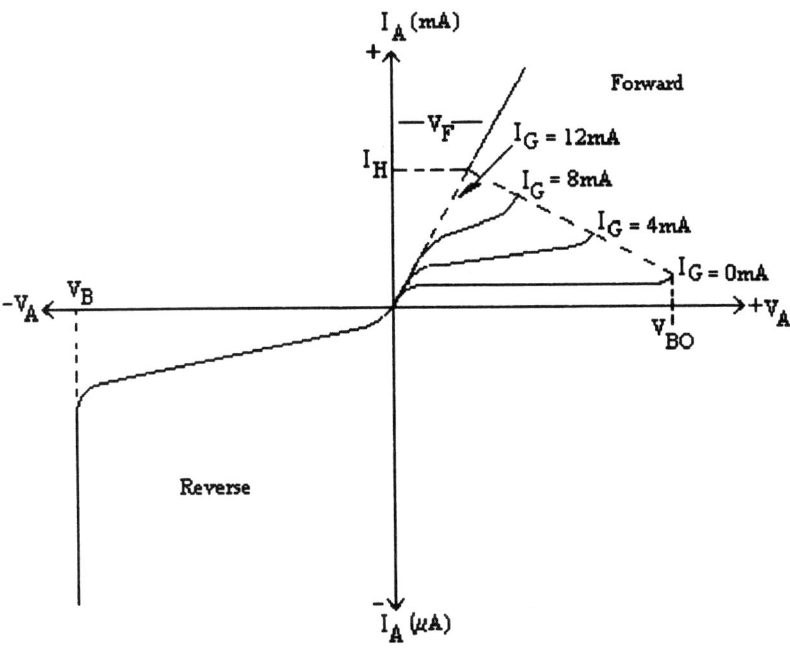

Fig. 9.1-0

A typical set of anode characteristics for the thyristor is shown in Fig. 9.1-1.

Both V_{BO} and V_B can be up to 250 V for thyristors.

At I_G = 12 mA blocking ceases; the thyristor from thereon acts as a normal *pn* junction diode.

V_B	=	reverse breakdown voltage
V_{BO}	=	breakover voltage
I_H	=	holding current
V_F	=	forward voltage drop across thyristor
V_A	=	anode voltage

Example 9.2

Using a two-transistor analogy for a thyristor, explain its principle of operation. Hence determine an expression for the forward gain of the device. State the condition for which the thyristor will switch from a blocking to a conducting state.

Solution

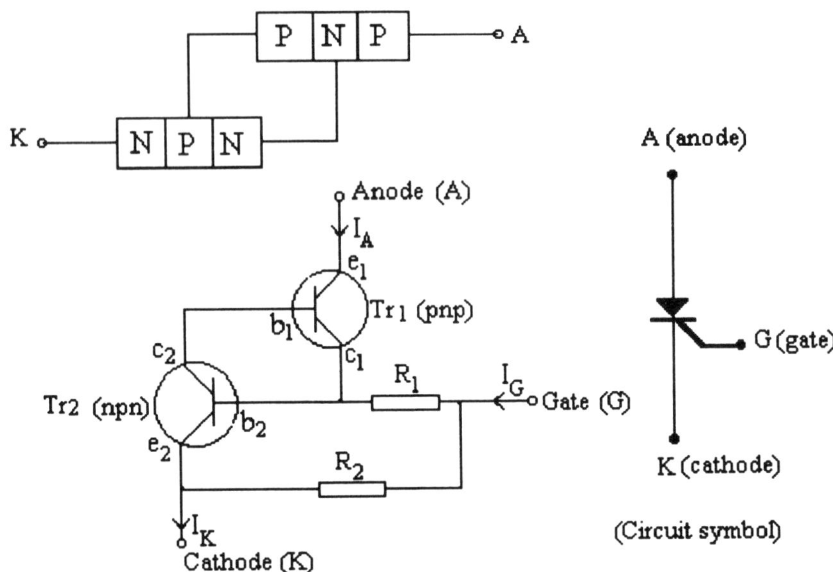

(Two-transistor representation of Thyristor)

Fig. 9.2-0

Fig. 9.2-0 shows the two-transistor-namely a pnp and an npn-representation of a thyristor, or SCR. The basic characteristics of the thyristor can be readily understood by referring to the two-transistor interconnection in Fig. 9.2-0 and the circuit shown in Fig.9.2-1.

(contd)

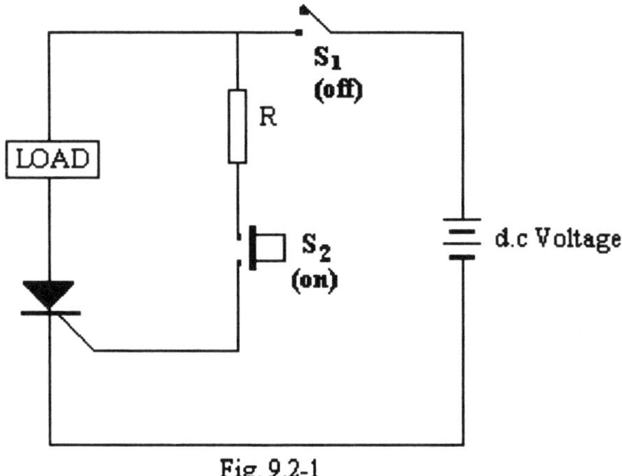

Fig. 9.2-1

Refer to Fig. 9.2-1

When S_1 is closed, the thyristor is "blocked" and acts (between anode and cathode) like an open switch.

Refer to Fig. 9.2-1

The above action is due to the fact that the base of Tr_2 is shorted to the cathode via R_1 and R_2 so that Tr_2 is cut off and passes negligible collector current into the base of Tr_1, which also is cut off and passes negligible current into the base of Tr_2. Consequently, under this condition both Tr_1 and Tr_2 are cut off and only small leakage current flows between the anode and cathode of the thyristor.

Refer to Fig.9.2-1

The thyristor can be turned on by applying a sufficiently large positive current to the gate by closing S_2. (S_1 remains closed.) The action of switching on the device can be explained as follows.

Refer to Fig.9.2-0

When the gate current is made sufficiently large and positive with respect to the cathode, gate current flows through R_1 and R_2 and the base-emitter junction of Tr_2. Thus, Tr_2 is biased on and its collector current feeds into the base of Tr_1. This base current is amplified by Tr_1 and fed back into the base of Tr_2, where it is again amplified and fed to Tr_1. A regenerative action thus takes place and both transistors switch rapidly to saturation.

(contd)

Under this condition the anode-to-cathode saturation voltage is the sum of Tr_1 saturation voltage and Tr_2 forward base-emitter voltage. As in the case of a normal silicon rectifier diode, the saturation voltage amounts to 1 or 2 V.

Once the thyristor has been turned on and is conducting in the forward direction, the gate loses all control and the thyristor stays on even with the gate current completely removed. Thus, only a short duration of a positive gate pulse is required to switch the device on. Considerable current is available between the gate and anode of the device; so a very small amount of gate power is needed to control very high external power loads.

Once the thyristor has self-latched into its conducting state, it can only be turned off by momentarily reducing its anode current to zero or below the *"minimum holding current"*; it follows that turn off is automatic in a.c. circuits near the zero crossing point at the end of each half-cycle.

A certain amount of internal capacitance exists between the anode and gate of the thyristor. Consequently, if a sharp rising voltage is applied to the anode, it is possible that the intrinsic capacitance can cause part of the anode rising signal to break through to the gate and thus trigger the device on. This "*rate-effect*" switch-on can be caused by the supply line transients and sometimes occurs at the instant of powering the intrinsic capacitance. *Rate effect* can be overcome by connecting a simple R-C filter network between the anode and cathode to limit the rate of rise to a safe value.

Refer to Fig. 9.2-0

$$I_{B1} = (1 - \alpha_1)I_A - I_{CBO1} \qquad \text{(Eq. 9.2-0)}$$

where

α_1 = common-base current gain of Tr_1

I_{CBO1} = leakage current of Tr_1

$$I_{C2} = \alpha_2 I_K + I_{CBO2} \qquad \text{(Eq. 9.2-1)}$$

where

α_2 = common-base current gain of Tr_2

I_{CBO2} = leakage current of Tr_2

(contd)

The cathode current I_K is the sum of the anode and gate currents, that is,

$$I_K = I_A + I_G$$

and $$I_{C2} = I_{B1}$$

Equating Eqs. 9.2-0 and 9.2-1 we get,

$$\alpha_2 I_K + I_{CBO2} = (1 - \alpha_1)I_A - I_{CBO1}$$

Substituting for I_K, we get

$$\alpha_2 I_A + \alpha_2 I_G + I_{CBO2} = (1 - \alpha_1)I_A - I_{CBO1}$$

$$\alpha_2 I_A - I_A(1 - \alpha_1) = -\alpha_2 I_G - I_{CBO1} - I_{CBO2}$$

$$I_A(1 - \alpha_1) - \alpha_2 I_A = \alpha_2 I_G + I_{CBO1} + I_{CBO2}$$

$$\therefore \quad I_A = \frac{\alpha_2 I_G + I_{CBO1} + I_{CBO2}}{1 - (\alpha_1 + \alpha_2)}$$

Analyzing the equation for I_A, we see that when $(\alpha_1 + \alpha_2) = 1$, the anode current I_A tends to infinity.

In practice, this equation corresponds to the conducting state of the device. Usually, the critical condition occurs by increasing the current gain α_2; this is increased when the gate current is injected into junction J_2 (refer to Fig. 9.1-0). Thus by increasing the current density, the thyristor is triggered into conduction.

The α current gain can be converted to the β current gain by substituting

$$\alpha = {}^{\beta}/(\beta + 1)$$

Example 9.3

(a) *Draw a circuit diagram showing how a single thyristor can be used to control a unidirectional load current. Assume a sinusoidal supply voltage.*

Discuss the merits and limitations of the method of gate control used.

(b) *Fig. 9.3-0 shows a thyristor employing a.c. phase shift control on its gate. Sketch the waveforms showing supply voltage, firing angle and load current.*

Draw the phasor diagram for the phase-shifting circuit, neglecting the loading effect of the gate circuit, and hence determine the firing angle ϕ when $R = 1.0\ k\Omega$. Prove any formula used.

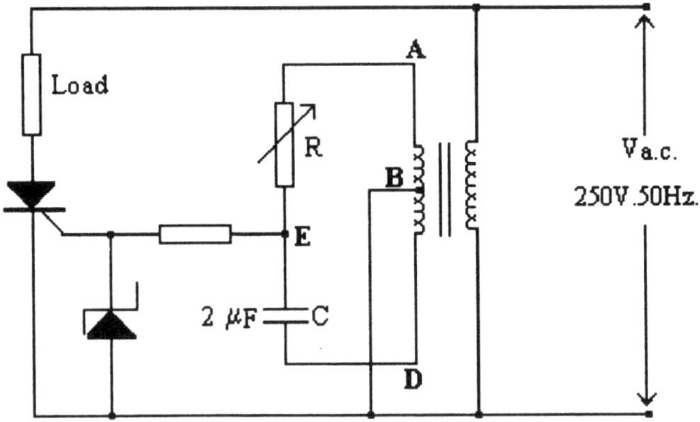

Fig. 9.3-0. Thyristor with a.c. phase shift control

Solution

(a) Fig. 9.3-1 shows the basic circuit diagram of a single thyristor controlling a unidirectional load.

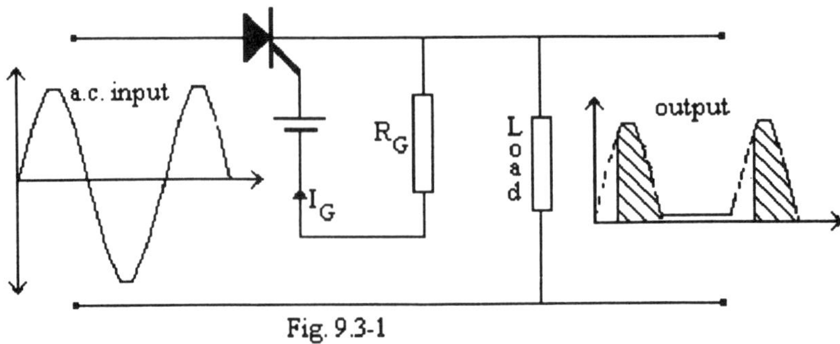

Fig. 9.3-1

(contd) *(a)*

Refer to Fig. 9.3-1

On the positive half-cycle of the input voltage the thyristor will be triggered at some voltage of V_{BO} (the breakdown voltage) corresponding to the gate current I_G.

On the negative half-cycle the anode voltage is reduced to zero, and the thyristor returns to its blocking state. The output voltage will be pulses, each pulse being a fraction of one half-cycle; that is, each pulse will be less than 180°.

The main advantage of this method of control is its simplicity. The gate resistor R_G limits the gate-cathode power from exceeding the maximum value.

The thyristor can be triggered only in the first half of the positive cycle of the a.c. input, which in effect makes the maximum triggering angle 90°.

(b) In the circuit of Fig. 9.3-0 the voltage phase at point B shifts through 180° as R is varied from maximum to zero.

Waveforms of the supply voltage (V_S), gate-cathode voltage and load current (I_L) for a given firing angle ϕ are shown in Fig. 9.3-2. By varying R from zero to maximum, ϕ can be varied from 0° to 180°.

At 180° the thyristor will not be triggered and the circuit is said to be fully phased-back.

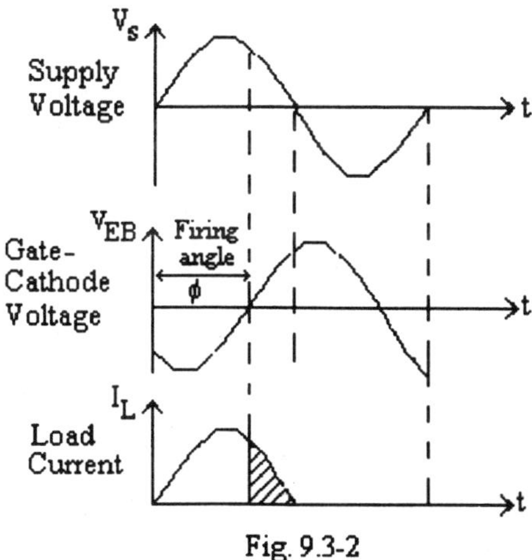

Fig. 9.3-2

(contd)

(b)

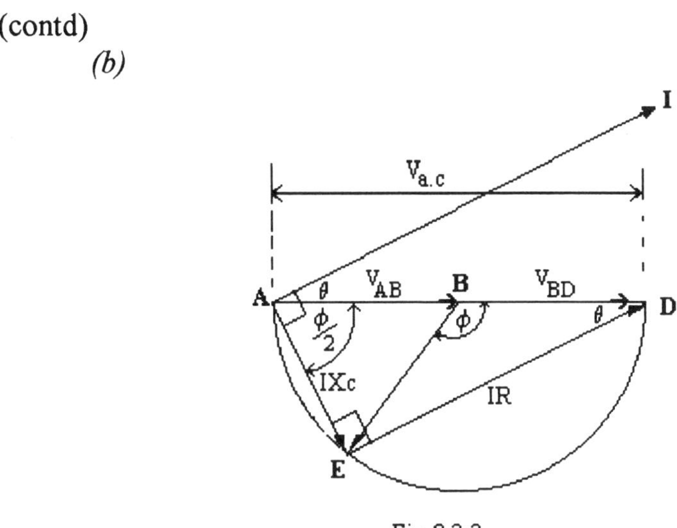

Fig. 9.3-3

Fig. 9.3-3 shows the phasor diagram of the *phase-shift*ing circuit (neglecting the loading effect of the gate circuit).

The current I leads IX_c by 90° and it is in-phase with R.
B is the midpoint of Va.c. since it is the center tap of the transformer.
By geometry, the locus of point E is a semicircle of diameter AD.
In the phasor diagram BD and BE are radii of the semicircle, making BDE and ABE isosceles triangles.
The firing angle ϕ can vary from 0° to 180° by varying R.

<u>Refer to the Phasor Diagram in Fig. 9.3-3</u>

$$\angle ABE = (180^o - \phi)$$

$$\angle BAE = \angle BEA = \phi/2 \quad \text{(isosceles triangle)}$$

Therefore in triangle AED,

$$\tan \phi/2 = IR/IX_c = \omega CR$$

$$\therefore \quad \phi/2 = \tan^{-1}(\omega CR)$$

$$\phi = 2\tan^{-1}(2\pi \times 50 \times 2 \times 10^{-6} \times 10^3)$$

$$= 2\tan^{-1} 0.628 = \underline{\mathbf{64.26^o}}$$

The function of the Zener diode is to protect the gate-cathode from excessive forward voltage on the positive half-cycle and to shunt the current from the cathode-gate circuit in the negative half-cycle.

Example 9.4

A thyristor has the following relationship between the breakdown voltage V_{BO} (volts) and the d.c. gate current I_G (mA):

$$V_{BO} = 480 - 20I_G$$

If the thyristor is employed in the circuit shown in Fig. 9.4-0, derive an expression for the mean load current and hence calculate its maximum and minimum values and corresponding values of I_{Gmax}

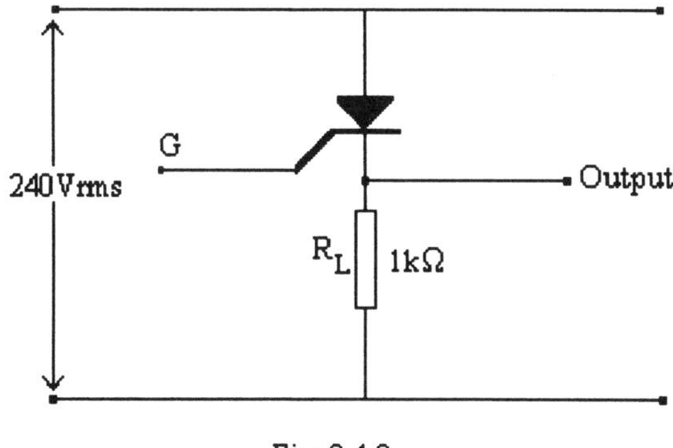

Fig. 9.4-0

Solution

Mean load current is

$$I_{d.c.} = \frac{1}{2\pi}\int_{\omega t}^{\pi} \frac{V_{max}\sin\omega t \; d(\omega t)}{R_L} \quad \text{amps}$$

$$= \frac{1}{2\pi}\int_{\omega t}^{\pi} \frac{240\sqrt{2}\sin\omega t \; d(\omega t)}{1000} \quad \text{amps}$$

$$= \frac{240\sqrt{2}}{2\pi \times 1000}\int_{\omega t}^{\pi} \sin\omega t \; d(\omega t) \quad \text{amps}$$

$$= \frac{120\sqrt{2}}{\pi \times 1000}\Big[-\cos\omega t\Big]_{\omega t}^{\pi} \quad \text{amps}$$

$$= \frac{120\sqrt{2}}{\pi}\Big[(-\cos\pi) - (-\cos\omega t)\Big] \text{ mA}$$

$$= \frac{120\sqrt{2}}{\pi}\Big[1 + \cos\omega t\Big] \text{ mA} \quad \textbf{(Eq.9.4-0)}$$

Equation 9.4-0 is the expression for the mean load current.

(contd)

Maximum load current is taken when $\omega t = 0$ or $\cos \omega t = 1$

$$\therefore \quad I_{max} = \frac{120\sqrt{2}}{\pi} \times 2 = \underline{\mathbf{108\ mA}}$$

$$I_{min} = \frac{120\sqrt{2}}{\pi} = \underline{\mathbf{54\ mA}}$$

Thyristor breakover occurs when

$$V_{BO} = V_{max} \sin \omega t$$

$$= 480 - 20I_G$$

$$\therefore \quad I_{Gmax} = \underline{\mathbf{24\ mA}}$$

Example 9.5

Draw the circuit diagram of a phase-controlled inverse parallel controller. Such a controller is used to supply a resistance furnace of 10 Ω resistance from a 400 Vr.m.s supply. Calculate the power supply to the furnace if the delay angle is 90°.

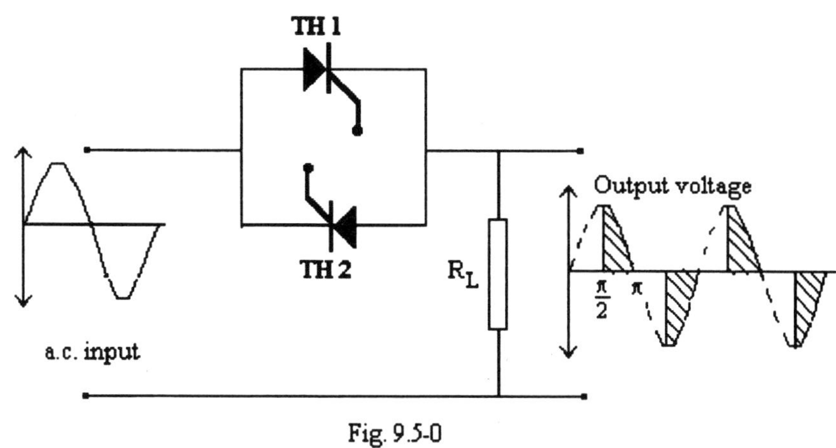

Fig. 9.5-0

The circuit diagram of Fig. 9.5-0 shows a phase-controlled inverse parallel a.c. controller employing two thyristors. The thyristors are so connected that firing occurs in both halves of the input supply.
The waveforms at the gates of TH1 and TH2 will be exactly 180° out of phase. On the positive half-cycle of the input voltage, TH1 fires at $\pi/2$ rad, or 90°, while on the negative half-cycle, TH2 fires at an angle of $(\pi + \pi/2)$ rad, or (180° + 90°), giving the output waveform shown.

The thyristor may be fired anywhere in the positive half-cycle of the anode supply. In this example, pulses in one direction trigger one thyristor while pulses in the opposite direction trigger the other. A pulse train is generated and synchronized to the anode supply, at the predetermined firing angle, to maintain a continuous firing signal.

Generally, pulses have fast rise times and are thus ideal for firing thyristors. Pulse powers can be high to ensure firing of thyristors under all conditions. This condition also limits the output power of the firing circuit, since the rating of gate power is determined by the average power supplied.
Ideal outputs for firing thyristors include, among others, astable multivibrators.

(contd)

Since an output is obtained between π/2 and π then, R.M.S voltage across R_L is;

$$V_{r.m.s} = \left[\frac{1}{\pi}\int_{\pi/2}^{\pi} V_m^2 \sin^2\omega t \; d(\omega t)\right]^{1/2}$$

Let $\omega t = \theta$.

Then $$V_{r.m.s} = \left[\frac{1}{\pi}\int_{\pi/2}^{\pi} V_m^2 \sin^2\theta \, d\theta \right]^{1/2}$$

$$= \left[\frac{V_m^2}{\pi}\int_{\pi/2}^{\pi} \sin^2\theta \, d\theta \right]^{1/2}$$

$$= \left[\frac{V_m^2}{\pi}\int_{\pi/2}^{\pi} \frac{1 - \cos 2\theta}{2} d\theta\right]^{1/2}$$

$$= \left[\frac{V_m^2}{2\pi}\left(\theta - ½ \sin 2\theta\right)_{\pi/2}^{\pi}\right]^{1/2}$$

$$= \left[\frac{V_m^2}{2\pi}\left(\pi - ½ \sin 2\pi\right) - \left(\frac{\pi}{2} - ½ \sin 2\,\frac{\pi}{2}\right)\right]^{1/2}$$

$$= \left[\frac{V_m^2}{2\pi} \text{ x } \frac{\pi}{2}\right]^{1/2}$$

Over a complete cycle, $$V_{r.m.s} = \left[2\left(\frac{V_m^2}{2\pi} \text{ x } \frac{\pi}{2}\right)\right]^{1/2}$$

$$= \left[\frac{V_m^2}{2}\right]^{1/2}$$

$$= \frac{V_m}{\sqrt{2}}$$

Power supplied to the furnace = V^2/R

$= (400/\sqrt{2})^2 \text{ x } 1/10$

$= 80000 \text{ x } 1/10$ = **<u>8000 watts</u>**

or **<u>8 kW</u>**

Example 9.6

The circuit diagram of Fig. 9.6-0 shows a thyristor with a fixed bias connected across a capacitor C, which is allowed to charge up via a large-value resistor R.

The bias is arranged so that the thyristor fires when the voltage across the capacitor is 150 V and stops conducting when the capacitor voltage falls to 75 V.

Derive an expression for the power dissipated in the load R_L in each operation and calculate its value when $R_L = 100\ \Omega$, stating any assumptions made.

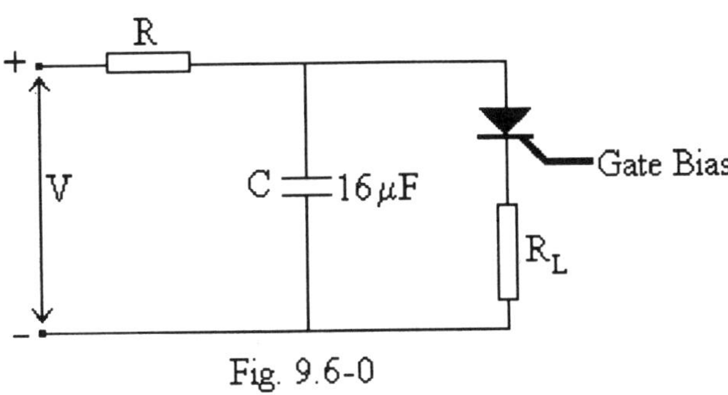

Fig. 9.6-0

Solution

Energy stored in capacitor $= \frac{1}{2} CV_1^2$ joules (V_1 = voltage across C)

Capacitor discharges through R_L to V_2 in t sec.

$\therefore$ Energy loss by capacitor $= \frac{1}{2} C(V_1^2 - V_2^2)$ joules

Power P dissipated = joules/sec

Power dissipated in t sec, P $= \frac{1}{2} C(V_1^2 - V_2^2)/t$ (Eq. 9.6-0)

It is assumed that the voltage drop across the thyristor is negligible and switching actions are instantaneous.

(contd)

The time taken for capacitor C to discharge from 150 V (V_1) to 75 V (V_2) is

$$V_2 = V_1 e^{-t/CR_L}$$

$$75 = 150e^{-t/CR_L}$$

$$^{75}/_{150} = e^{-t/CR_L}$$

$$^{1}/_{2} = e^{-t/CR_L}$$

$$2 = e^{t/CR_L}$$

$$\therefore \quad t = CR_L \log_e 2$$

$$= 16 \times 10^{-6} \times 100 \log_e 2$$

$$= 1.6 \times 10^{-3} \times 0.693$$

$$= 1.109 \text{ mS} \qquad \approx \underline{\mathbf{1.11\ mS}}$$

Substituting the value of t into Eq. 9.6-0, we get

$$P = ½ \times \frac{16 \times 10^{-6}(150^2 - 75^2)}{1.11 \times 10^{-3}}$$

$$= \frac{8 \times 10^{-3} \times 16875}{1.11}$$

$$= \underline{\mathbf{121.62\ W}}$$

Example 9.7

The circuit of Fig. 9.7-0 used a four-layer (pnpn) diode to generate a sawtooth waveform for operation at 500 Hz. The diode is rated at $V_{BO} = 100$ V, $I_A = 20$ A and $I_H = 0.02$ A. The breakdown voltage V_{BO} occurs at one-third the supply voltage V_S. To ensure turn-off, R is taken at (V_S/I_H + 10 kΩ). Determine the value of

resistor R

capacitor C

diode resistance R_D.

Sketch the shape of the output waveform.

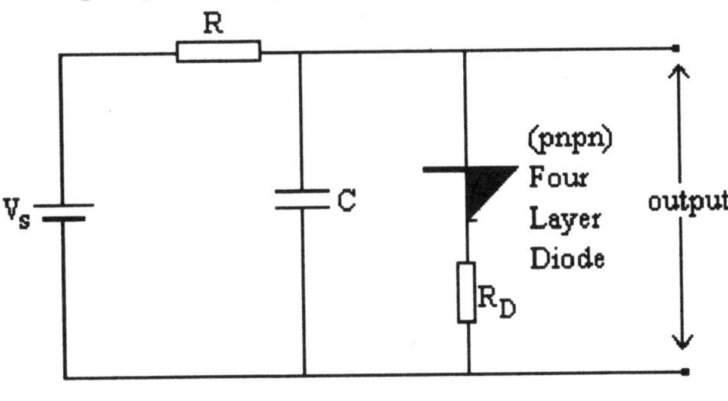

Fig. 9.7-0

[Note Basics: The silicon *pnpn* or four-layer diode consists of three pn junctions in series as shown in Fig. 9.7-1.

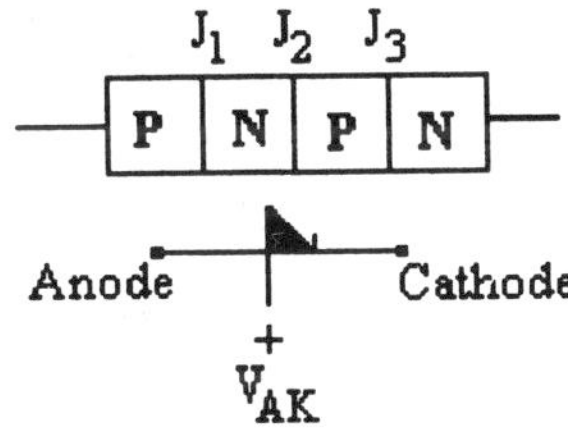

Fig. 9.7-1

The symbol is a half-diode or numeral 4 with the apex pointing in the direction of positive current flow. In the cross-section representation, for negative anode-to-cathode voltage V_{AK}, junction J_2 is forward biased but junctions J_1 and J_3 are reverse biased. As a result, this diode is a reverse-blocking diode. The characteristic for negative V_{AK}, the off state, resembles that of a reverse-biased diode.

Solution

In the circuit of Fig. 9.7-0 the capacitor C charges up to V_{BO} and discharges through the diode and its resistance R_D.

(contd)

$$V_{BO} = (1/3)\ V_s$$

$$\therefore \quad V_s = 3\ V_{BO}$$

$$= 3 \times 100 \qquad = \underline{\mathbf{300\ Volts}}$$

$$R = V_s/I_H + 10\ k\Omega$$

$$= 300/0.02 + 10\ k\Omega$$

$$= 15\ k\Omega + 10\ k\Omega \qquad = \underline{\mathbf{25k\Omega}}$$

Now $V_c = V_o - V_o e^{-t/CR}$

i.e. $V_{BO} = V_s - V_s e^{-t/CR}$

$$= V_s(1 - e^{-t/CR})$$

$$V_{BO}/V_s = 1 - e^{-t/CR}$$

$$e^{-t/CR} = 1 - V_{BO}/V_s$$

$$= 1 - 100/300 = 2/3$$

$$e^{t/CR} = 3/2$$

$$t/CR = \log_e 1.5 = 0.4$$

But $t = 1/f$

$$\therefore \quad 1/fCR = 0.4$$

$$1/C = 0.4Rf$$

$$\therefore \quad C = 1/(0.4 \times 25 \times 10^3 \times 500)$$

$$= 2 \times 10^{-7} \times 10^6 \qquad = \underline{\mathbf{0.2\ \mu F}}$$

$$R_D = V_{BO}/I_A$$

$$= 100/20 \qquad = \underline{\mathbf{5\ \Omega}}$$

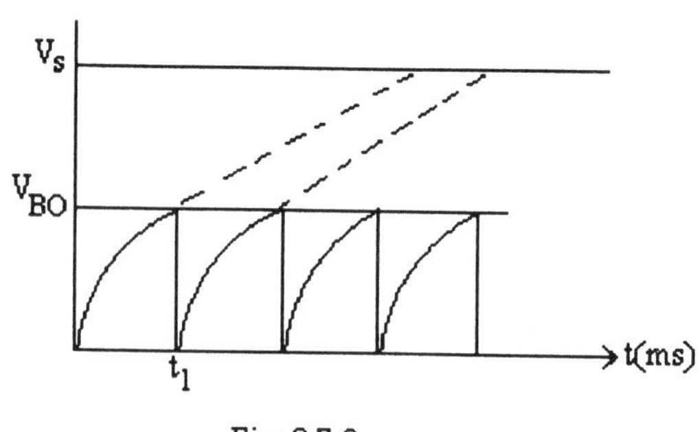

Fig. 9.7-2

The output approximates a sawtooth.

Example 9.8

The high power gains, low leakage currents and high current carrying capacity of thyristors (SCRs) make them ideal for use in a variety of alarm systems. Briefly describe the following practical alarm systems, with associated circuits, suitable for use in residential and industrial premises.

1. *a remote-operated alarm*
2. *make-to-operate burglar alarm*
3. *break-to-operate burglar alarm*
4. *tamper-proof burglar alarm*
5. *a delayed self-latching burglar alarm*

Solution

1. *Remote-Operated Alarm*

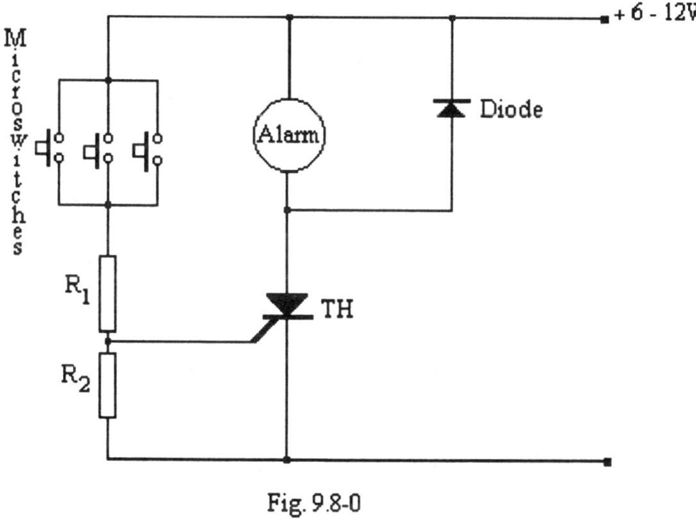

Fig. 9.8-0

This is the simplest alarm circuit, shown in Fig. 9.8-0.
The circuit is nonlatching and activates when any of the microswitches are closed. Standby current may be designed for a few milliamperes by choice of suitable values for R_1 and R_2. With a suitable range of standby current, the effect of cable resistance can be reduced, and the micro-switches connected in parallel may be positioned a distance away.

(contd)

2. *Make-To-Operate Burglar Alarm*

The remote-operated alarm (Fig. 9.8-0) may be converted to a simple self-latching multiinput burglar/fire alarm by the inclusion of a latching resistor R_3 and a reset button as shown in Fig. 9.8-1.

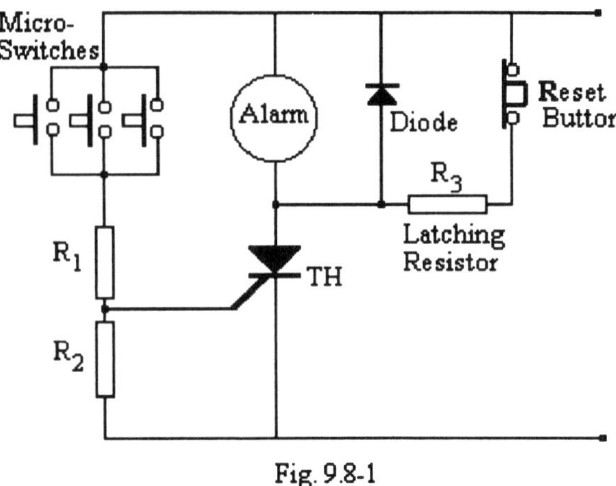

Fig. 9.8-1

The circuits of Figs. 9.8-0 and 9.8-1 may be utilized for different operation in the home and industry by simply replacing the microswitches with other options. For example:

1. If the switches are of the microswitch or reed type, the alarm can be made to activate when a door or window is opened.

2. If the switches are of the pressure pad type, the alarm can be made to operate when someone stands on a mat or a vehicle passes over the pressure pad.

3. If thermostat switches are used, the circuit can function as an automatic fire alarm.

The circuit can be designed to operate from a 12-Vd.c. supply, drawing very low current when the alarm is initially switched on to standby.

(contd)

3. *Break-To-Operate Burglar Alarm*

The alarm circuit in Fig. 9.8-1, though useful as a burglar alarm, can be disabled by cutting the cable linking the microswitches to R_1 or by cutting the supply lines.

A more reliable burglar alarm circuit is shown in Fig. 9.8-2. If any of the switches are briefly disturbed or if their connecting leads are cut, the circuit will activate and self-latch in the on state until reset by depressing the reset button.

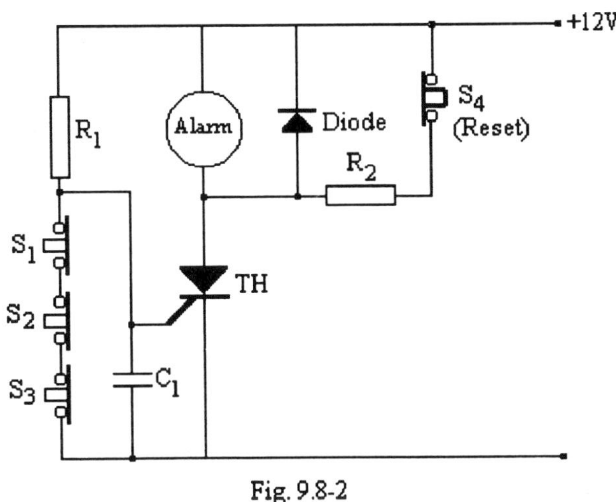

Fig. 9.8-2

Capacitor C_1 prevents false alarms due to inadvertent bouncing of contacts, vibration or other shock conditions.

The alarm will only activate if the contacts are made to open for a period in excess of a few milliseconds.

The bleed current or standby current can be reduced by the use of one or two transistor amplifiers to increase the gate sensitivity of the thyristor. By reducing the standby current, the life of the supply battery is extended. Figure 9.8-3 shows the inclusion of a single-transistor amplifier in the circuit.

(contd)

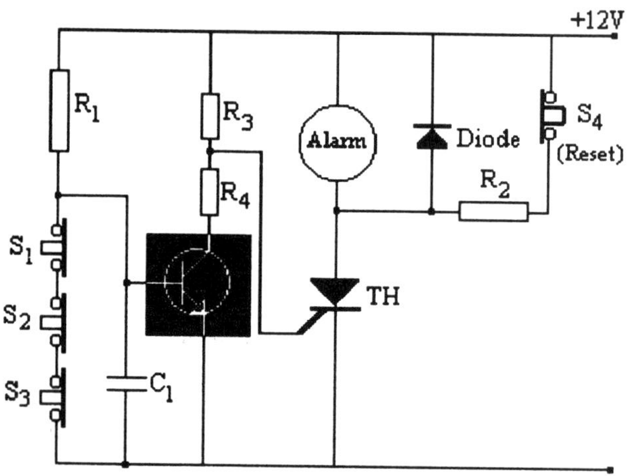

Fig. 9.8-3. Single-transistor amplifier break-to-operate burglar alarm

4. *Tamper-Proof Burglar Alarm*

The preceding circuits are not fully tamper-proof; they can be disabled by connecting a jumper lead across the switches S_1, S_2 and S_3. Figure 9.8-4 shows a circuit in which this deficiency can be overcome. The circuit combines both break-to-operate and make-to-operate switches. An intruder is unlikely to differentiate which alarm leads are for break-to-operate or make-to-operate. Consequently, if the wrong leads are broken or cut, the circuit will instantly activate and self-latch, and the alarm will only stop sounding when the reset button is depressed.

(contd)

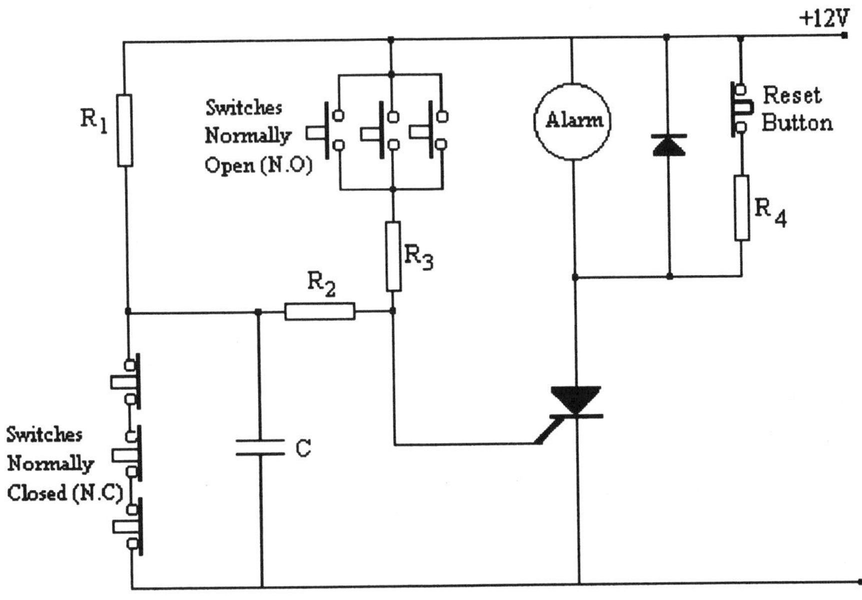

Fig. 9.8-4. A simple tamper-proof alarm

To extend the life of the battery, a two-transistor amplifier may be included in the circuit as shown in Fig. 9.8-5

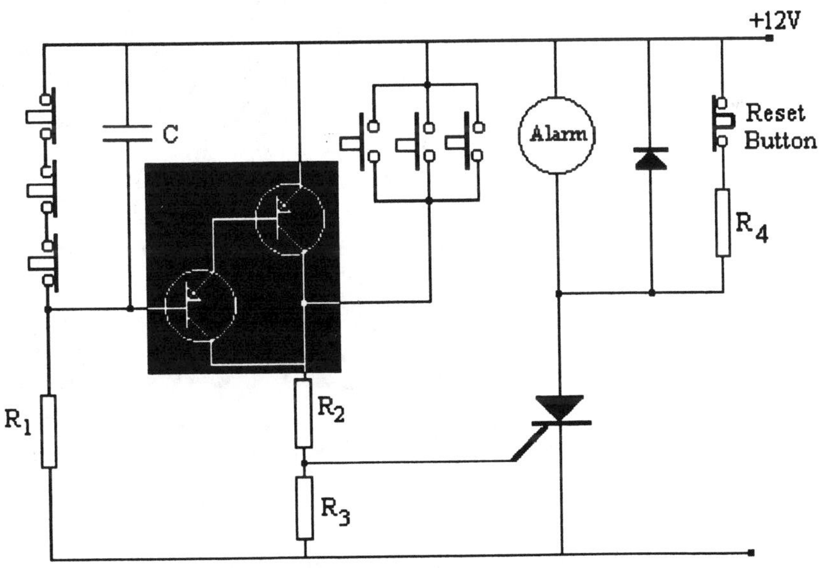

Fig. 9.8-5. A simple burglar-proof alarm with a two-transistor amplifier

This circuit (Fig. 9.8-5) will take a standby current far below that in Fig. 9.8-4. It can be designed around components to take a standby current of approximately 80% less than its counterpart in Fig. 9.8-4.

5. *Delayed Self-Latching Burglar Alarm*

The alarm circuits described so far have no provision for the owner or the last person to leave the secured premises without activating the alarm. That is, when the last person leaving the protected premises turns the alarm on to standby, it becomes impossible for him or her to leave the premises without activating the alarm.

This facility can be achieved by fitting the alarm with a delayed circuit which ensures that the alarm does not activate and goes into self-latching mode until after a predetermined time from the instant the system was initially put into standby condition.

Fig. 9.8-6 shows a simple burglar alarm with a delayed self-latching facility, giving the last person sufficient time to leave the secured premises without activating the alarm.

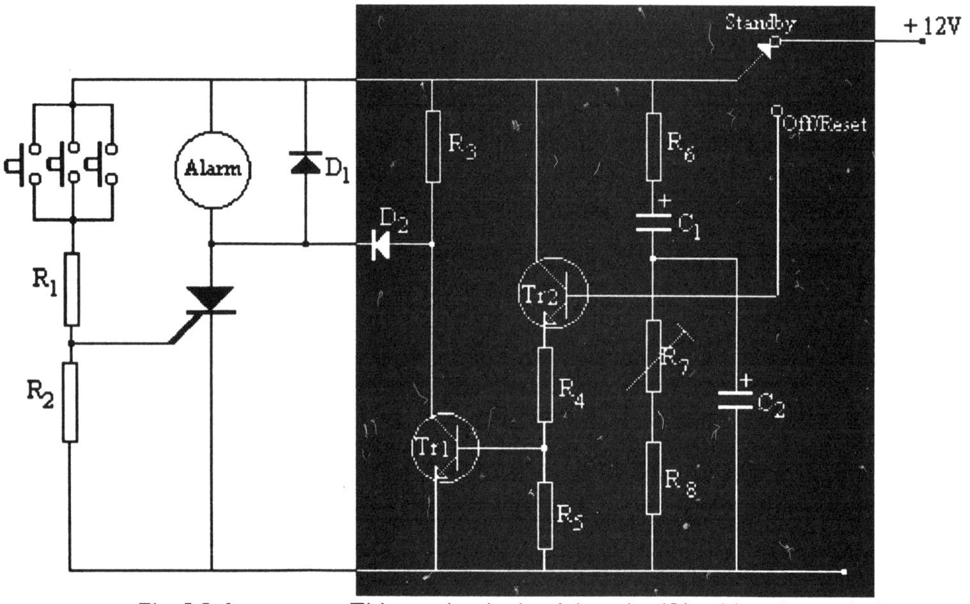

Fig. 9.8-6

This section is the delayed self-latching time switch which replaces the latching resistor and reset button

An alternative to the circuit in Fig. 9.8-6 is the *relay time switch*.

This device is connected across one of the premises' exit doors and is arranged to disable the door alarm switch for a brief period as the owner or the last person leaves the protected premises.

CHAPTER 10

LOGIC ELEMENTS AND NETWORKS

Introduction

The nature of digital logic embraces the very wide field of the logic-decision making process in electronics, which forms the theoretical basis of computers, digital clocks, electronic calculators, banking machines and all other digital electronic devices.

Philosophically, we are all by nature familiar with logic, for logic devices serve us in almost every sphere of life. Our minds continuously use logic in decision making.

The processing of information in digital form requires special circuits, and the efficient design of digital circuits requires a special numbering system and even a special algebraic mathematics called **Boolean algebra**.

In digital logic circuits there are only two voltage levels. These two voltage levels are referred to as *Logic 1 and Logic 0 states*. Since there are only two states, digital logic is said to be in **binary code**.

The significance of the binary nature of digital electronics lies in the fact that logic circuits carry out all decision-making and memory functions by using no more than the two logic states; numbers and letters of the alphabet or other information used in the process can be encoded by using the two logic states.

In digital logic there are three basic elements. The AND gate, the OR gate and the NOT gate (sometimes referred to as the INVERTER).

By interconnecting a number of these gates into circuits, they can perform addition, subtraction, multiplication, division and various complex operations and can even run a whole computer.

Each basic logic operation is indicated by a symbol and its function by a *truth table* that indicates all possible input combinations and the corresponding output.

The AND gate

The AND gate is a device whose output is a *logic 1* when all, and only all, of its inputs are at *logic 1.*

The Boolean algebra equation of an AND gate with two inputs A and B and output X is

$$\mathbf{A \cdot B = X}$$

This means that A and B must be logic 1 if X is to be logic 1.

The OR gate

The OR gate is a device whose output is a *logic 1* if either of its inputs are at *logic 1.*

The Boolean algebra equation of an OR gate with two inputs A and B and output X is:

$$\mathbf{A + B = X}$$

This means that X is at logic 1 if either A or B is at logic 1.

The NOT gate or INVERTER

The NOT gate or INVERTER has only a single input; it does not perform a decision-making function that is dependent on a combination of inputs. Instead, **the NOT gate or INVERTER simply converts a logic 1 at its input to a logic 0 at its output or conversely a logic 0 at its input to a logic 1 at its output.**

The Boolean algebra equation for a NOT gate or INVERTER with input A and output X is

$$\mathbf{X = \overline{A}}$$

The bar above the A is read as "NOT A."

The examples that follow will strengthen the student's understanding and knowledge of logic elements and associated networks.

Example 10.1

(a) *Prepare a graphical table of the following logic symbols in common use: AND, OR, NOT, NAND, NOR.*

(b) *Draw and briefly explain the action of diode logic circuits which perform the following functions:*

(i) *a two-input AND gate*

(ii) *a two-input OR gate*

Clearly indicate the polarities in each logic circuit and construct truth tables for the two-input AND and OR circuits.

Solution

(a) Logic Symbols in Common Use

Symbols in common use	Function
A, B → [&] → $X = A \& B$ A, B → [•] → $X = A \cdot B$	AND
A, B → [+] → $X = A + B$	OR
A → ▷○ → $X = \overline{A}$	NOT
A, B → [&]○ → $X = \overline{A \& B}$ A, B → [•]○ → $X = \overline{A \cdot B}$	NAND
A, B → [+]○ → $X = \overline{A + B}$	NOR

(contd)

(b) *(i)* Diode **AND** Gate

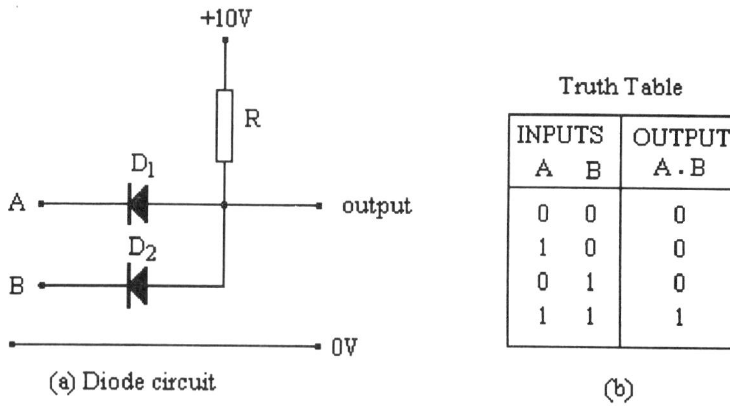

Truth Table

INPUTS A	INPUTS B	OUTPUT A . B
0	0	0
1	0	0
0	1	0
1	1	1

(b)

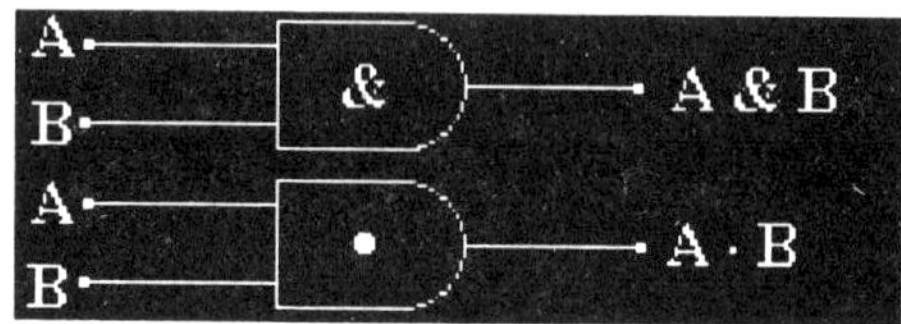

(c) Symbols in common use

Fig. 10.1-0. A two-input **AND** gate

Consider Fig.10.1-0(a)

If the inputs are in the form of positive voltage pulses (with respect to ground, referred to as positive logic), then inputs A and B will reverse bias both diodes, no current will flow through R and hence, there will be a positive output logic 1. Diodes D_1 and D_2 are said to be cut off.

When A and B are at '0' V the output is 0 V; both diodes D_1 and D_2 are conducting. If input A goes to 10 V, diode D_1 is reverse biased (cut off) but at the same time D_2 still conducts and holds the output at 0 V. Similarly, if B goes to 10 V while A is still at 0 V, the output still remains at 0 V. However, when inputs A and B are raised to 10 V both diodes are reverse biased (cut off), no current flows through resistor R and the output goes to +10 V, that is, a positive logic 1.

As indicated in the truth table of Fig.10.1-0(b), an output appears only when there are inputs at A and B, that is, when both inputs A and B are at logic 1.

(contd)

(b) (ii) Diode **OR** Gate

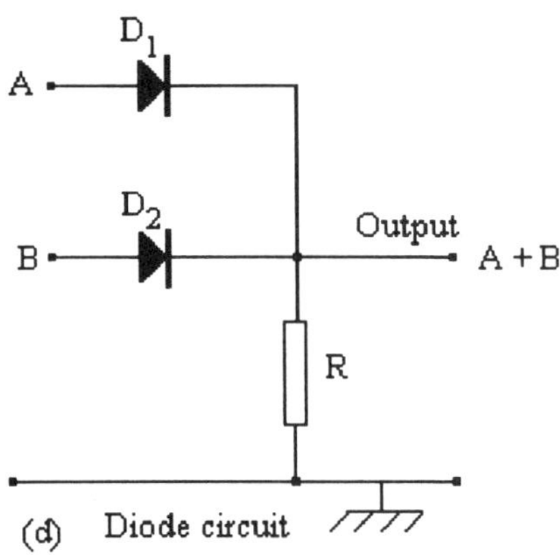

(d) Diode circuit

Truth Table

INPUTS A	B	OUTPUT A+B
0	0	0
1	0	1
0	1	1
1	1	1

(e)

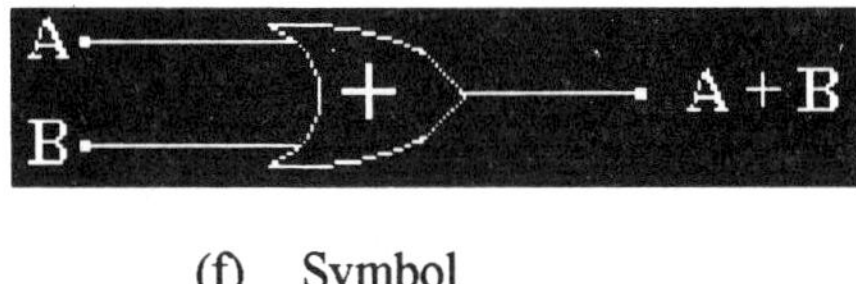

(f) Symbol

Fig. 10.1-1

Refer to Fig. 10.1-1

The symbol for an **OR** gate is shown in Fig. 10.1-1(f). As indicated in the truth table, the output is 1 when input A *or* B is 1 or both inputs are at logic 1. With no inputs [zero voltage; Fig. 10.1-1(d)], no current flows and the output is zero (logic 0)

An input of + 10 V (logic 1) at either terminal A *or* B or both forward biases the corresponding diode, current flows through the resistor R and the output voltage rises to nearly 10 V (logic 1).

Example 10.2

Draw and explain the action of the following logic circuits:

(i) *a transistor* ***NOT*** *gate*

(ii) *a two-input diode-transistor* ***NOR*** *gate*

Construct the truth table for the ***NOT*** *and the two-input* ***NOR*** *gates and sketch typical response curves for each gate.*

Solution

(i) Transistor **NOT** Gate

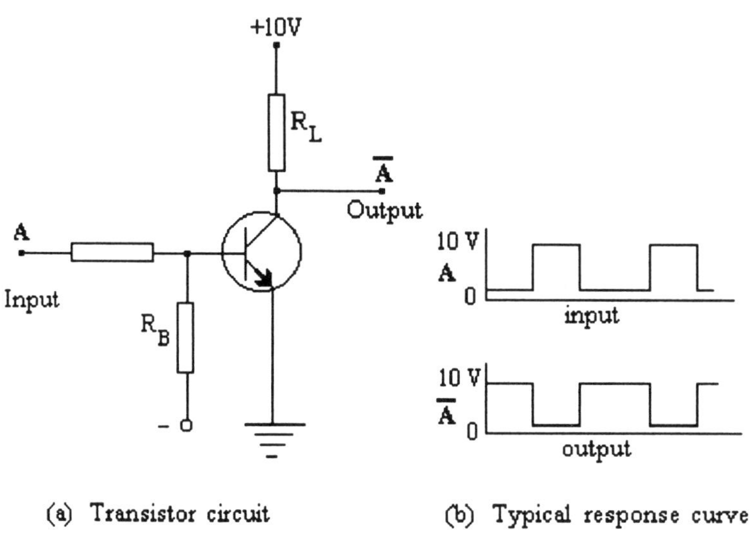

(a) Transistor circuit

(b) Typical response curve

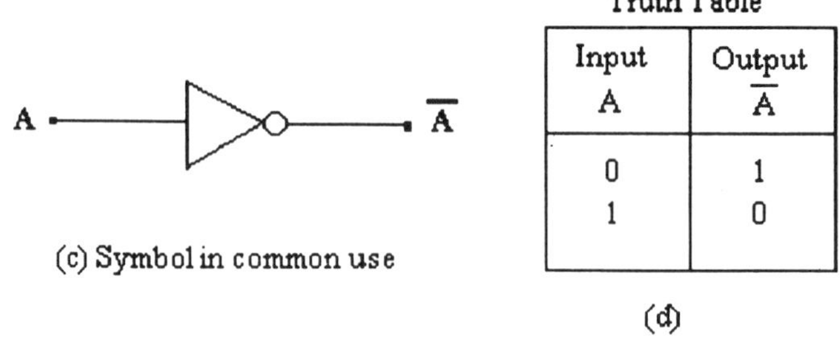

(c) Symbol in common use

Truth Table

Input A	Output $\overline{A}$
0	1
1	0

(d)

Fig.10.2-0. A transistor NOT gate

(contd) - Refer to Fig. 10.2-0(a)

(i) Transistor NOT Gate

With zero input (0) the transistor is cut off and held open by the negative base voltage and the output is +10 V (logic 1).
A positive input voltage (logic 1) forward biases the base-emitter junction and forces the transistor to change state. That is, the transistor is now switched ON, collector current flows and the output voltage drops to approximately zero volt (logic 0).

As indicated in the truth table, the **NOT** gate is an **INVERTER**.
The **NOT** gate can only accommodate one input. The output is the complement of the single input.

(ii) Diode-Transistor **NOR** Gate

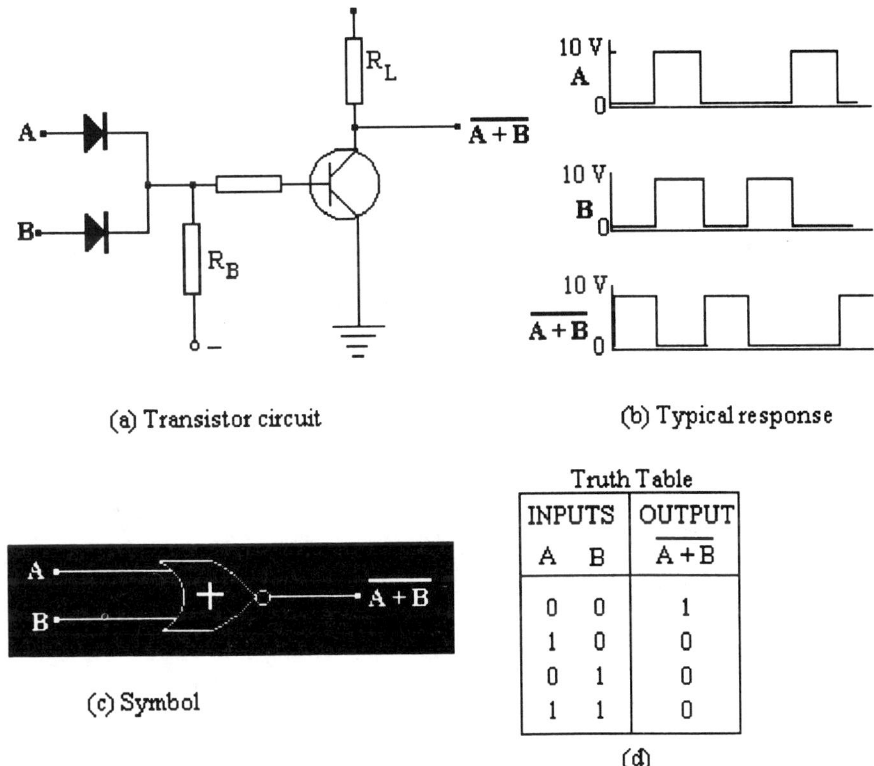

Truth Table

INPUTS		OUTPUT
A	B	$\overline{A+B}$
0	0	1
1	0	0
0	1	0
1	1	0

Fig. 10.2-1. A diode-transistor NOR gate

Refer to Fig. 10.2-1

In the NOR circuit, with no input the transistor is cut off by the negative base voltage and the output is +10 V (logic 1).
Put another way; *in a NOR gate an output is present when there is zero input.*

A positive input at terminal A or B raises the base potential which forward biases the base-emitter junction and drives the transistor to the ON state. With the transistor now turned on, the output voltage is nearly zero (logic 0).

From the truth table the output is 1 when both inputs are at zero (0); otherwise the output is zero (0)

Alternatively, in a NOR gate the output is logic 1 when all inputs are at zero (0); otherwise the output is zero (0).

Example 10.3

(a) *Define positive logic and hence explain the consequence of positive and negative logic levels.*

(b) *State the main disadvantages of diode-resistor logic.*

Solution

(a) *Positive logic* may be defined as the voltage used to represent binary 1 level which is higher than that used to represent binary 0 level.

As an example, refer to Table 10.3-0, where H represents the high-voltage levels and Z the zero-voltage levels of a two-input gate.

Table 10.3-0

INPUTS		OUTPUT
A	B	X
Z	Z	Z
H	Z	Z
Z	H	Z
H	H	H

Positive Logic Notation

In positive logic notation

$$H = 1$$

and $$Z = 0$$

Negative Logic Notation

In negative logic notation

$$H = 0$$

and $$Z = 1$$

If we rewrite the voltage levels in Table 10.3-0 using positive logic notation, we obtain Table 10.3-1 and for negative logic notation, we obtain Table 10.3-2.

(contd) - *(a)*

The result of rewriting Table 10.3-0 in positive logic notation is shown in Table 10.3-1.

The result of re-writing Table 10.3-0 in negative logic notation is shown in Table 10.3-2.

Table 10.3-1

INPUTS		OUTPUT
A	B	X
0	0	0
1	0	0
0	1	0
1	1	1

Table 10.3-2

INPUTS		OUTPUT
A	B	X
1	1	1
0	1	1
1	0	1
0	0	0

Comparing Table 10.3-1 with the truth table of the two-input AND gate in Example 10.1 we note that by using positive logic notation the gate generates the AND function of the input.
In the negative logic notation the gate generates the OR function of the truth table of Example 10.1 for the two-input OR gate.

The consequences of positive and negative logic may be summarized as follows:

*An **AND** gate in positive logic notation operates as an **OR** gate in negative logic notation.*

(b) The main disadvantage of diode-resistor logic is the drop in gate voltage due to the diode forward resistance.
As a result, high logic level falls and low logic level rises, which limits the number of gates that can be connected in cascade.

Further disadvantages are low switching speed and high power dissipation in the load resistor.

Example 10.4

(a) *With the aid of diagrams, explain the function of a diode-transistor **NAND** gate and construct the truth table for a two-input gate. Sketch a typical response curve.*

(b) *Show how **NAND** gates may be used to perform the function of a two-input **OR** gate.*

Solution

(a) **NAND** Gate

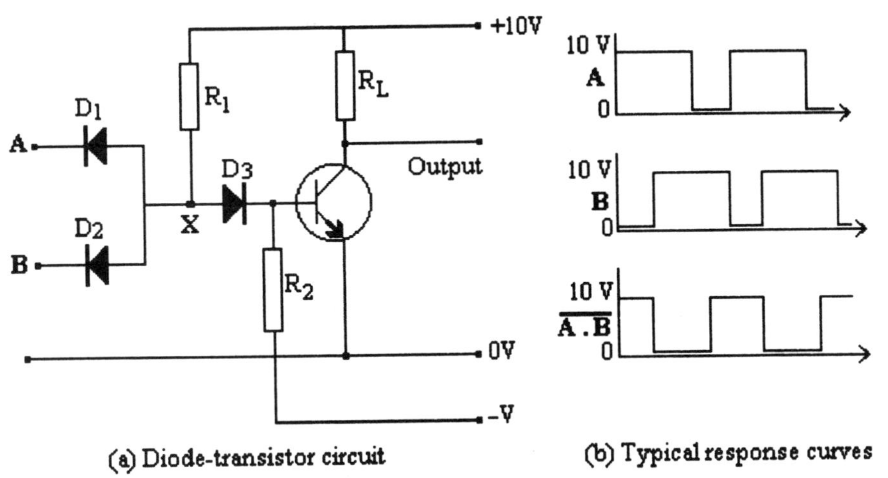

(a) Diode-transistor circuit

(b) Typical response curves

Truth Table

INPUTS		OUTPUT
A	B	$\overline{A.B}$
0	0	1
1	0	1
0	1	1
1	1	0

(d)

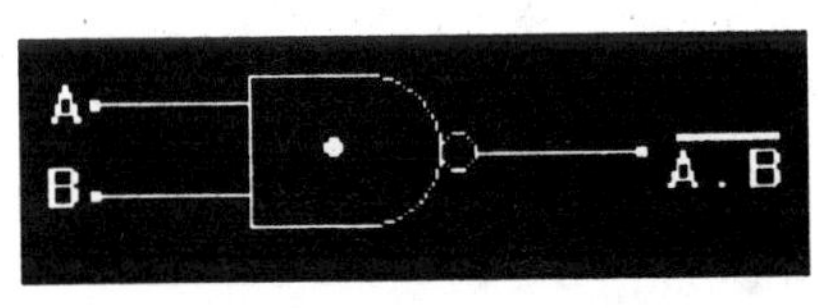

(c) Symbol in common use

Fig. 10.4-0. A diode-transistor **NAND** gate

(contd)

(a) The truth table of Fig.10.4-0(d) defines the function of the NAND gate.

Refer to Fig.10.4-0(a)

With positive inputs at A and B (i.e., A and B are at logic 1) the diodes are reverse biased and no current flows through D_1 and D_2. That is, D_1 and D_2 are blocking. However, current flows through R_1 and D_3, thus forward biasing the transistor sufficiently to saturate it. That is, the transistor is switched ON and conducting, and hence the output voltage is approximately 0 V (logic 0).

If either A or B has a zero input, the corresponding diode conducts and the voltage at junction X is V_D volts (where V_D is the voltage drop across the forward-biased diode (D_1 or D_2), resulting in current flowing through D_3 and R_2, making the transistor base voltage negative, that is, $-V_D$ volts, which cuts it off. Consequently, the output voltage goes to +10 V (logic 1).

In practice, a second diode is usually placed in series with the base of the transistor, which will further ensure that the transistor will cut off.

(b) ***NAND** gates performing the function of a two-input **OR** gate*

The desired logic is shown in Fig. 10.4-1(a) and the desired function is defined by the truth table in Fig.10.4-1(b).

Comparing the truth table in Fig.10.4-1(b) with the truth table of the NAND gate in Fig.10.4-0(d), we note that if each input were inverted (replaced by its complement), the NAND gate would produce the desired result as indicated in the truth table.

(contd) - *(b)*

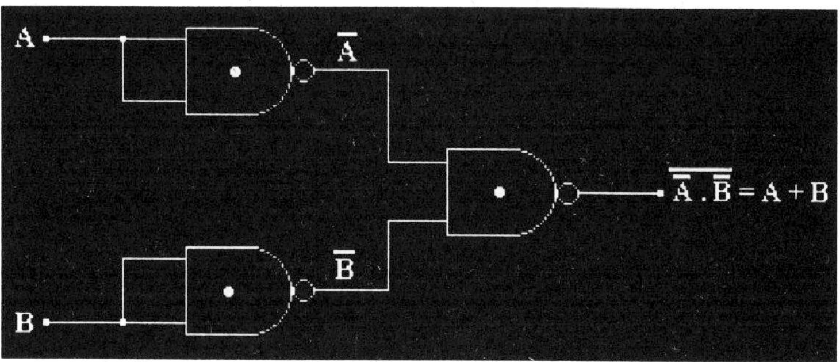

(a) NAND gates

Truth Table

INPUTS A	B	Function f	OUTPUT $\overline{A}$	$\cdot \overline{B}$	OUTPUT $\overline{\overline{A} \cdot \overline{B}}$
0	0	0	1	1	0
1	0	1	0	1	1
0	1	1	1	0	1
1	1	1	0	0	1

(b)

Fig.10.4-1. Combination of NAND gates to function as an OR gate

To provide simple inversion, both terminals of a NAND gate are tied together; the truth table is the desired NAND realization.
The combination of the NAND gates is equivalent to an OR gate in that it performs the same logic operation.

The Boolean expression for the function **f** is defined as follows:

$$\mathbf{f} = A + B = \overline{\overline{A} \cdot \overline{B}}$$

Example 10.5

(a) *Show how simple logic gates may be used to satisfy Truth Table 10.5-0.*

(b) *From Truth Table 10.5-1 give a Boolean expression for the output Z.*

(c) *Using Boolean algebra, simplify the expression determined in (b); that is, simplify the expression for the output Z.*

(d) *Sketch the logic diagram for the simplified expression obtained in (c).*

Truth Table 10.5-0

INPUTS		OUTPUT
A	B	Z
0	0	0
1	0	1
0	1	1
1	1	0

Truth Table 10.5-1

INPUTS			OUTPUT
A	B	C	Z
0	1	1	0
0	1	0	0
1	0	1	1
1	1	0	1
0	0	0	0
1	0	0	0
1	1	1	1
0	0	1	0

Solution

[Note: In the past the AND sign in $A \cdot B$ has been retained to focus attention on the logic expression. However, the simpler equivalent form AB will be used whenever convenient.]

(a) From Truth Table 10.5-0, an output 1 is required for the input combinations shown in **rows 2 and 3.**

Therefore

$$Z = A\overline{B} + \overline{A}B$$

The required logic circuit is shown in Fig. 10.5-0

(contd) - *(a)*

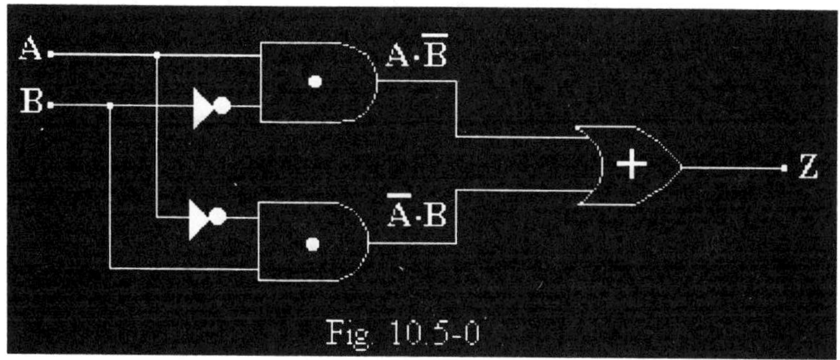

Fig. 10.5-0

(b) From Truth Table 10.5-1 the Boolean expression for the output Z is

$$Z = A\overline{B}C + AB\overline{C} + ABC$$

(c) The simplified expression is as follows:

$$\begin{aligned} Z &= A\overline{B}C + AB(C+\overline{C}) \\ &= A\overline{B}C + AB \qquad \textbf{[Note: } C+\overline{C} = 1+0 = 1\textbf{]} \\ &= A(B + \overline{B}C) \\ &= AB + AC \\ &= A(B + C) \end{aligned}$$

(d) The logic diagram for the simplified expression Z = A (B + C) is shown in Fig.10.5-1

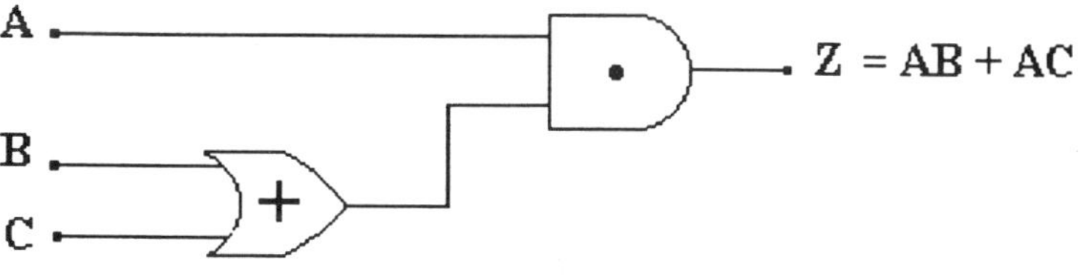

Fig. 10.5-1

Example 10.6

(a) *State DeMorgan's theorems for two quantities A and B. Hence draw the logic diagrams using NAND gates only equivalent to two-input OR and AND gates.*

(b) *Define the cumulative law, associative law and distributive law associated with logic circuit equations.*

Solution

(a) DeMorgan's theorems state

(i) that a NOR gate $(\overline{A+B})$ is equivalent to an AND gate with NOT circuits in the inputs $(\overline{A}\cdot\overline{B})$

(ii) that a NAND gate $(\overline{A\cdot B})$ is equivalent to an OR gate with NOT circuits in the inputs $(\overline{A}+\overline{B})$

In general, DeMorgan's theorems state the following:

To obtain the inverse of any Boolean function, invert all variables and replace all **OR**s by **AND**s and all **AND**s by **OR**s.

Alternatively DeMorgan's theorems state that the logic complement of a function is obtained if we

(i) logically invert each term in the expression

(ii) interchange the "dots" with plusses and vice-versa.

For example

$$\overline{A\cdot B\cdot C} = \overline{A}+\overline{B}+\overline{C}$$

$$\overline{A+B+C} = \overline{A}\cdot\overline{B}\cdot\overline{C}$$

(contd)

(a) *Combination of NAND gates equivalent to two-input OR gate*

The desired function is $f = A + B$

Applying the general form of DeMorgan's theorems

$$f = A + B = \overline{\bar{A} \cdot \bar{B}}$$

This suggests a NAND gate with NOT inputs

Since $\overline{A \cdot A} = \bar{A}$, a NAND gate with inputs tied together performs the NOT operation.

The logic circuit is shown in Fig. 10.6-0.

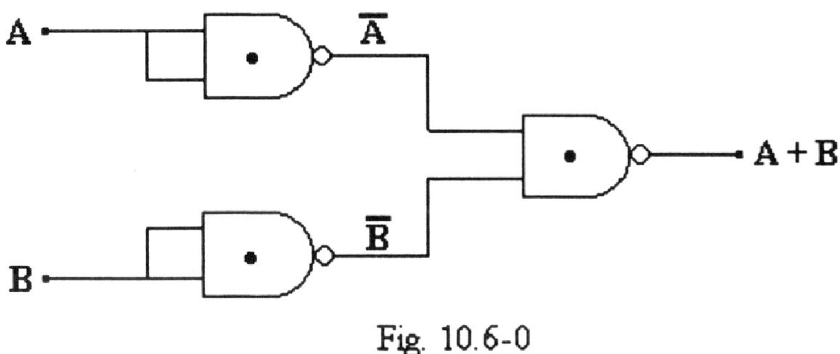

Fig. 10.6-0

For the AND function, $f = AB = \overline{\overline{AB}}$.

The logic function is shown in Fig. 10.6-1.

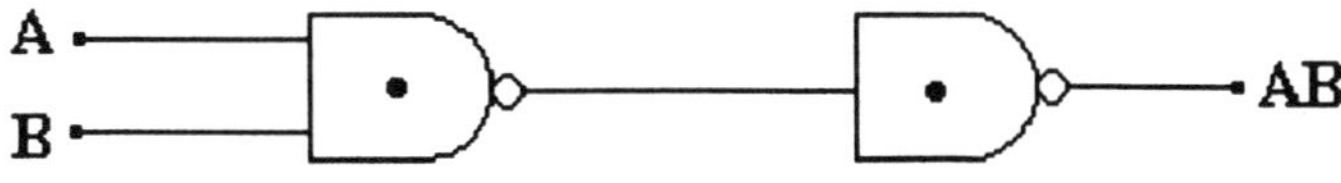

Fig. 10.6-1

(b) *Cumulative Law*

This law states that the order in which terms or variables appear in an equation is irrelevant. For example

$$A + B = B + A$$

$$A \cdot B = B \cdot A$$

Associative Law

This law states that the order in which identical functions are performed is irrelevant. For example

$$A + B + C = (A + B) + C = A + (B + C)$$

$$A \cdot B \cdot C = (A \cdot B) \cdot C = A \cdot (B \cdot C)$$

[Note: Care must be taken in combining terms together in brackets. For example $A + B \cdot (C + D) \neq (A + B) \cdot (C + D)$; however, $A + B \cdot (C + D) = A + (C + D) \cdot B$.]

Distributive Law

This is expressed in two forms:

(i) *Product of sums* expression, e.g.

$$A + (B \cdot C \cdot D + \cdots) = (A + B) \cdot (A + C) \cdot (A + D)$$

(ii) *Sum of products* expression, e.g.

$$A \cdot (B + C + D + \cdots) = A \cdot B + A \cdot C + A \cdot D$$

Note:

Useful Expressions of OR gate	Useful Expressions of AND gate
$A + 0 = A$	$A \cdot 0 = 0$
$A + 1 = 1$	$A \cdot 1 = A$
$A + A = A$	$A \cdot A = A$
$A + \overline{A} = 1$	$A \cdot \overline{A} = 0$
	$\overline{\overline{A}} = A$

Example 10.7

(a) *With the aid of a circuit diagram explain the operation of a resistor-transistor **NOR** logic element.*

(b) *Show in logic form how one or more such elements may be used together to produce a logic **AND** function, a logic **OR** function and a logic **NOT** function.*

Solution

(a) Using Positive Logic — Using Negative Logic

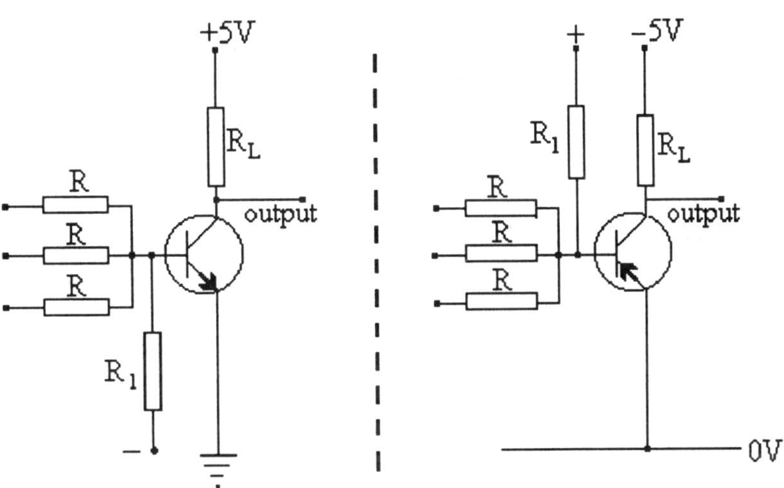

Fig. 10.7-0. RTL **NOR** gate

With no input the transistor is cut off by the negative base voltage and the output is at +5 V (Binary 1). If any input goes sufficiently positive so as to saturate the base-emitter junction, the transistor is switched ON, and the output voltage drops to 0 V (Binary 0). Hence the circuit performs the **NOR** function using positive logic.

When all inputs are at 0 V (Binary 0), the base-emitter junction is held positive, the transistor is cut off and the output voltage is +5 V (Binary 1). If any input goes to -5 V and the values of R_1 and R are such that the base voltage is made sufficiently positive to saturate the transistor, the output voltage goes to 0 V (Binary 0). Hence the circuit performs the **NOR** function using negative logic.

(contd)

(b) **AND** Function - Refer to Fig. 10.7-1

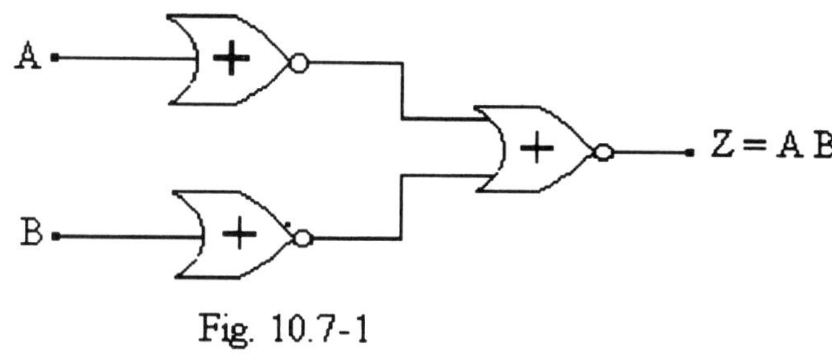

Fig. 10.7-1

$$Z = AB$$

$$\therefore \quad \overline{Z} = \overline{AB}$$

$$= \overline{A} + \overline{B}$$

$$\text{Also } Z = \overline{\overline{Z}} = \overline{\overline{A} + \overline{B}}$$

OR Function - Refer to Fig.10.7-2

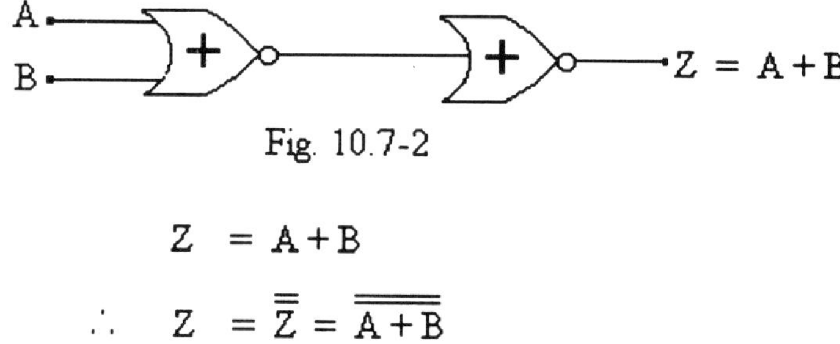

Fig. 10.7-2

$$Z = A + B$$

$$\therefore \quad Z = \overline{\overline{Z}} = \overline{\overline{A + B}}$$

NOT Function - Refer to Fig.10.7-3

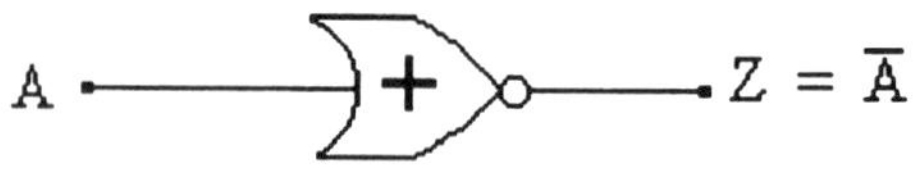

A $\quad Z = \overline{A}$

Fig. 10.7-3

The NOT function is simply a one-input NOR gate.

$$Z = \overline{A}$$

Example 10.8

The circuit diagram of Fig. 10.8-0 is that of a DTL (diode-transistor logic) ***NAND*** *gate with the following values:*

$$V_{cc} = 5\ V$$

$$R_B = R_L = 5\ k\Omega$$

$$\beta = 30$$

$$V_{ce}(sat) = 0.3\ V \quad at\ V_{BE} \geq +0.7\,V$$

The forward voltage drop of each diode is assumed to be 0.7 V.

If input A is 5 V and input B is 0.3 V, determine the output state of the transistor. If input B is raised to 2 V, calculate the maximum possible no-load collector current and the required minimum base current to provide saturation.

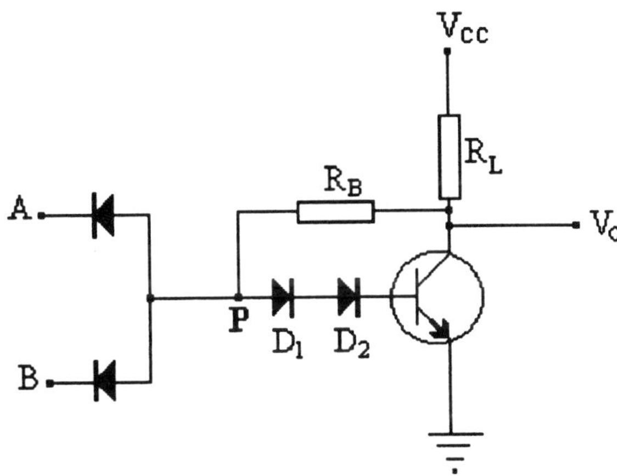

Fig. 10.8-0

Solution

The lowest input voltage dictates the output state of the NAND gate.

Input $A = 5\ V$ *and* $B = 0.3\ V$

For $B = 0.3$ V, $\quad V_P = 0.3 + 0.7 = 1.0$ V

Assume current flows through series diodes D_1 and D_2, then

$$V_{BE} = V_P - 2(0.7)$$

$$= 1.0 - 1.4 = -0.4\ V$$

(contd)

Since $V_{BE} = -0.4$ V the transistor is cut off.

$$\therefore \quad \mathbf{V_o} = V_{CC} - I_C R_L$$

$$= 5 - 0 = \mathbf{5\ V}$$

The output state of the transistor is at Binary 1

Input B = 2 V

When input B = 2 V

$$V_P = \text{input B} + 0.7\ \text{V}$$

$$= 2 + 0.7 \qquad = 2.7\ \text{V}$$

The critical value for switching the transistor ON is

$$V_P = 2 \times 0.7 + 0.7 \qquad = 2.1\ \text{V}$$

The critical value (2.1 V) is exceeded (2.7 V). Hence the transistor is switched ON.

$$\therefore \text{ Base current flows and } I_B = (V_{CC} - 2.7)/R_B$$

$$= (5 - 2.7)/(5 \times 10^3) \qquad = \underline{0.46\ \text{mA}}$$

However, the maximum possible no-load collector current is

$$I_C = V_{CC}/R_L$$

$$= 5/(5 \times 10^3) \qquad = \underline{\mathbf{1.0\ mA}}$$

This value of collector current (1.0 mA) will require a minimum base current of only

$$I_B = I_C/\beta$$

$$= (1.0 \times 10^{-3})/30 \qquad = \underline{\mathbf{0.033\ mA}}$$

Therefore, the transistor is in saturation and

$$\mathbf{V_o} = \mathbf{V_{ce}(sat)} \qquad = \underline{0.3\ \text{V}}$$

Example 10.9

(a) *With the aid of logic diagrams for the following two circuits construct a truth table to show that circuit 1 and circuit 2 are equivalent.*

Circuit 1: Inputs B and C to an ***OR*** *gate whose output is* ***AND*** *gated with input A to give an output X.*

Circuit 2: Two ***AND*** *gates with separate inputs B and C, and shared input A whose outputs enter an* ***OR*** *gate to give output Y.*

(b) *Draw a logic diagram of the following expression. Using Boolean algebra, simplify the expression and then draw the single logic element which it represents.*

$$X = A[\overline{\overline{B} + \overline{A}(C + \overline{B}\,\overline{C})}]$$

Solution

(a)

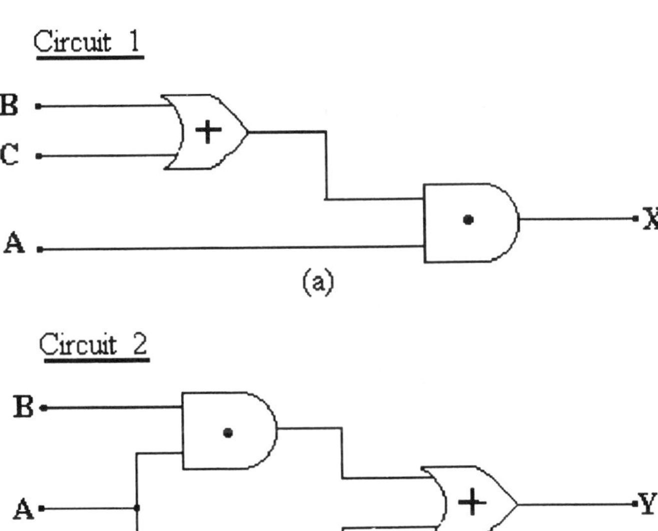

(b)

Fig. 10.9-0

(contd) - *(a)*

The logic diagrams for Circuit 1 and Circuit 2 are shown in Fig. 10.9-0(a) and 10.9-0(b) and the corresponding truth tables are shown in Table 10.9-0.

Truth Table 10.9-0
For Logic Circuits (a) and (b) of Fig. 10.9-0

Inputs			Input variables			Output (a)	Output (b)
A	**B**	**C**	**B+C**	**AB**	**AC**	**X=A(B+C)**	**Y=AB+AC**
0	0	0	0	0	0	0	0
0	0	1	1	0	0	0	0
0	1	0	1	0	0	0	0
0	1	1	1	0	0	0	0
1	0	0	0	0	0	0	0
1	0	1	1	0	1	1	1
1	1	0	1	1	0	1	1
1	1	1	1	1	1	1	1

From the truth table it can be seen that the X and Y columns are identical. Therefore, **X = Y**.

(b) To simplify the expression by using Boolean algebra, proceed as follows:

$$
\begin{aligned}
X &= A\left[\overline{\overline{\overline{B} + \overline{A}(C + \overline{B}\,\overline{C})}}\right] \\
&= A\left\{B\left[A + \overline{C}(B + C)\right]\right\} \\
&= AB\left[A + B\overline{C} + \overline{C}C\right] \\
&= AB\left[A + B\overline{C}\right] \\
&= A\,A\,B + A\,B\,B\,\overline{C} \\
&= A\,B + A\,B\,\overline{C} \\
&= A\,B\,(1 + \overline{C}) \\
&= A\,B
\end{aligned}
$$

Note
$\overline{\overline{A}\cdot\overline{B}} = A + B$ $\overline{\overline{A} + \overline{B}} = A\cdot B$ A double negative converts an AND to OR and vice versa
Note double negatives $\overline{\overline{B}}\,\overline{\overline{C}} = B + C$ $\overline{\overline{B}} + \overline{\overline{A}} = B\,A$
Refer to Example 10.6 Useful Expressions $\overline{C}\,C = 0$ $A\cdot A = A$ $B\cdot B = B$

Hence function X is an **AND** gate. The logic circuit is shown in Fig. 10.9-1

(contd)

(b)

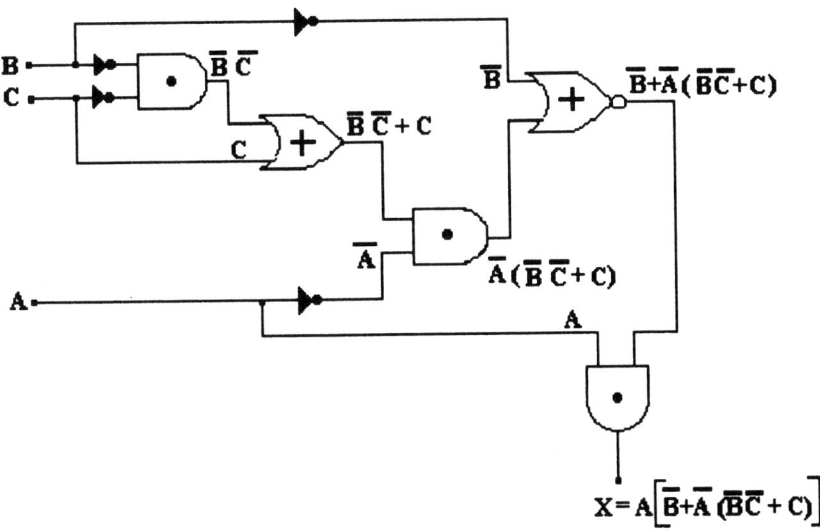

Fig. 10.9-1. The logic circuit to obtain X

Example 10.10

With the aid of a circuit diagram, describe the operation of a transistor-transistor logic (TTL) NAND gate.

Solution

The emitter-base junction is essentially a **pn** junction diode. Thus the input diodes A and B of diode-transistor logic (DTL) gate in Fig. 10.8-0 can be replaced by a multiemitter transistor. Refer to Fig. 10.10-0.

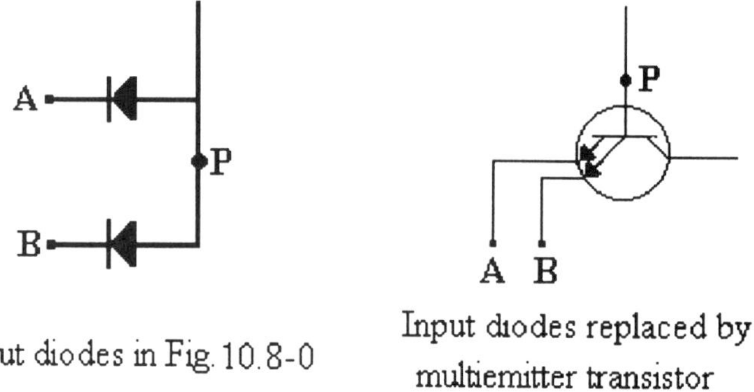

Fig. 10.10-0

The transistor-transistor logic NAND gate is now drawn in Fig.10.10-1.

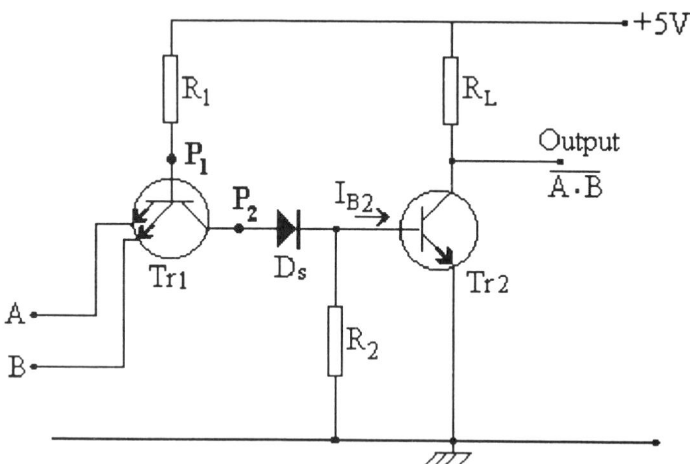

Fig.10.10-1. A basic TTL gate derived from a DTL gate

(contd)

Refer to Fig.10.10-1

The collector-base junction of Tr_1 provides an offset voltage equal to that of one series diode (approximately 0.7 V).

If we assume that the inputs A and B are High (Binary 1), then all the base-emitter diodes of Tr_1 are reverse biased.

However, the base-collector junction of Tr_1 is forward biased and current flows through R_1 across the base-collector junction, assuming diode D_S is sufficiently forward biased to switch **ON** Tr_2.
Since Tr_2 is **ON**, the output is Low, and the emitter-base junction of Tr_2 and diode D_S are forward biased. Therefore, the voltage at point $\mathbf{P_2}$ (V_{P2}) must be approximately 0.7 V + 0.7 V = 1.4 V and the base-collector junction of Tr_1 must be forward biased to supply sufficient base current to Tr_2.
As a result of the preceding, the voltage at point $\mathbf{P_1}$ (V_{P1}) is about 1.4 V + 0.7 V = 2.1 V.
This condition requires both inputs A and B to be High (Binary 1), that is, greater than 1.4 V.

If any of the inputs go to level 0 (Binary 0), the corresponding base-emitter diode becomes forward biased (V_{P1} = 0.3 + 0.7 = 1.0 V).
Under this condition, D_S cannot be forward biased and the current I_{B2} is **insufficient** to switch **ON** Tr_2. Therefore, Tr_2 is **cut off** and the output voltage is High (Binary 1).

Thus the circuit performs the **NAND** function. In other words the output is **NOT (A · B)** and this is a **NAND** gate.

Example 10.11

Analyze the logic circuit shown in Fig.10.11-0 and hence construct the truth table to show that this circuit can be replaced by a single NAND gate.

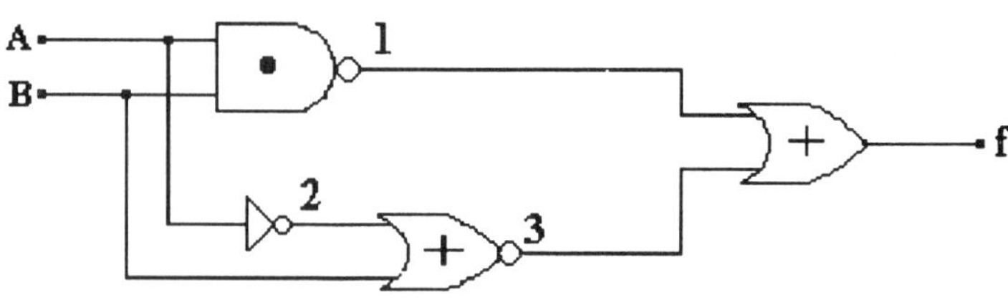

Fig. 10.11-0

Solution

We will now consider the outputs at points 1, 2 and 3.

Point 1: Output at NAND gate $= \overline{A\,B}$

Point 2: Output at NOT gate $= \overline{A}$

Point 3: Output at NOR gate $= \overline{\overline{A} + B}$

The above suboutputs are indicated in the diagram of Fig.10.11-1.

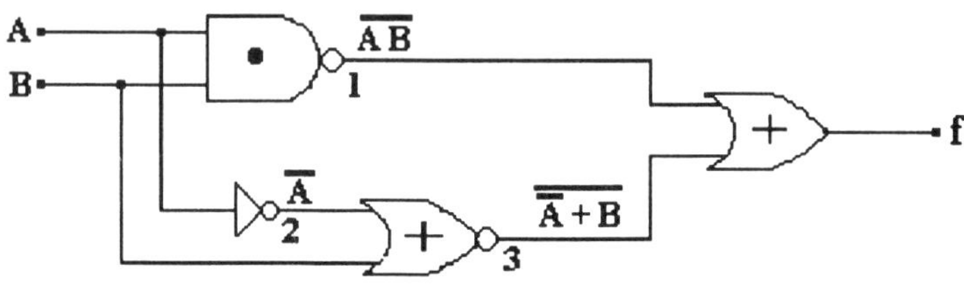

Fig. 10.11-1

The overall function can be simplified as shown hereunder.

$$
\begin{aligned}
f &= \overline{A\,B} + \overline{\overline{A} + B} \\
&= (\overline{A} + \overline{B}) + A\,\overline{B} && \text{(DeMorgan's Rule)} \\
&= \overline{A} + \overline{B}\,(1 + A) && \text{(Distribution)} \\
&= \overline{A} + \overline{B} && (1 + A = 1) \\
&= \overline{A \cdot B} && \text{(DeMorgan's Rule)}
\end{aligned}
$$

Hence f $= \overline{A \cdot B}$ is a NAND gate.

(contd)

Table 10.11.0 shows the construction of the truth table

Truth Table 10.11-0

1		2	3	4
A	B	$\overline{AB}$	$\overline{\overline{A}+B}$	f
0	0	1	0	1
0	1	1	0	1
1	0	1	1	1
1	1	0	0	0

The truth table demonstrates that the circuit of Fig.10.11-0 or 10.11-1 can be replaced by a single NAND gate.

Consider the logic gate of the single two-input NAND gate of Fig.10.11-2 and its supporting truth table in Table 10.11-1

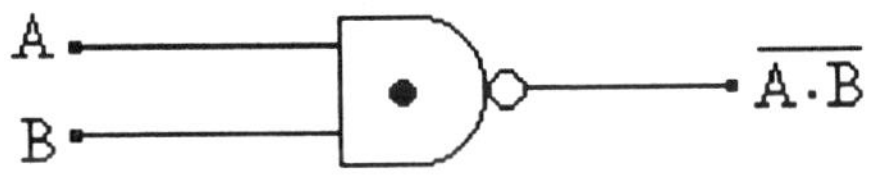

A

B

$\overline{A \cdot B}$

Fig. 10.11-2. Two-input NAND gate

Truth Table 10.11-1

1		2
A	B	$\overline{A \cdot B}$
0	0	1
0	1	1
1	0	1
1	1	0

Comparison of columns **1** and **2** of Table 10.11-0 and columns **1** and **2** of Table 10.11-1 verifies that function **f** is a **single NAND gate.**

CHAPTER 11

TUNED CIRCUITS, RESONANCE, and the Q-FACTOR

INTRODUCTION

A tuned circuit basically consists of an inductor and capacitor connected either in series or in parallel; the resistance of the inductor is assumed to be negligible and the capacitor lossless.

At a particular frequency, known as the *resonance frequency*, the tuned circuit appears similar to a pure resistive circuit.

As a result of resonance the applied e.m.f E and current I are in phase and hence the power factor is unity.

SERIES TUNED CIRCUIT

Fig. 11.0-0. Series Circuit

Inductive reactance,	X_L	=	$2\pi f L$
Capacitive reactance,	X_C	=	$1/(2\pi f C)$
Effective reactance,	X_T	=	$X_L - X_C$

Consider the case when $X_L > X_C$

When X_L is greater than X_C, the effective reactance is inductive and the current I lags the applied e.m.f E.

Refer to Fig. 11.0-1

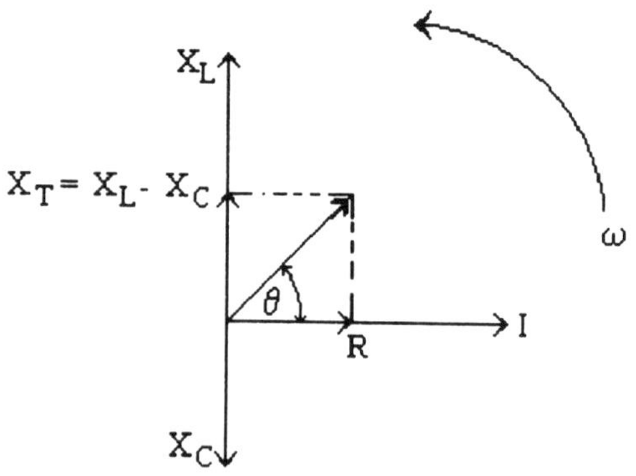

Fig. 11.0-1. Phase relationship when $X_L > X_C$

(contd)

Consider the case when $X_C > X_L$

When X_C is greater than X_L, the effective reactance is capacitive and the current I leads the applied e.m.f E.

Refer to Fig.11.0-2

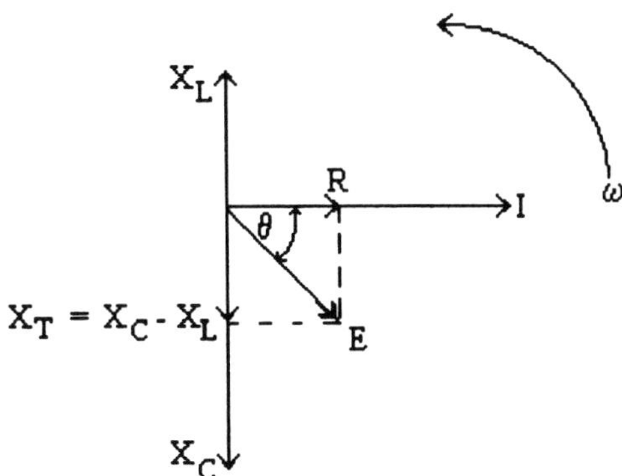

Fig. 11.0-2. Phase relationship when $X_C > X_L$

Consider the case when $X_L = X_C$

When $X_L = X_C$, the circuit is tuned to the *Resonance frequency*, f_o.

At the resonance frequency, $X_L = X_C$

$$2\pi f_o L = 1/(2\pi f_o C)$$

$$f_o^2 = 1/(4\pi^2 LC)$$

$$\therefore \quad f_o = 1/[2\pi\sqrt{(LC)}]$$

where L is in henrys and C is in farads

Circuit Impedance Z

The impedance of the tuned circuit varies with frequency and is given by

$$Z = \sqrt{[R^2 (\omega L - 1/\omega C)^2]} \text{ ohms}$$

At the resonance frequency, the effective reactance $X_T = 0$. Therefore

$$Z = R \text{ ohms}$$

(contd)

Response Curves

Fig. 11.0-3 shows the current-frequency response curves for a low value of R and a high value of R.

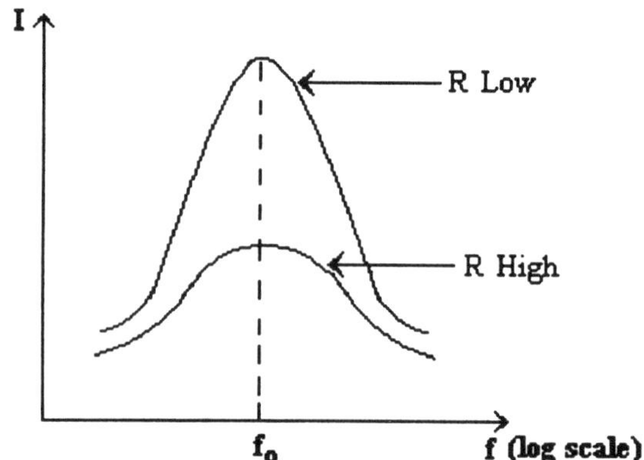

Fig. 11.0-3. Current-frequency curves

Control of the response curves can be achieved by selecting suitable values of R.

PARALLEL TUNED CIRCUIT

Fig. 11.0-4 shows a two-branch parallel circuit.

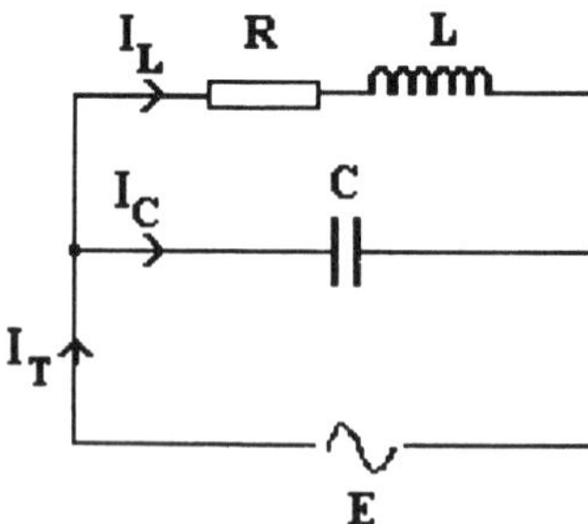

Fig. 11.0-4. Two-branch parallel circuit

(contd)

Consider the case of low frequencies ($I_L > I_C$)

At low frequencies, I_L is large due to X_L being low and this current lags the applied e.m.f E. The total current is

$$I_T = \sqrt{(I_L^2 + I_C^2)}$$

Consider the case of high frequencies ($I_C > I_L$)

At high frequencies, the capacitive branch takes the larger current since at high frequencies X_C is small.

Consider the case at the resonance frequency (f_0)

At resonance, the total current I_T is in phase with the e.m.f. The resonance frequency f_0 is given by

$$f_0 = \frac{1}{2\pi}\sqrt{\left(\frac{1}{LC} - \frac{R^2}{L^2}\right)}$$

At resonance, the impedance, termed the *dynamic resistance* R_D, is given by:

$$R_D = L/CR \text{ ohms}$$

Response Curve

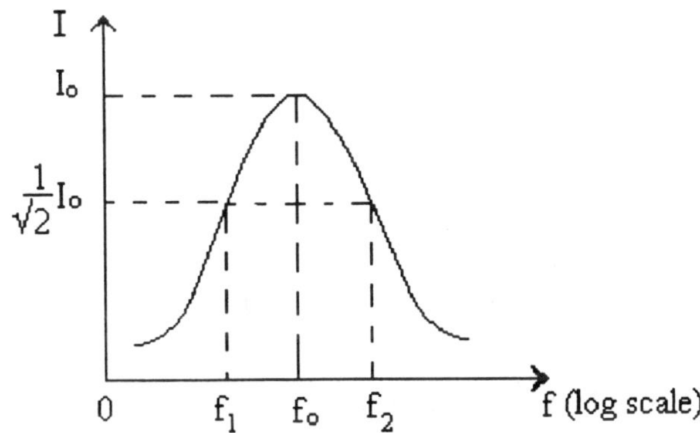

Fig. 11.0-5. Impedance-frequency curve

Refer to Fig. 11.0-5

The band of frequencies f_1 to f_2 is called the *band width*.

At f_0, the current I_0 is maximum. The points where the current is $I_0/\sqrt{2}$, corresponding to f_1 and f_2, are called the half-power points.

(contd)

Q-FACTOR

The Q-factor is given as

$$Q = 2\pi \frac{\text{maximum energy stored}}{\text{energy dissipated per cycle}}$$

This equation is the fundamental expression for the Q-factor and may be applied to circuits of all kinds.

Q-factor of an inductor

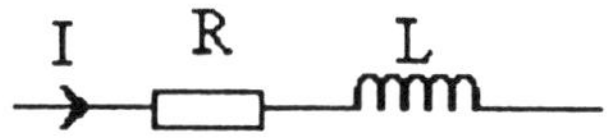

I R L

Fig. 11.0-6

The energy dissipated in the inductor is given by the product of the average power in the resistor and the period T or 1/f.

Refer to Fig. 11.0-6

Average power $= \left(\frac{I_{max}}{\sqrt{2}}\right)^2 R$

Period T $= 1/f$

Maximum energy stored in magnetic field $= \frac{1}{2} L I_{max}^2$

$$\therefore \quad Q = 2\pi \frac{\frac{1}{2} L I_{max}^2}{\left(\frac{I_{max}}{\sqrt{2}}\right)^2 R \times \frac{1}{f}}$$

$$= 2\pi \frac{\frac{1}{2} L I_{max}^2}{\frac{(I_{max})^2}{2} R \times \frac{1}{f}}$$

$$= \frac{2\pi f L}{R} = \frac{\omega L}{R}$$

That is, the Q-factor of an inductor is the ratio of its reactance/resistance.

Q-factor of a capacitor

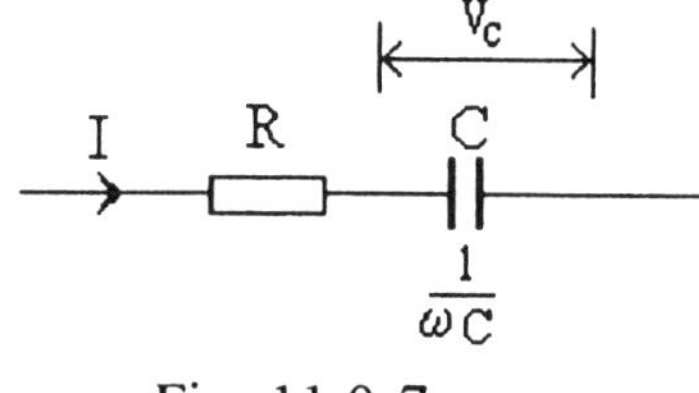

Fig. 11.0-7

(contd)

Refer to Fig. 11.0-7

Maximum energy stored $= \frac{1}{2} C V_{max}^2$

$= \frac{1}{2} I_{max}^2/\omega^2 C$

Energy dissipated per cycle $= \left(\frac{I_{max}}{\sqrt{2}}\right)^2 R \times {}^1/f$

$$\therefore \quad Q = 2\pi \frac{\frac{1}{2} I_{max}^2 / \omega^2 C}{\frac{(I_{max}^2)}{2} R \times \frac{1}{f}}$$

$$= \frac{2\pi f}{\omega^2 CR} = \frac{1}{\omega CR}$$

The Q-factor of a capacitor is the ratio of its reactance/resistance.

Q-factor of a series resonance RLC circuit

At resonance, the energy stored in the magnetic field is equal to the energy stored in the electric field.

$$\therefore \quad \frac{1}{2} CV_{max}^2 = \frac{1}{2} LI_{max}^2$$

Then $$Q_o = \frac{2\pi \times LI^2}{I^2R/f_o}$$

$$= \omega_o L/R = 1/\omega_o CR$$

That is, the Q-factor of a series resonance circuit is the same as that of the inductor if the losses in the capacitor are zero.

Now from the above

$$\omega_o L = {}^1/\omega_o C$$

$$\therefore \quad \omega_o = {}^1/\sqrt{(LC)}$$

Substituting for ω_o in the expression for $Q = \omega_o L_{/R}$, we get

$$Q = {}^L/R\sqrt{(LC)}$$

$$= \mathbf{{}^1/R \sqrt{({}^L/C)}}$$

(contd)

The current at resonance in the tuned series circuit is

$$I_0 = {}^{E}/R$$

$$\therefore \quad V_L = I_0\omega_0 L$$

$$= ({}^{E}/R)\,\omega_0 L = \mathbf{QE}$$

$$\text{Also} \quad V_C = {}^{I_0}/\omega_0 C$$

$$= {}^{E}/R\omega_0 C = \mathbf{QE}$$

Q-factor of a parallel resonance circuit

$$Q = 2\pi \frac{\frac{1}{2} L I_{max}^2}{\left(\frac{I_{max}}{\sqrt{2}}\right)^2 R \times \frac{1}{f_0}}$$

$$= \frac{\boldsymbol{\omega_0 L}}{\mathbf{R}}$$

The dynamic impedance of a parallel resonance circuit is given by

$$R_D = {}^{L}/CR \text{ ohms}$$

Multiplying top and bottom by ω_0, we get

$$R_D = {}^{\omega_0 L}/\omega_0 CR$$

$$\text{Now} \quad {}^{1}/\omega_0 CR = Q$$

$$\therefore \quad \mathbf{R_D = Q\omega_0 L}$$

$$\text{Also} \quad {}^{\omega_0 L}/R = Q$$

$$\therefore \quad \mathbf{R_D = {}^{Q}/\omega_0 C}$$

Consider the tuned parallel resonance circuit shown in Fig. 11.0-8.

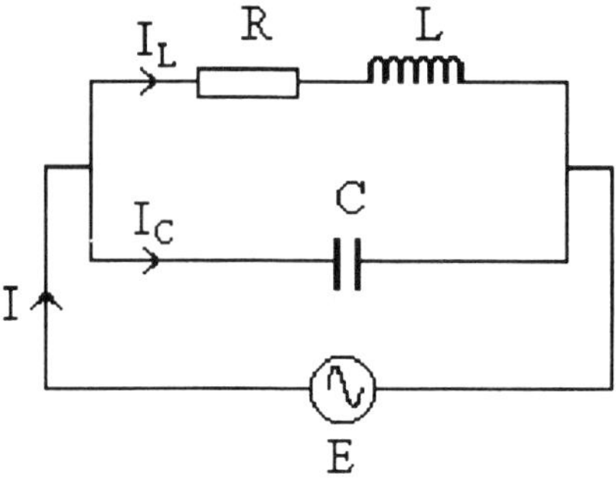

Fig. 11.0-8

(contd)

Refer to Fig. 11.0-8

The e.m.f $E = I R_D$

$$I_c = E/X_c$$

$$= \frac{I R_D}{1/\omega_0 C}$$

$$= \frac{I L \omega_0 C}{CR} \qquad (R_D = L/CR)$$

$$= \frac{I \omega_0 L}{R}$$

$$\therefore \quad \mathbf{I_c = Q\,I}$$

Similarly

$$I_L = \frac{E}{\sqrt{[R^2 + (\omega_o L^2)]}}$$

$$= \frac{E}{\omega_o L \sqrt{[1 + (\frac{R}{\omega_o L})^2]}}$$

$$= I R_D \frac{1}{\omega_o L \sqrt{[1 + (\frac{R}{\omega_o L})^2]}}$$

$$= I \frac{L}{CR} \frac{1}{\omega_o L \sqrt{[1 + (\frac{R}{\omega_o L})^2]}}$$

$$\frac{Q I}{\sqrt{[1 + \frac{1}{Q^2}]}}$$

$$\therefore \quad \mathbf{I_L \approx Q\,I}$$

Effect of source resistance, R_s on the Q-factor - Series R, L and C

The source resistance R_S will add to the resistance of the tuned circuit and hence reduces the Q-factor.

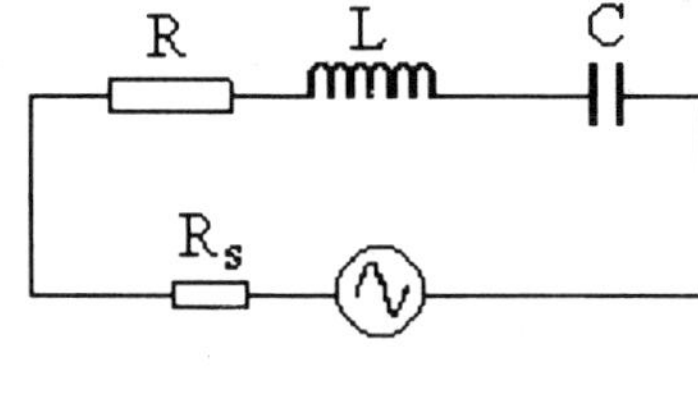

Fig. 11.0-9

(contd)

Refer to Fig. 11.0-9

$$Q = \frac{\omega_o L}{R}$$

$$Q_{eff} = \frac{\omega_o L}{R}\left(\frac{1}{1+\frac{R_s}{R}}\right)$$

$$= \frac{Q}{1+\frac{R_s}{R}}$$

Consider the circuit shown in Fig. 11.0-10 - **Parallel R, L and C**

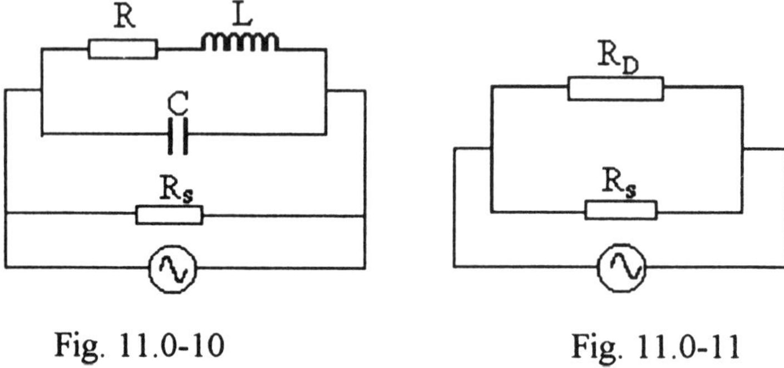

Fig. 11.0-10 Fig. 11.0-11

At resonance the circuit of Fig. 11.0-10 can be represented by its dynamic impedance, R_D shown in Fig. 11.0-11

$$R_{D(eff)} = \frac{R_D R_s}{R_D + R_s}$$

$$\therefore \quad Q_{eff} = \frac{R_{D(eff)}}{\omega_o L}$$

(Substitute for $R_D = Q\omega_o L$)

$$Q_{eff} = \frac{Q\omega_o L R_s}{(Q\omega_o L R_s)\omega_o L}$$

$$= \frac{Q R_s}{R_D + R_s}$$

$$= \frac{Q}{1+\frac{R_D}{R_s}}$$

Note:

This introduction has summarized the basic theory and expressions employed in series and parallel tuned circuits, resonance, Q-factor and effect of source resistance. It is hoped that the solution to past examination papers which follows will help to reinforce the student's knowledge and understanding of the subject matter.

Example 11.1

(a) *An inductor is tuned to series resonance at 1500 kHz by a capacitor of 1600 pF. Determine the new frequency of resonance when the capacitor is increased by 4025 pF.*

(b) *A tuned circuit comprises an inductor and a capacitor in series and is tuned to resonate at a frequency of 1.0 MHz. Determine the new frequency of resonance if the value of the inductor is increased by 21%.*

Solution

(a)

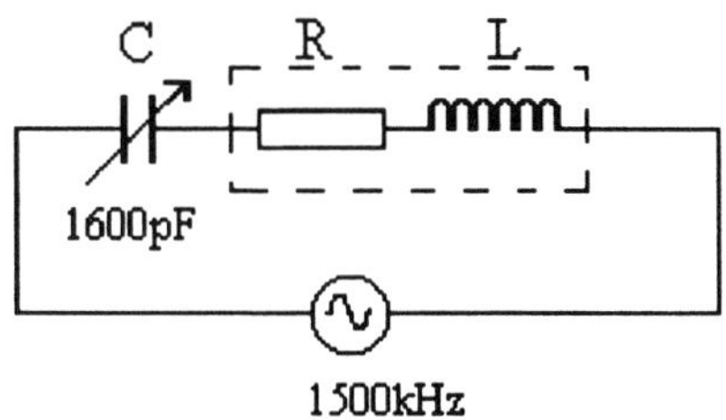

Fig.11.1-0. Series tuned circuit

$$\text{Resonance frequency} = \frac{1}{2\pi\sqrt{LC}}$$

$$\therefore \quad L = \frac{1}{f_o^2 4\pi^2 C}$$

$$= \frac{1}{(1500 \times 10^3)^2 \times 4 \times 3.14^2 \times 1600 \times 10^{-12}}$$

$$= \frac{10^{12}}{225 \times 10^4 \times 10^6 \times 4 \times 3.14^2 \times 16 \times 10^2}$$

$$= \frac{1}{225 \times 4 \times 3.14^2 \times 16} = \underline{\mathbf{7.043\ \mu H}}$$

Now C is increased by 4025 pF.

$$\therefore \quad C = 1600 + 4025 = 5625\text{pF}$$

At resonance $\quad f_o = {}^1/[2\pi\sqrt{(LC)}]$

(contd) *(a)*

$$f_o = \frac{1}{2\pi\sqrt{[7.043 \times 10^{-6} \times 5625 \times 10^{-12}]}}$$

$$= \frac{1}{2\pi\sqrt{[7.043 \times 5625]} \times 10^{-9}}$$

$$= \frac{10^9}{2 \times 3.14 \times 199} \text{ Hz}$$

$$= \frac{10^6}{2 \times 3.14 \times 199} \text{ kHz} \qquad = \underline{\mathbf{800\ kHz}}$$

(b)

Resonance frequency, f_o = 10^6 Hz = $^1/[2\pi\sqrt{(LC)}]$

Let C = 1

and L = 1

∴ f_o = $^1/2\pi\sqrt{(1 \times 1)}$

L is now increased by 21%

∴ L = 1.21

New frequency of resonance f_o' = $^1/(2\pi\sqrt{1.21})$

$$\therefore \quad \frac{f_o'}{f_o} = \frac{\dfrac{1}{2\pi\sqrt{1.21}}}{\dfrac{1}{2\pi\sqrt{1}}}$$

$$= \frac{2\pi\sqrt{1}}{2\pi\sqrt{1.21}}$$

$$f_o' = \frac{f_o}{\sqrt{1.21}}$$

$$= \frac{10^6}{\sqrt{1.21}} \qquad = \underline{\mathbf{0.909\ MHz}}$$

$$= \underline{\mathbf{909\ \ kHz}}$$

Example 11.2

(a) *A tuned circuit is resonant at 600 kHz. Determine the capacitance required to tune the circuit to resonate at 300 kHz.*

(b) *If one section of a two-gang variable-tuning capacitor is used to tune the aerial circuit to the incoming signal, calculate the maximum and minimum values of the capacitor required to cover the frequency range 500 kHz to 1500 kHz when tuned with an inductance of 150 μH. Hence specify the range of the second section required to tune the frequency converter oscillator over the range 1000 kHz to 2000 kHz if the same value of inductance is used to tune the oscillator circuit.*

Solution

(a)

$$f_o = {}^1/[2\pi\sqrt{(LC)}]$$

$$600\text{ kHz} = {}^1/[2\pi\sqrt{(LC)}]$$

To simplify calculation

$$\text{Let } {}^1/(2\pi\sqrt{L}) = \text{constant} = k$$

$$\therefore \quad 600\text{ kHz} = {}^k/\sqrt{C}$$

For resonance at 300 kHz

$$300\text{ kHz} = {}^k/\sqrt{C_1}$$

$$\therefore \quad \frac{600\text{ kHz}}{300\text{ kHz}} = \frac{k/\sqrt{C}}{k/\sqrt{C_1}}$$

$$2 = \frac{\sqrt{C_1}}{\sqrt{C}}$$

$$\therefore \quad \sqrt{C_1} = 2\sqrt{C}$$

$$\mathbf{C_1 = 4C}$$

Therefore the new value of capacitance must be 4 times the original value.

(contd)

(b) Inductive reactance, $X_L = 2\pi f L = \omega L$

Capacitive reactance, $X_C = 1/(2\pi f C) = {}^1/\omega C$

X_L is proportional to frequency (f)

X_C is inversely proportional to frequency (f)

The maximum value of C corresponds to the minimum value at the resonance frequency.

$$f_o = \frac{1}{2\pi\sqrt{LC}}$$

$$f_o^2 = \frac{1}{4\pi^2 LC}$$

$$\therefore \quad C_{max} = \frac{1}{4\pi^2 f_o^2 L}$$

$$= \frac{1}{4\pi^2 \times (500 \times 10^3)^2 \times 150 \times 10^{-6}}$$

$$= \underline{\mathbf{676.16\,pF}}$$

$$C_{min} = \frac{1}{4\pi^2 \times (1500 \times 10^3)^2 \times 150 \times 10^{-6}}$$

$$= \underline{\mathbf{75.13\,pF}}$$

The range of frequency is 500 to 1500kHz.

$\therefore$ Frequency ratio = 1500 : 500 = **3 : 1**

Capacitance ratio required to tune to 1500kHz is C_{max}/C_{min}.

C_{max} = 676.16 pF

C_{min} = 75.13 pF

$\therefore$ Ratio = **9 : 1**

Second Section - Frequency Converter Oscillator

Capacitance for 1000 kHz = [(500/1000)/2] x 676.16 = **169.04pF**

Capacitance for 2000 kHz = [(1000/2000)/2] x 169.04 = **42.26pF**

$\therefore$ Capacitance range is **42 pF to 169 pF**

Example 11.3

(a) *Give an expression to obtain the Q-factor of a parallel resonance circuit.*

(b) *A parallel resonance circuit, consisting of a 50 μH inductor of 5 Ω resistance with a capacitor is tuned to 5/2π MHz and is connected in series to a 1.0 mA constant-current generator. Calculate*

  (i) *the current flowing in the inductor*
  (ii) *the dynamic resistance of the circuit*
  (iii) *the voltage developed across the circuit*

Solution

(a)

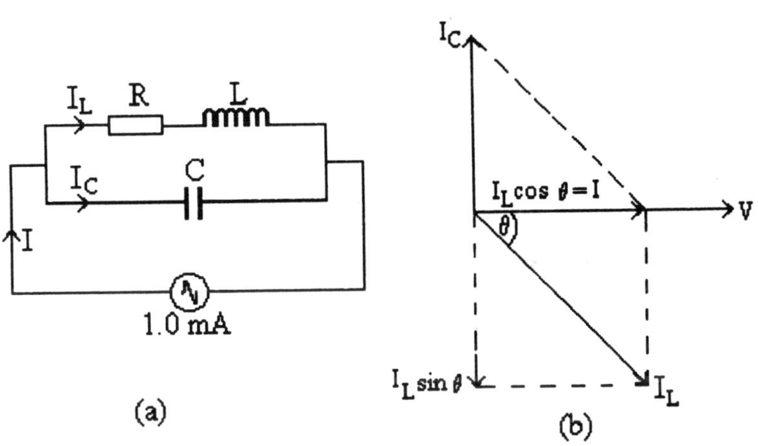

Fig. 11.3-0

At resonance, the Q-factor is given by

$$Q = 2\pi \frac{\text{maximum energy stored}}{\text{energy dissipated per cycle}}$$

$$= 2\pi \frac{\frac{1}{2} L I_{max}^2}{\left(\frac{I_{max}^2}{2}\right) R \times \frac{1}{f_o}} = \frac{\omega L}{R}$$

(contd)

(a) Refer to Fig. 11.3-0(b)

Alternatively, at resonance the Q-factor is given by

$$Q = \frac{\text{current through the capacitor}}{\text{total circuit current}}$$

$$= \frac{I_C}{I} = \frac{I_L \sin\theta}{I_L \cos\theta} = \tan\theta = \frac{X_L}{R}$$

$$\therefore \quad Q = \frac{2\pi f_o L}{R} = \frac{\omega_o L}{R}$$

(b) (i)

$$Q = \frac{2\pi f_o L}{R}$$

$$= \frac{2\pi \times 5 \times 10^6 \times 50 \times 10^{-6}}{2\pi \times 5} = 50$$

∴ Current through inductor is

$$I_L = Q\,I = 50 \times 1.0 \times 10^{-3} = \mathbf{\underline{50\ mA}}$$

(ii) The maximum circuit current occurs at resonance which will be when Z is a minimum.

$$\therefore \quad \omega_o L = {}^1/\omega_o C$$

and $\quad f_o = {}^1/[2\pi\sqrt{(LC)}]$

The circuit impedance at the resonance frequency is $Z = {}^L/CR$ ohms. This is purely resistive and is called the *dynamic impedance* R_D.

$$R_D = {}^L/CR$$

$$= {}^{\omega_o L}/\omega_o CR$$

$$= {}^Q/\omega_o C = Q\omega_o L$$

$$= 50 \times 2\pi f_o L$$

$$= \frac{50 \times 2\pi \times 5 \times 10^6 \times 50 \times 10^{-6}}{2\pi}$$

$$= \mathbf{\underline{12.5\ k\Omega}}$$

(iii) The voltage developed across the tuned circuit is

$$V = I\,R_D$$

$$= 1.0 \times 10^{-3} \times 12.5 \times 10^3 = \mathbf{\underline{12.5\ V}}$$

Example 11.4

(a) *Explain the effect of source resistance on the Q-factor of*

(i) a series resonance circuit

and (ii) a parallel resonance circuit

(b) *The collector circuit of a transistor amplifier consists of a 100 pF capacitor in parallel with a coil of Q-factor 100. The circuit resonates at a frequency of $10^6/2\pi$ Hz. If the transistor has an output resistance of 25 kΩ, determine the Q-factor of the tuned circuit.*

Solution

(a) (i) This subject matter has been dealt with in the introduction and will briefly be repeated.

In a series resonance circuit, the source resistance R_S will add to the resistance of the tuned circuit, thus changing the Q-factor.

Consider the case of the Q-factor with no source resistance.

$$Q = \omega_o L/R$$

Consider the case of the Q-factor with the source resistance taken into account. The effective Q-factor (Q_{eff}) will now be

$$Q_{eff} = \omega_o L/(R + R_S)$$

The source resistance has the effect of reducing the Q-factor.

(ii) The parallel resonance circuit

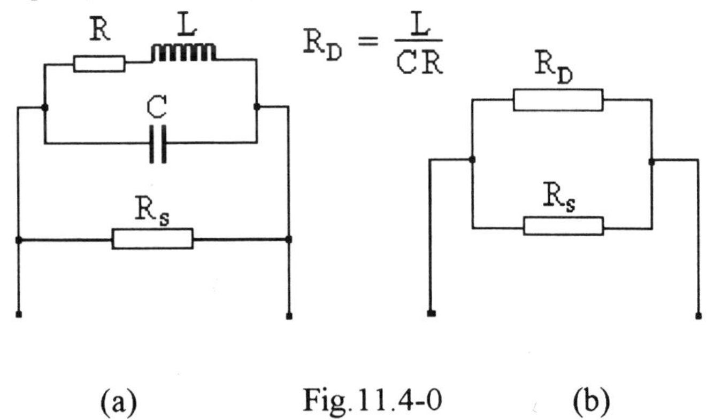

(a) Fig.11.4-0 (b)

(contd)

(a) (ii) Refer to Fig.11.4-0 which shows a parallel resonance circuit shunted by the source resistance R_S.

At resonance, a parallel tuned circuit has a dynamic resistance R_D. *Consider the case of the Q-factor with the source resistance taken into consideration. The dynamic resistance will now be*

$$R_D(\text{eff}) = R_D R_S/(R_D + R_S)$$

$$R_D = Q\omega_o L$$

$$Q_{eff} = \frac{Q R_S}{R_D + R_S} = Q/(1 + R_D / R_S)$$

As a result of the source resistance, the Q-factor will be reduced.

(b)

$$R_D = Q\omega_o L = \frac{Q}{\omega_o C}$$

$$= \frac{100}{2\pi \times \frac{10^6}{2\pi} \times 100 \times 10^{-12}}$$

$$= \underline{\mathbf{1.0\ M\Omega}}$$

$$Q_{eff} = \frac{Q}{1 + \frac{R_D}{R_s}}$$

$$= \frac{100}{1 + \frac{10^6}{25 \times 10^3}} = \underline{\mathbf{2.44}}$$

Example 11.5

(a) *If the voltage gain at the extreme ends of the bandwidth of an amplifier is 3 dB down on the maximum gain, show that these frequencies are the 'half-power points'.*

(b) *The load of a tuned collector amplifier consists of a capacitor of 2 μF in parallel with a coil of inductance 10 mH and Q-factor of 14.15 at the resonance frequency. The amplifier has an a.c. input resistance of 700 Ω and a current gain of 152. Assuming that the shunting effect of the transistor is negligible, calculate*

(i) *the frequency at which the amplifier voltage gain is maximum*

(ii) *the voltage gain at this frequency*

(iii) *the amplifier bandwidth*

State the effect on the bandwidth if the tuned circuit is shunted by the a.c. input resistance of a second transistor stage.

Solution

(a)

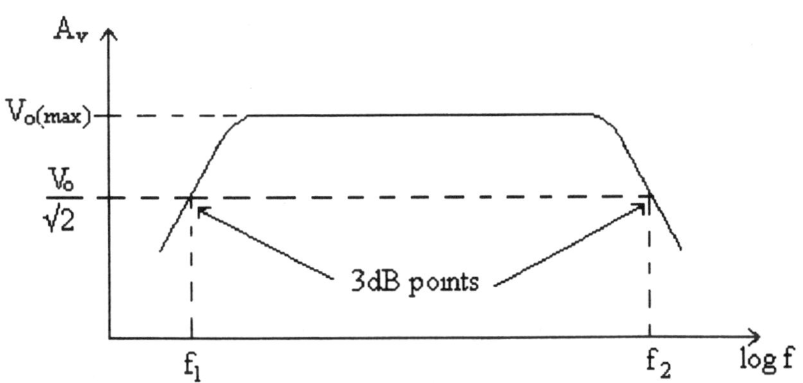

Fig.11.5-0. Bandwidth

Refer to Fig. 11.5-0

At the extreme ends of the bandwidth, the voltage gain is 3 dB down on the maximum voltage gain.

(contd)

(a) Now $V_o/V_{(max)} = {}^1/\sqrt{2}$; i.e., $V_{o(max)} = \sqrt{2}V_o$

Output power P_o is proportional to V_o^2.

That is $P_o \propto V_o^2$

$P_o = kV_o^2$

$\therefore \quad P_o/P_{o(max)} = kV_o^2/k(\sqrt{2}V_o)^2$

$= kV_o^2/2kV_o^2$

$= {}^1/2$

$\therefore \quad \mathbf{P_o = {}^1/2\ P_{o(max)}}$

This is half the power at resonance. Hence the frequencies at the extremes of the bandwidth (f_1- f_2) are referred to as **'the half-power points'**.

Alternatively

The power developed at resonance frequency, f_o is

$P_o = I_o^2 R$

At f_1 and f_2 (Fig.11.5-0), the power developed is

$P = (I_o/\sqrt{2})^2 R$

$= I_o^2 R/2$

$= \mathbf{P_o/2}$

That is, the power at resonance is referred to as the **'half-power points'**.

(b) (i) The amplifier voltage gain is maximum at the resonance frequency (f_o).

At resonance, $f_o = {}^1/[2\pi\sqrt{(LC)}]$

$= {}^1/[2\pi\sqrt{(10 \times 10^{-3} \times 2 \times 10^{-6})}]$

$= 10^4/(2\pi\sqrt{2})$

$=$ **<u>1.126 kHz</u>**.

(contd) *(b)*

(ii) Voltage gain, $A_v = V_{out}/V_{in}$

$V_{in} = I_{in} R_{in}$

$= 700I_{in}$

$V_{out} = I_o R_D$

$= A_i I_{in} \times R_D$

$R_D = Q \omega_o L$

$= 14.15 \times 2\pi \times 1.126 \times 10^3 \times 10 \times 10^{-3}$

$= 1000\ \Omega$

$\therefore$ $V_{out} = A_i \times 1000\ I_{in}$

$= 152 \times 10^3\ I_{in}$

Voltage gain, $A_v = V_{out}/V_{in}$

$= \dfrac{152 \times 10^3\ I_{in}}{700I_{in}}$ $=$ **217.14**

(iii) Bandwidth $= f_o/Q$

$= \dfrac{1.126 \times 10^3}{14.15}$ $=$ **79.58 Hz**.

Example 11.6

A series resonance circuit is connected to a supply E of 1.0 V, internal resistance 5 Ω and frequency 1.0 MHz.
Calculate the values of L and C such that the voltage across C is 50 V at resonance frequency. Assume the resistance of L to be negligible compared with the source resistance.

Solution

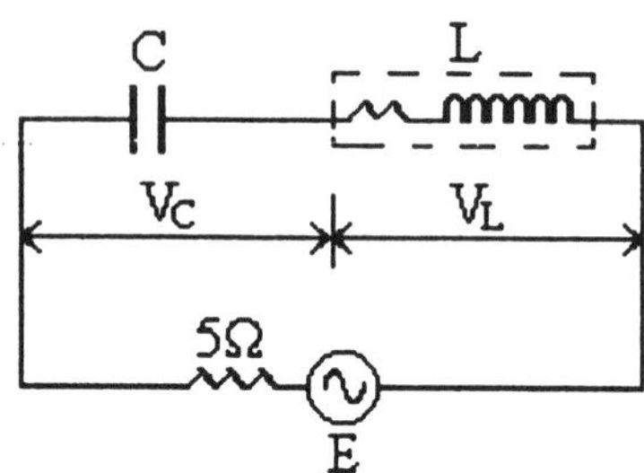

Fig. 11.6-0

At resonance

$$V_L = 2\pi f_o L \times I_o$$
$$= \omega_o L \times E/R$$
$$= \frac{\omega_o L}{R} \times E = \mathbf{QE} \qquad \text{..(Eq. 11.6-0)}$$

and

$$V_C = \frac{1}{2\pi f_o C} \times I_o$$
$$= \frac{I_o}{\omega_o C}$$
$$= \frac{E}{R} \times \frac{1}{\omega_o C} = \mathbf{QE} \qquad \text{...... (Eq. 11.6-1)}$$

Also
$$Q = \frac{\omega_o L}{R} = \frac{1}{\omega_o C R} \qquad \text{...... (Eq. 11.6-2)}$$

$$\therefore \quad \omega_o^2 = \frac{R}{LCR}$$

$$\therefore \quad \omega_o = \frac{1}{\sqrt{LC}} \qquad \text{.......... (Eq. 11.6-3)}$$

Substituting for ω_o in Eq. 11.6-2, we obtain

$$Q = \frac{L}{R\sqrt{LC}} = \frac{1}{R}\sqrt{L/C} \qquad \text{.................. (Eq. 11.6-4)}$$

(contd)

From Eq. 11.6-1

$$V_c = QE$$

$$\therefore \quad Q = V_{c/E} = 50$$

From Eq. 11.6-4

$$Q = (^1/R)\,[\sqrt{(^L/C)}]$$

$$\therefore \quad 50 = {^1/5} \times \sqrt{(^L/C)}$$

$$\text{and} \quad \sqrt{L} = 50 \times 5 \times \sqrt{C} \qquad \text{.......... (Eq. 11.6-5)}$$

From Eq. 11.6-3

$$\omega_o = 2\pi f = {^1/\sqrt{(LC)}} \qquad \text{...... (Eq. 11.6-6)}$$

Substituting Eq.11.6-5 into Eq.11.6-6, we get

$$2\pi f = \frac{1}{\sqrt{(LC)}}$$

$$= \frac{1}{50 \times 5 \times \sqrt{C} \times \sqrt{C}}$$

$$= \frac{1}{250C}$$

$$\therefore \quad C = \frac{1}{2\pi \times 10^6 \times 250} \approx \underline{\mathbf{637\ pF}}$$

From Eq. 11.6-6

$$\omega_o = \frac{1}{\sqrt{(LC)}}$$

$$\therefore \quad L = \frac{1}{4\pi^2 \times 10^{12} \times 637 \times 10^{-12}}$$

$$= \underline{\mathbf{39.8\ \mu H}}$$

Example 11.7

(a) *A coil of inductance 100 μH has a Q-factor of 60 at a frequency of 1.0 MHz. Determine the required value of a shunting capacitor to cause resonance at this frequency and hence calculate the dynamic impedance of this resonance circuit.*

(b) *If this dynamic impedance becomes the load to an amplifier which has an internal resistance of 60 kΩ, determine the bandwidth of the device.*

Solution

(a) At resonance

$$f_o = \frac{1}{2\pi\sqrt{LC}}$$

$$\therefore \quad C = \frac{1}{4\pi^2 \times 10^{12} \times 100 \times 10^{-6}} = \underline{\mathbf{254\ pF}}$$

Dynamic impedance

$$R_D = Q\omega_o L$$

$$= 60 \times 2\pi \times 10^6 \times 100 \times 10^{-6} = \underline{\mathbf{37.68\ k\Omega}}$$

(b) The effective load (R_{eff}) on the amplifier is R_D in parallel with its internal impedance (R_{int}). That is

$$R_{eff} = \frac{R_D \times R_{int}}{R_D + R_{int}}$$

$$= \frac{37.68 \times 60}{37.68 + 60} = \underline{23.14\ k\Omega}$$

$$Q_{eff} = \frac{R_{eff}}{\omega_o L}$$

$$= \frac{23.14 \times 10^3}{2\pi \times 10^6 \times 100 \times 10^{-6}} = \underline{36.8}$$

Amplifier bandwidth $= f_o/Q_{eff}$

$$= 10^6/36.8 = \underline{\mathbf{27.17\ kHz}}$$

Example 11.8

(a) *A tuned-collector transistor amplifier operates at 500 kHz and has a 3 dB bandwidth of 10 kHz. The transistor is operated in the common-emitter configuration and has R_{IN} = 30 kΩ, R_{out} = 30 kΩ* and g_m = *35 mS. The collector-tuned circuit employs a capacitor tap of such ratio that the effective collector load into which the transistor works is 15 kΩ. If the power loss in the collector-tuned circuit is 6 dB, calculate the power gain of the stage.*

(b) *If the transistor is unilateralized, calculate the power gain which can be obtained.*

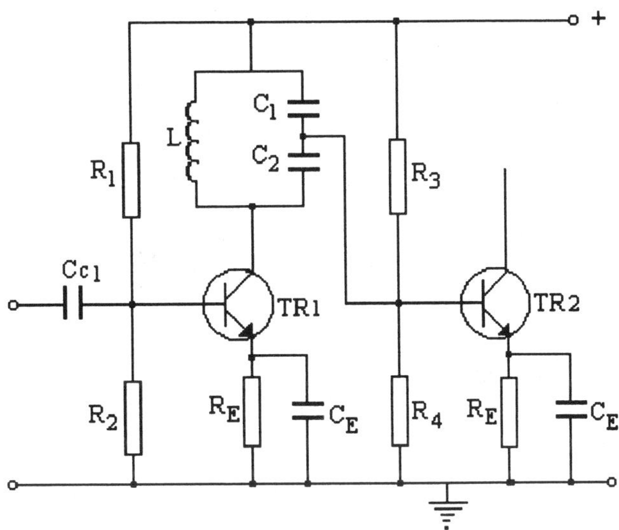

Fig. 11.8-0. Tuned-collector transistor amplifier

Solution
(a)

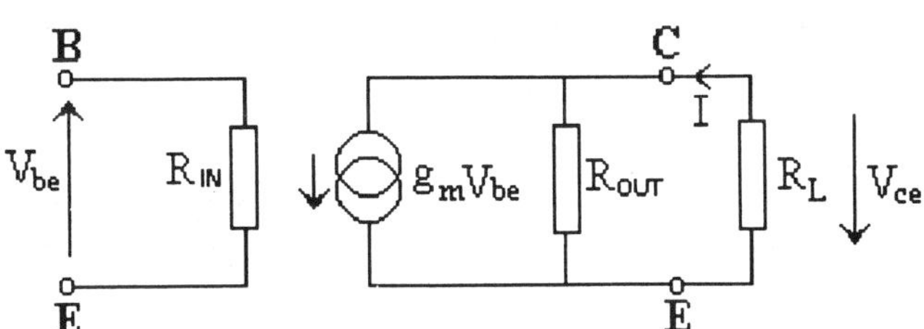

Fig. 11.8-1 Simplified equivalent circuit of unilateralized transistor tuned amplifier

(contd)

(a) The collector current gain $\delta I_c = h_{fe}\delta I_b$.

This current may be expressed in terms of the mutual conductance g_m of the transistor.

$$g_m = \delta I_c/\delta V_{be} \quad (V_{ce} \text{ constant})$$

Refer to Fig. 11.8-1

$$P_{IN} = \frac{V_{be}^2}{R_{IN}}$$

$$P_{OUT} = I^2 R_L$$

$$I = \frac{g_m V_{be} R_{OUT}}{R_{OUT} + R_L}$$

$$\therefore \quad P_{OUT} = \left(\frac{g_m V_{be} R_{OUT}}{R_{OUT} + R_L}\right)^2 R_L$$

$$\text{Power gain, } A_p = \frac{P_{OUT}}{P_{IN}} = \frac{(g_m V_{be} R_{OUT})^2}{(R_{OUT} + R_L)^2} R_L \times \frac{R_{IN}}{V_{be}^2}$$

$$= \frac{g_m^2 R_{OUT}^2 R_L R_{IN}}{(R_{OUT} + R_L)^2} \quad \text{.......... (Eq. 11.8-0)}$$

$$= \frac{35^2 \times 10^{-6} \times 30^2 \times 10^6 \times 15 \times 10^3 \times 1 \times 10^3}{(30 + 15)^2 \times 10^6}$$

$= \underline{8167}$

Now convert 8167 into decibel units, which would be

dB = 10 log 8167 = <u>39.1dB</u>

The tuned circuit introduces a power loss of 6 dB.

Therefore, overall gain = (39.1 – 6) dB = <u>33.1dB</u>

= log (33.1/10) = log 3.31

= $\log^{-1}$ 3.31 = <u>2041.7</u>

That is, **<u>overall gain = 2041.7 or 33.1 dB</u>**.

(contd)

(b) Maximum power transfer is obtained when the source resistance is equal to the load resistance. In our case, the maximum gain is obtained when the output resistance R_{OUT} of the transistor (source resistance) is equal to the load resistance R_L.

Refer to Eq. 11.8-0

For max. Power, $R_{OUT} = R_L$

$$\therefore \quad A_{p(max)} = \frac{g_m^2 R_{OUT}^3 R_{IN}}{(2 R_{OUT})^2}$$

$$= \frac{g_m^2 R_{OUT} R_{IN}}{4}$$

$$= \frac{35^2 \times 10^{-6} \times 30 \times 10^3 \times 1 \times 10^3}{4}$$

$$= \underline{9187.5}$$

$$= \underline{39.63 \text{ dB}}$$

Tuned circuit losses = 6 dB

Therefore, $A_{p(max)}$ = (39.63 – 6) dB = 33.63 dB

= **2306.7**

Example 11.9

A Class A r.f. amplifier stage has a load circuit comprising an inductor of 100 μH in parallel with a loss-free capacitor of 200 pF. The effective Q-factor of the inductor is 56.25. Calculate the frequency of resonance and the frequencies of the upper and lower 3 dB points, f_1 and f_2.
Three such stages are connected in tandem. If the output voltage at resonance is 10 V, calculate the output voltage at f_1 and f_2.
Calculate the value of the resistor that would have to be connected in series with the inductor to make the 3 dB bandwidth of the single-stage amplifier equal to 40 kHz.

Solution

$$\text{Resonance frequency, } f_o = 1/[2\pi\sqrt{(LC)}]$$
$$= 1/[2\pi\sqrt{(100 \times 10^{-6} \times 200 \times 10^{-12})}]$$
$$= 1/[2\pi\sqrt{(2 \times 10^{4})} \times 10^{-9}]$$
$$= 10^{7}/(2\pi\sqrt{2})$$
$$= 0.1126 \times 10^{7} \qquad = \underline{\mathbf{1.126\ MHz}}$$

$$\text{3 dB bandwidth} = \frac{\text{resonance frequency}}{\text{effective Q-factor}} = f_o/Q$$
$$= \frac{1.126 \times 10^{6}}{56.25} \qquad = \underline{20.018\ \text{kHz}}$$

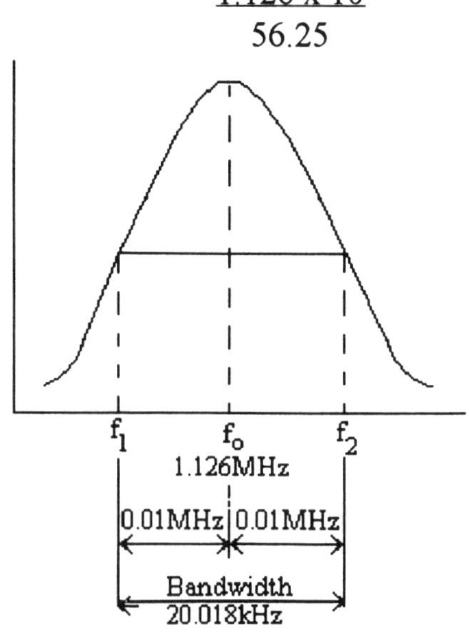

Fig. 11.9-0

(contd)

Refer to Fig.11.9-0

Lower 3 dB point, $f_1 = (1.126 - \frac{20.018 \times 10^{-3}}{2})$ MHz

$= 1.126 - 0.010009 \quad =$ **1.115991 MHz**

Upper 3 dB point, $f_2 = 1.126 + 0.010009 \quad =$ **1.136009 MHz**

Output voltage at resonance = 10 V

∴ For three stages, output at the lower 3 dB point, f_1 and upper 3 dB point $f_2 = {}^{10}/(\sqrt{2})^3$

$= {}^{10}/(2 \times \sqrt{2})$

$= 5 \times 0.707 \quad =$ **3.536 V**

Required bandwidth = 40 kHz

$Q = f_0/B$

$= \frac{1.126 \times 10^6}{40 \times 10^3} \quad = 28.15$

Total resistance $= \frac{2\pi f_0 L}{Q}$

$= \frac{2\pi \times 1.126 \times 10^6 \times 100 \times 10^{-6}}{28.15}$

$= 25.12\ \Omega$

Initial Q = 56.25

Initial resistance $= \frac{2\pi f_0 L}{Q}$

$= \frac{2\pi \times 1.126 \times 10^6 \times 100 \times 10^{-6}}{56.25}$

$= 12.57\Omega$

Resistance to be added = Total resistance - Initial resistance

$= 25.12 - 12.57 \quad =$ **12.55 Ω**

CHAPTER 12

MEMORY CIRCUITS

INTRODUCTION

The basic decision-making elements of digital logic were previously introduced in Chapter 10. The logic gates made decision based on the input data provided to them. Digital logic elements–memory elements–are constructed from AND and OR gates, even though their performance characteristics are quite different.

In practice, the role of memory elements in digital logic circuits is so fundamentally important that they are regarded as an independent group. **The basic characteristics and some applications of these memory elements (called flip-flops) are the subject of this chapter.**

AND and OR gates can be connected to provide memory; that is, they can remember if a signal of logic 1 or 0 level has been connected to their inputs and make the fact available at the outputs.

A memory is normally used to store information for some period of time, after which it is discarded. To implement this type of memory, two NOR gates are required; the circuitry is simple and very important. **It is called an S-R (Set-Reset) flip-flop or S-R latch and it is the most basic form of the class of circuits called flip-flops from which other circuits are developed**. Any difficulty in analyzing the S-R flip-flop lies in the fact that the outputs are connected back to the inputs; thus the signal goes through the circuit and is fed back to the input lines. As a result, the input signal has multiple effects. Feedback connection is essential in giving a logic circuit its memory ability.

All basic flip-flops are available in integrated circuit (IC) form with only the connecting pins accessible. Therefore, only the type (the truth table) and pin assignment need to be known to use an I C containing one or more flip-flops. This chapter consists of brief descriptions dealing with a few particular aspects of the subject only. No attempt has been made to produce a complete textbook on memory elements; however it is hoped that the description will be useful in understanding the basic structures of memory elements.

Example 12.1

Describe the operation of a basic S-R flip-flop using cross-connected NOR gates.

Solution

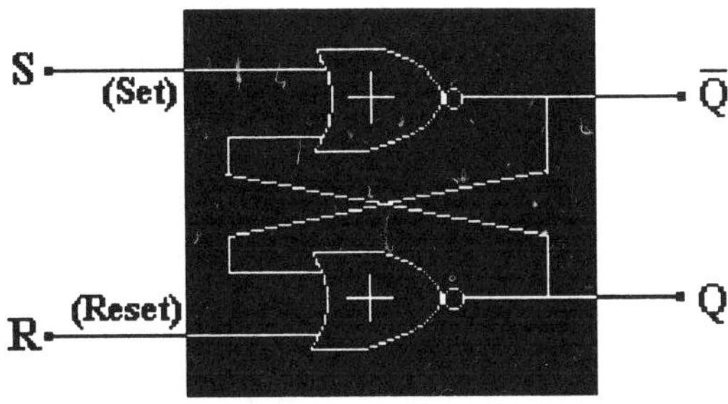

Fig. 12.1-0 Basic S-R flip-flop or latch

To understand the operation of the S-R flip-flop or latch, begin by applying a logic 1 level at the Set input S, while holding the Reset input R at logic 0. [Remember: a NOR gate output is a logic 0 whenever one or both of its inputs are at logic 1.]
With a logic 1 input at S, the output at $\overline{Q}$ goes to the logic 0 state. However logic 0 at $\overline{Q}$ is connected to the lower gate, thus making the lower gate have two logic 0 inputs. Hence its output Q goes to logic 1. The logic 1 at Q is fed back to the input of the upper gate, and the upper gate now has two logic 1 inputs. Being a NOR gate, it needs only one high input to maintain its output at the logic 0 state. Consequently, the logic 1 at S can now become a logic 0 and the output at $\overline{Q}$ will remain unchanged.

We have traced the original input through the circuit in a figure 8 pattern. If the logic 1 at S goes to logic 0, the figure 8 pattern can be traced again, where it can be observed that all conditions still remain the same. Note that the S input was used to activate (or set) the circuit. As a result, the gates have latched themselves in certain states, and as long as the R (Reset) input remains at logic 0, nothing will change. Thus, this device is called a latch.

(contd)

To reset the circuit, input R is raised to logic 1. Tracing a figure 8 pattern beginning at R will give the same results as before, except that now Q will become a logic 0 and $\overline{Q}$ a logic 1. The circuit has been Reset and its memory erased.

The S-R flip-flop or latch has the property that whenever the Set input is raised to a logic 1, the Q output will become latched in a Q = 1 state, and whenever the Reset input is raised to a logic 1, Q will become a logic 0. If the inputs are raised alternately, Q and $\overline{Q}$ will each alternate between logic 1 and logic 0 states.

Only momentary input is required to produce a complete transition; that is, very short pulses can be used to trigger the flip-flop.

Attempting to set and reset simultaneously creates an undesirable state. This operating condition is unacceptable in practice and actual circuits are designed to avoid this state ambiguity. However, it does not change the basic function of the circuit, which is that of remembering a logic1 or 0 at its inputs.

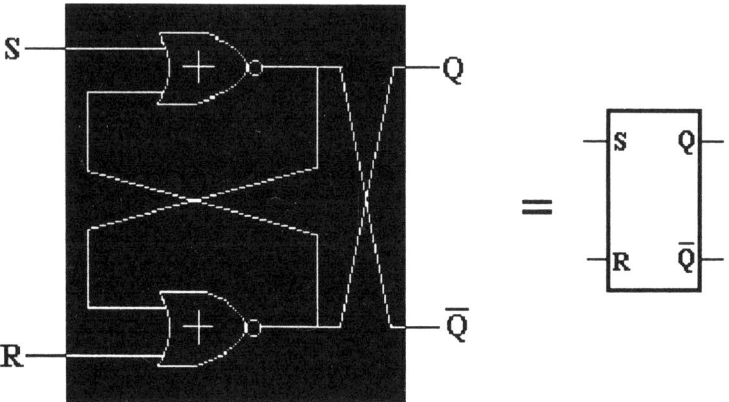

(a) Functional representation of the S-R flip-flop or S-R latch as a simple memory element

(b) Standard flip-flop symbol as it is used in logic diagrams

Fig.12.1-1. Basic flip-flop symbol

Example 12.2

An S-R flip-flop incorporating two AND gates (known as a gated S-R flip-flop) is shown in Fig. 12.2-0.

Briefly explain the functions of the Clock, Preset and Clear signals.

Solution

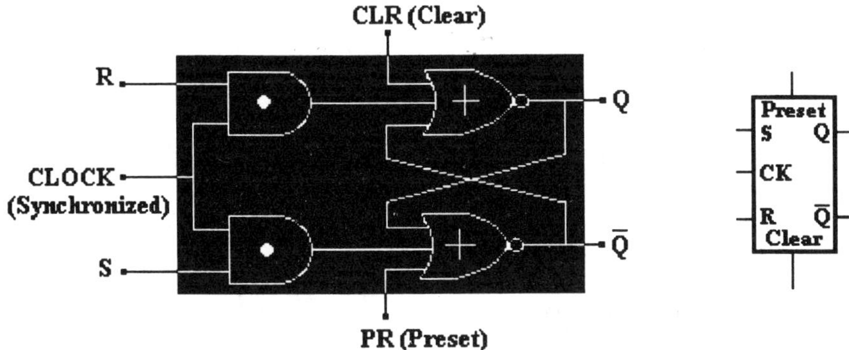

Fig.12.2-0. Clocked S-R flip-flop or S-R latch

The three basic control signals common to most flip-flops are **Clock, Preset** and **Clear**.

<u>Refer to Fig.12.2-0</u>

Two gates are connected to the inputs of the flip-flop and a clock signal is connected so that it can **enable** or **disable** both gates simultaneously. These gates block the R and S inputs from producing a change of state (logic level) of the flip-flop while the clock signal is low.

When the clock signal is raised to logic 1 level, any logic 1 signal on the R and S inputs is *gated* in (permitted to enter). The clock can then return to logic 0 again to disable the input gates and prevent other signals from entering the flip-flop or latch. The clock signal creates what is called a *gate*. Unless the gate is open, the state of the S-R latch cannot be changed by the S and R inputs.

Note: *The synchronizing signal from the clock (CK) is frequently desiginated* **E,** *the enabling terminal.*

(contd)

To gain a better understanding of this *gate*, consider a calculator circuit that uses several flip-flops (memory elements) for the storage of numbers. To carry out addition, a number is first entered by depressing keys on the calculator keyboard. A clock signal connected to all the flip-flops is then raised to a logic 1 level, and the numbers are *gated* into the flip-flops, after which the flip-flops are disabled again by placing the clock in a logic 0 state. The clock signal is maintained at logic 0 so that the flip-flops are prevented from receiving additional numbers while calculations are being processed by other circuits.

Thus, **a clock signal can be used to clock or gate data into both inputs of the S-R flip-flop or latch.**

Another fundamental purpose of a clock signal is to synchronize. In the calculator example, storage of data required several flip-flops. The same clock signal was used to enable and disable the data inputs to all flip-flops, thus entering the data into the flip-flops synchronously.

Preset and Clear are inputs used to Set or Reset a flip-flop without including the data and clock inputs.

In other words, Preset and Clear can be used to Set and Reset the flip-flop when the clock signal is low.

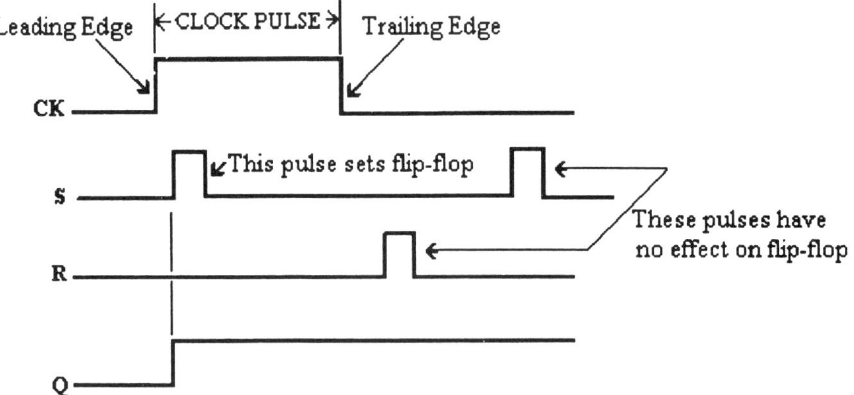

Fig. 12.2-1. Pulse waveforms for an S-R flip-flop

Example 12.3

(a) *Show in block diagrams the following:*

(i) *the derivation of a D-latch from an S-R flip-flop*

(ii) *an S-R latch from NAND gates*

Briefly describe their operation and explain the term "race condition" as applied to the NAND gate latch

(b) *Briefly describe the operation of a D flip-flop.*

Solution

(a) (i) **D-LATCH FROM S-R FLIP-FLOP**

The D-latch is shown in Fig.12.3-0. The unacceptable state of R = 1 and S = 1 simultaneously is avoided by the modified S-R flip-flop. Because an inverter is used to produce the R input, the inputs to the NOR gates will always be complementary; that is, when one input is at logic 1, the other must be at logic 0. In this manner the race condition is avoided. Thus, the D-latch offers a simple solution to the race problem of the S-R latch and is a very commonly used circuit.

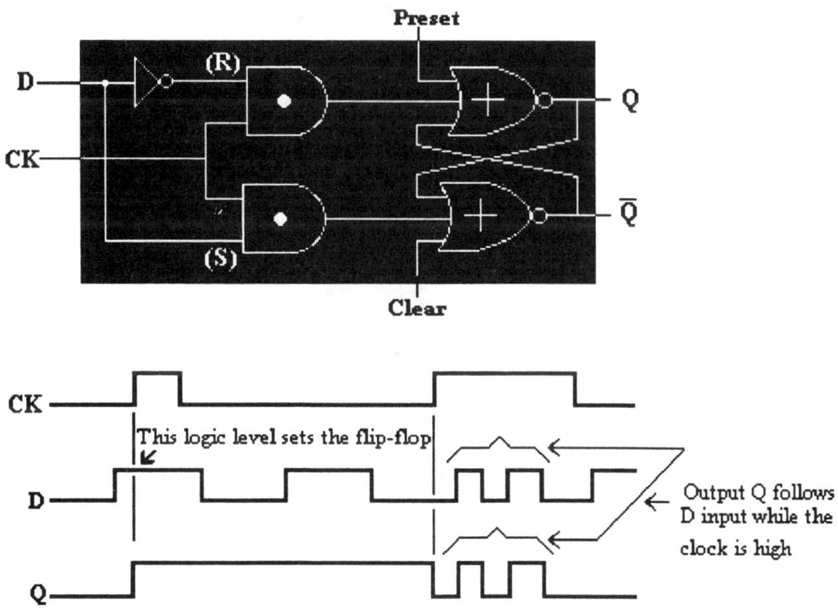

Fig. 12.3-0. D-Latch with timing signals

(contd)

Refer to Fig.12.3-0

As a result of connecting an inverter between the R and S terminals, the D-latch has only one data input terminal.

The latch is set at Q = 1 by clocking in a logic 1 and reset at Q = 0 by clocking in a logic 0.

To use the D-latch correctly, the desired clock signal, high or low input, is applied to the data input line, and the clock signal is then removed before the data input line signal is allowed to change. Immediately upon the removal of the clock signal, the circuit is latched and the data input line can change in any manner without affecting the outputs. This is illustrated in the timing diagram.

[Note: Alternatively, the D input can be applied prior to the clock signal.]

(ii) S-R LATCH FROM NAND GATES

The basic S-R latch circuit is constructed from NAND gates as shown in Fig.12.3-1. The operation is similar to that of the S-R flip-flop previously described. However, both types are in common use by engineering designers and it is important to be able to recognize them and know their differences. The two types have the same ability to store information. *The* **NAND** *gate implementation of the latch is set and reset using a logic 0 level instead of a logic 1 level.* This is indicated by the inversion bars on the $\overline{R}$ and $\overline{S}$ inputs in Fig. 12.3-1.

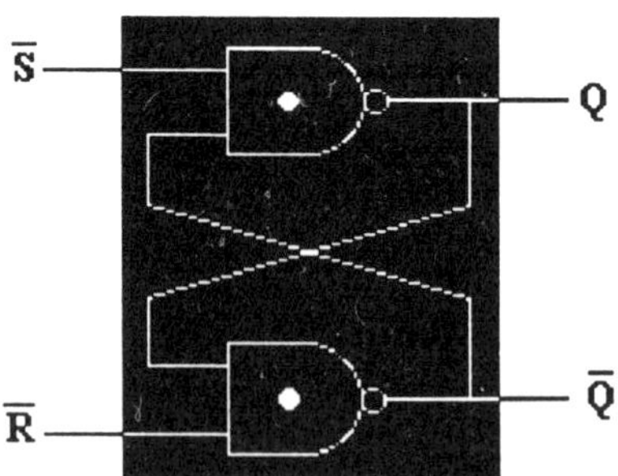

Fig.12.3-1. S-R latch from NAND gates

To store a logic 1 at Q (logic 0 at Q), the conditions at the input must be $\overline{S}$ = 0 and $\overline{R}$ = 1. This condition is memorised when the inputs return

(contd)

to $\overline{R} = 1$ and $\overline{S} = 1$.

The undesirable condition to be avoided in the NAND gate S-R latch is $\overline{R} = 0$ and $\overline{S} = 0$ simultaneously.

RACE CONDITION

The reason for avoiding the $\overline{R} = 0$ and $\overline{S} = 0$ input state in the NAND gate latch is to eliminate the danger of what is called a **race condition**. This condition is created by the following sequence.

(i) Assume that both inputs are at logic 0. The outputs are therefore both at logic 1.

(ii) As a result of (i), the NAND gates simultaneously have one of their inputs at logic 1, and they must change their output states to logic 0.

(iii) However, the logic 0 outputs will remove one logic 1 input from each NAND gate. Consequently, both gates must switch back again to logic 1 outputs.

(iv) These outputs reestablish two logic 1 inputs to each NAND gate and the outputs switch again to logic 0.

So long as the two gates change state simultaneously, this sequence continues indefinitely; both gates continuously switch back and forth between logic 0 and logic 1 outputs with high speed.

However, one of the gates will switch from one state to the other a little faster and cause the flip-flop to latch. The gates literally race each other to change states and whichever changes its output first will prevent the other gate from changing.

In this situation, the final output is unpredictable and therefore this sequence of input states should be avoided.

[Note: The race condition is also possible in the NOR gate latch.]

(contd)

(b) **THE D FLIP-FLOP**

The D flip-flop and the D-latch are functionally closely related. Both have a single data input and a logic 1 or 0 is used to set or reset them. The difference between the two is the manner in which the clock signal is used to gating-in data. The output latch takes place only at the instant when clock signal changes from a logic 0 to a logic 1 level, and at no other time. The essential feature of the D flip-flop is that it samples data present at the input only with the *rising edge* of the clock pulse shown in the timing diagram of Fig. 12.3-2.

This is known as **edge triggering** and is the characteristic of most flip-flops.

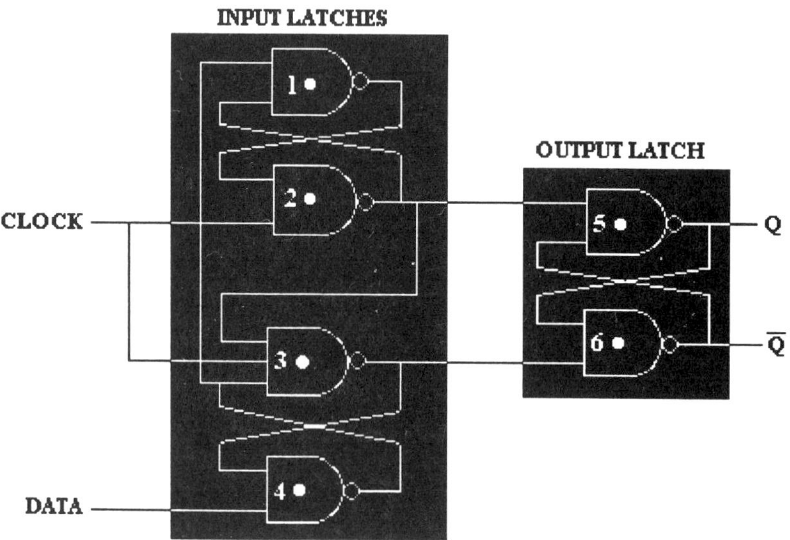

[Note: Preset and Clear inputs not shown in circuit.]

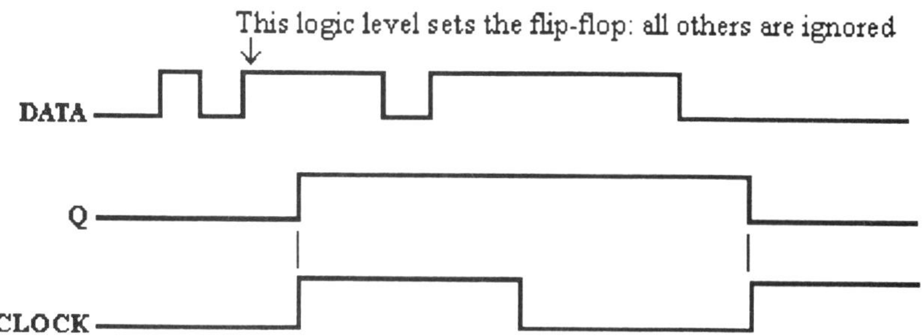

Fig. 12.3-2. D flip-flop with timing diagram

(contd)

Refer to Fig.12.3-2

The circuit shows that the D flip-flop consists of two interconnected latches (gates 1 and 2 and gates 3 and 4) and an output latch (gates 5 and 6). The input latches are connected so that when the clock signal goes from logic 0 to logic 1 (the leading edge of the clock pulse), it causes the input latches to latch in complementary states. That is, one always supplies a logic 1 and the other a logic 0 to the output latch.

Which way they latch is determined by the state of the data line at the leading edge of the clock pulse.

Once the clock signal is high (logic 1), it holds both input latches in their existing states and the data line will have no further effect.

When the clock signal goes low (logic 0) again, both input latches supply a logic 1 to the output latch and the data line can only affect the status gates 1 and 2.

Example 12.4

(a) *Briefly explain the principle of a master-slave S-R flip-flop.*

(b) *Show how the master-slave S-R flip-flop is structured to become a master-slave J-K flip-flop.*

Solution

(a) The basic principle of the master-slave relationship is simulated in Fig. 12.4-0. It consists of two S-R flip-flops connected by synchronizing switches S1 and S2. The switches are shown connected with S1 open and S2 closed; however, the switches can also be shown with S1 closed and S2 open.

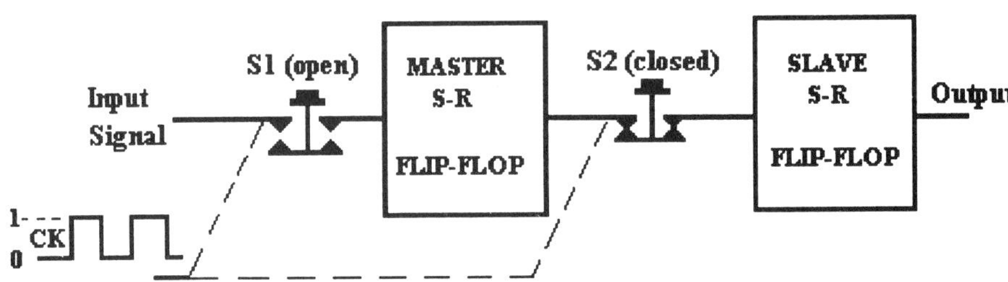

Fig. 12.4-0. Master-slave flip-flop

At level 0 of the clock (CK) signal, synchronous operating switch S1 is open and S2 closed, thus allowing data stored in the MASTER to be transferred to the SLAVE.

When the clock signal rises to level 1, S1 closes and S2 opens, permitting new data to be transferred into the MASTER while the previous data are retained by the SLAVE.

When the clock signal again falls to level 0, S1 opens and S2 closes, thus isolating the MASTER from the input, and once again the MASTER transfers data to the SLAVE.

In this type of master-slave S-R-flip flop, the output changes state when the clock signal changes from level 1 to level 0; that is, the output changes state on the **trailing edge** of the clock pulse. Because of this characteristic, this type of master-slave flip-flop is known as a **trailing edge triggered** flip-flop.

(contd) *(a)*

Figure 12.4-1 shows a typical block diagram of one form of master-slave flip-flop.

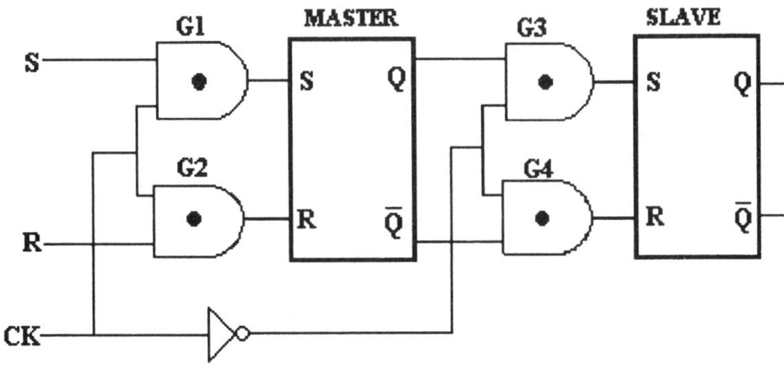

Fig. 12.4-1

In the block diagram AND gates G1 and G2 represent switch S1 and gates G3 and G4 are equivalent to switch S2 (refer to Fig.12.4-0); the INVERTER (NOT gate) provides the correct phase relationship between the switches.

(b) **MASTER-SLAVE J-K FLIP-FLOP**

Figure 12.4-2 shows a typical master-slave J-K flip-flop.
The master-slave principle is not limited to J-K flip-flops alone but is any arrangement in which the data are input into one S-R latch and subsequently transferred to the second.

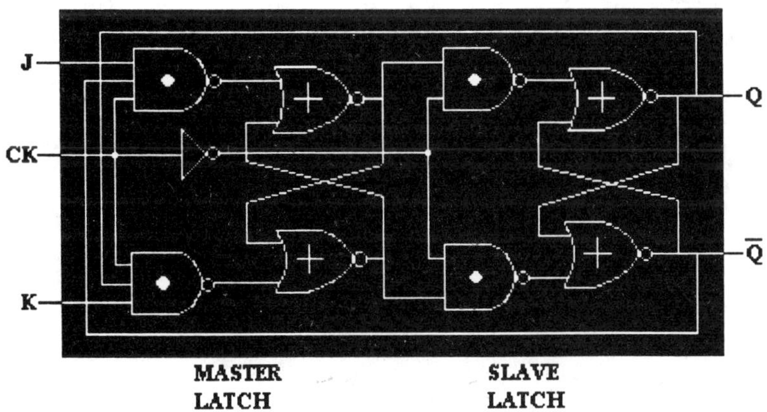

[Note: Preset and Clear inputs not shown in circuit]

Fig. 12.4-2. J-K flip-flop

(contd) *(b)*

The J-K flip-flop is the most commonly used flip-flop and also the most versatile and sophisticated. Like the S-R latch, it has two data inputs but has none of the shortcomings of the S-R latch. It cannot have an undefined output, and its latches are not subject to race conditions. Most versions of the J-K flip-flop are controlled by the trailing edge of the clock signal.

The master-slave J-K flip-flop is similar in structure to the master-slave S-R flip-flop with the exception that the outputs of the J-K flip-flop are fed back.

Figure 12.4-3 shows a typical block diagram of a J-K flip-flop.

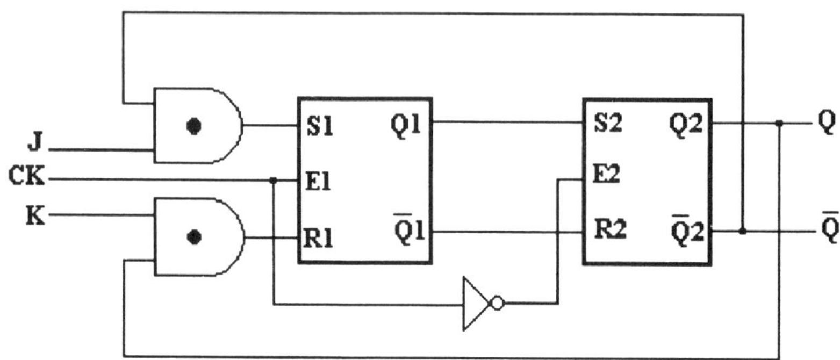

Fig.12.4-3. Typical J-K master-slave flip-flop

For greater flexibility, some versions of the J-K flip-flop include Preset (PR) and Clear (CLR) terminals. These terminals are normally held High (logic 1). If PR goes Low (logic 0), Q is forced to level 1, whereas if CLR goes Low ,Q is forced to level 0. The truth table is shown in Table 12.4-0. With 0 inputs at J and K, the clock (CK) has no effect and the flip-flop remains in its current state, Q_o.

When J = 1 and K = 0, the clock sets the flip-flop to Q = 1.

When J = 0 and K = 1, the clock resets the flip-flop to Q = 0.

When J = 1 and K = 1, the flip-flop toggles; that is, the output changes each time the clock (CK) goes Low (level 0).

Truth Table 12.4-0

J	K	Q
0	0	Q_o
1	0	1
0	1	0
1	1	Q_o

Example 12.5

Describe in general the basic principle of shift registers. Include in this description the sampling of data input with the clock pulses.

Solution

D flip-flops have been chosen for this exercise; however, J-K flip-flops would have served equally well.

Basically a shift register consists of a series of flip-flops interconnected in such a manner that the output of one flip-flop becomes the input of the next flip-flop as shown in Fig.12.5-0. [Note: FF1, FF2, FF3, and FF4 represent flip-flop 1, flip-flop 2, flip-flop 3 and flip-flop 4.]

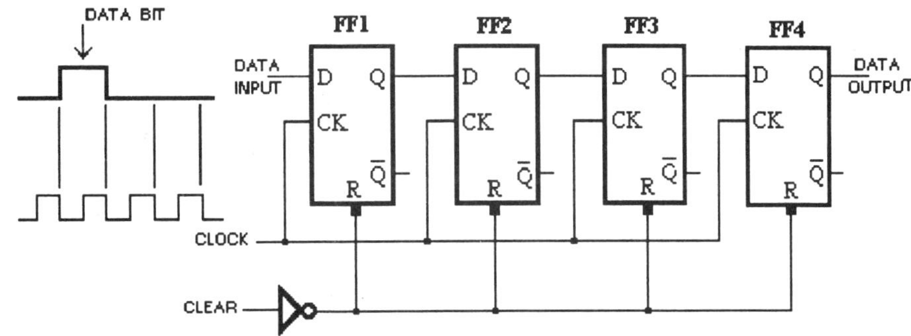

Fig.12.5-0. Basic shift register

Shift registers (flip-flops) are classified basically into two types:

(1) *Asynchronous* - [One flip-flop changes state and the change triggers a second flip-flop, which in turn will trigger a third flip-flop at the next change of state, then a fourth and so on.]

(2) *Synchronous* - [All flip-flops change state simultaneously.]

Asynchronous registers load their data one after another which is referred to as ***serial data loading.***

Synchronous registers load all their data simultaneously which is referred to as ***parallel data loading.***

Refer to Fig.12.5-0

There is a common clock signal and all flip-flops are set or reset *synchronously* (at the same time). Assume all flip flops are initially reset.

(contd)

During the time the flip-flops are clocked, a logic 1 data is supplied to the input of FF1 long enough to be sampled by the leading edge of one clock pulse.

Along with the clock pulse the logic 1 is stored at the output of FF1.

On the next clock pulse, the logic 1 in FF1 shifts to and sets FF2. Meanwhile, FF1 samples its input data again at the leading edge of the next pulse and observant of a logic 0 at the input loads a logic 0. On the third clock pulse, the logic 1 data moves to FF3 and on the fourth clock pulse, to FF4. If a fifth pulse is injected, the logic 1 is shifted out from FF4 and all flip-flops return to their initial state (logic 0).

It can be seen from the preceding that the logic 1 is simply being shifted through the flip-flops by the clock pulses. Both the logic 1 and logic 0 are called **bits** or **data bits**.

Consider storing a number in a shift register. Let each bit in the register represent 1 and let the combined number of 1's represent the number stored; for example, two bits will be contained in two of the four flip flops, and three bits in three flip-flops. Consequently, to enter a number to be stored in the shift register, four clock pulses must be supplied so that the appropriate logic level is present at the input of FF1 when the leading edge of each clock pulse occurs. This is shown in Fig.12.5-1.

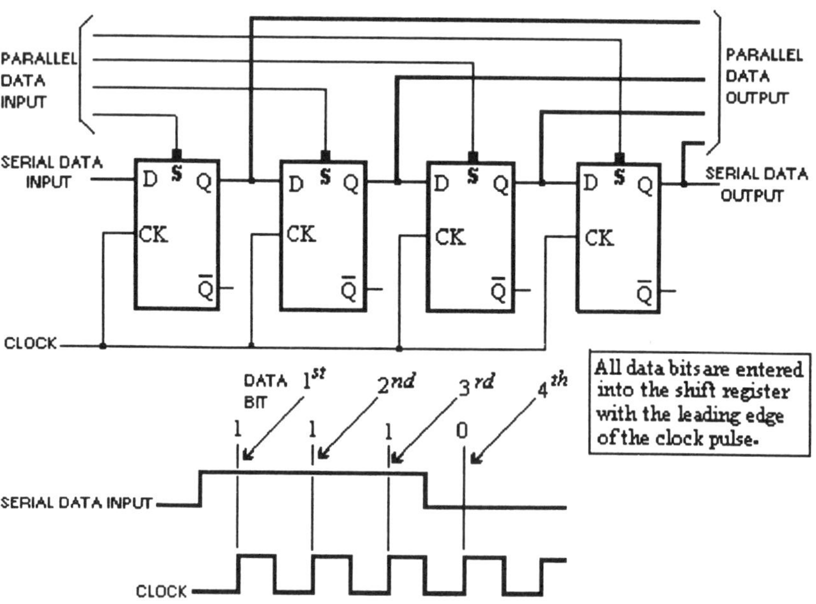

Fig.12.5-1. Methods of shifting data into and out of a shift register

(contd)

Refer to Fig.12.5-1

Assuming the flip-flops are all initially cleared (all flip-flops reset) and the four data bits are shifted through the register, the flip-flops will be in the following states at the end of the four clock pulses.

FF1 = 0 FF2 = 1 FF3 = 1 FF4 = 1

This can be written as 0111. If the number 2 was stored, it would be written as 0011; and the number 4 would be 1111. The four flip flops can can store five numbers: 0, 1, 2, 3 and 4.
Because the bits were loaded one after another and *shifted* or *ripple* through the register from one flip-flop to another, the sequence is referred to as *serial data loading* and the circuit is called *a 4-bit serially loaded shift register* [asynchronous register.]

Parallel loading the shift register is an alternative to serial loading. In this case, the input of each flip-flop is preset by a separate line. All the data bits are loaded simultaneously by setting the flip-flops to logic 1 through the preset input. Since the flip flops change state concurrently the register is said to be *synchronous.*
The data on the four preset inputs is called *parallel data.*

Just as the data bits can be loaded into the shift register in parallel or serially, they can also be read out all at the same time or one at a time.

Parallel read out is accomplished by simultaneously sampling data at the outputs of each flip-flop.

Serial read out is accomplished by shifting data through the register by the flip-flops and sampling at the output of FF4.

A shift register loaded serially and read out in parallel operates as a *serial-to-parallel converter.*
If the data are loaded in parallel and shifted out serially, the shift register is a *parallel-to-serial converter.*

Example 12.6

Briefly explain how RAM and ROM memories are stored in an integrated circuit (I C) for use in a computer or other relevant device.

Solution

RAM (Random Access Memory/Read And Write Memory)

In a digital computer, instructions and numbers are stored in an organized arrangement of elements called 'memory cells'. Each cell is capable of storing one bit of data.

A 64-bit memory has an 8 x 8 matrix of cells as shown in Fig.12.6-0.

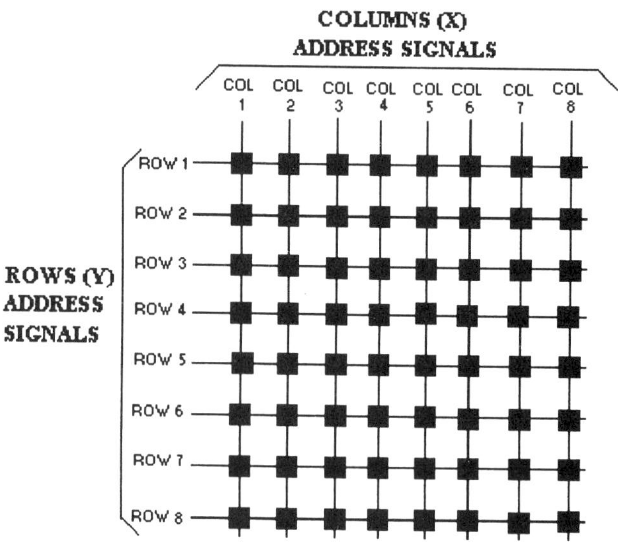

Fig.12.6-0. A typical arrangement of a 64-bit RAM

RAM is an integrated circuit memory into which data bits can be written and then read out again. On each RAM chip there are as many cells as there are bits. Each memory cell has an individual address located at the intersection of a ROW and a COLUMN. To access any one memory cell, one ROW address line and one COLUMN address line are enabled and the corresponding cell at the intersection of these two lines is then activated. The cell circuit is basically a flip-flop, though a capacitive element is also used. To write data into a typical RAM, the data bit location address and write-enabled signal must be supplied and then data can be entered. At any subsequent time, the computer can retrieve (read) the data after supplying the bit address and the read-enable signal.

(contd)

A typical I C memory of a general m x n RAM is shown in Fig. 12.6-1.

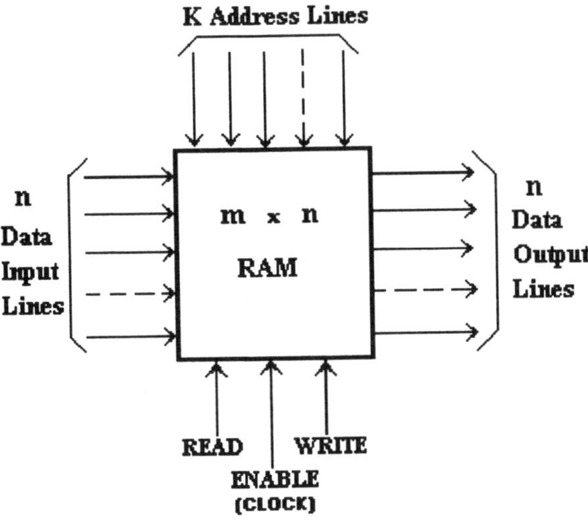

Fig. 12.6-1. General m x n RAM

The K lines may be designated as

$2^k = m$ (words) whose n (bits) are carried by the n parallel input and output lines.

ROM (Read Only Memory)

In ROM the data can be read out but not written over. All data is programmed in during manufacture. The computer can read the data at any address but cannot change or alter the stored data.
ROM also consists of a matrix of memory cells. The cell circuits are very simple and may consist of a single transistor or even a diode.
A ROM may also be considered as a kind of decoder that decodes input addresses into a set of outputs.

There are three types of ROMs: The manufacturer ***programmable ROM, a PROM*** that is programmable by the user, and an ***erasable PROM*** that can be erased and programmed several times.

(contd)

An I C memory of a typical 64 bit ROM is shown in Fig.12.6-2.

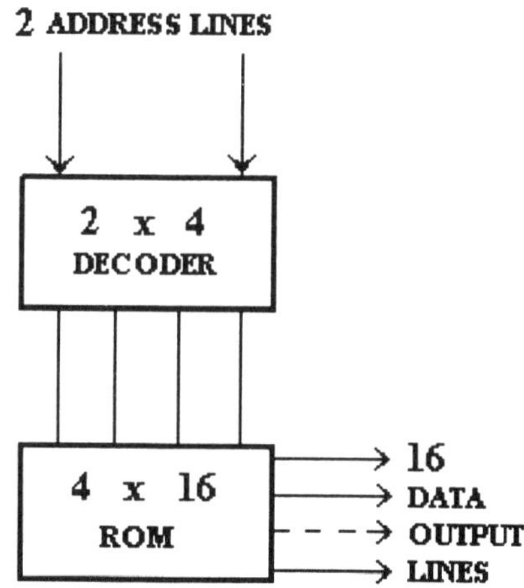

Fig.12.6-2. A typical 4 x 16 ROM

The 64 bits of the 4 x 16 ROM are stored in 64 memory cells arranged in $2^k = 2^2 = 4$ words x 16 bits available at the 16 output lines. The 2 x 4 decoder translates the coded address to the 16 output lines. The address is coded as a k-bit binary number.

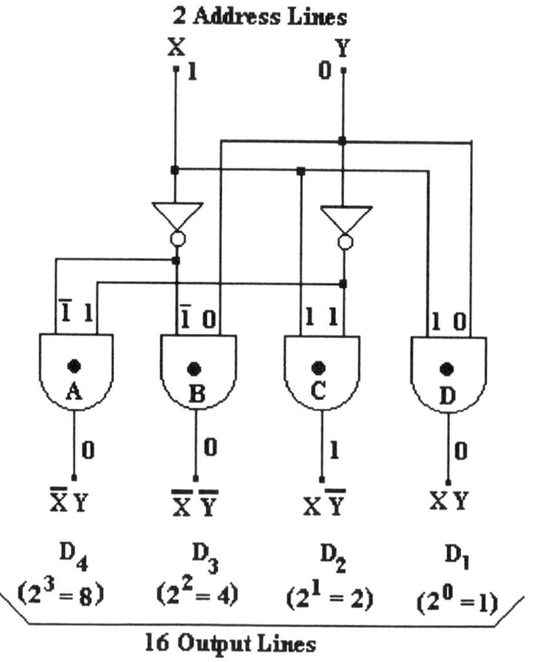

Fig. 12.6-3. A typical 2 x 4 decoder

(contd)

<u>Refer to Fig.12.6-3</u>

In the 2 x 4 decoder, the two address lines are X = 1 and Y = 0; that is, XY = 10 (read as "one zero").
With this address (XY = 10), an output will appear at D2 which specifies that 2 bits (2^1) connected to the 16 output lines is to be read.

APPENDIX

Trigonometrical Identities

$$\sin(90^\circ - \theta) = \cos\theta = \sin(90^\circ + \theta) \tag{1}$$

$$\cos(90^\circ - \theta) = \sin\theta = -\cos(90^\circ + \theta) \tag{2}$$

$$\tan(90^\circ - \theta) = \cot\theta = -\tan(90^\circ + \theta) \tag{3}$$

$$\sin^2\theta + \cos^2\theta = 1 \tag{4}$$

$$1 + \tan^2\theta = \sec^2\theta \tag{5}$$

$$1 + \cot^2\theta = \operatorname{cosec}^2\theta \tag{6}$$

$$\sin(A + B) = \sin A\cos B + \cos A\sin B \tag{7}$$

$$\sin(A - B) = \sin A\cos B - \cos A\sin B \tag{8}$$

$$\cos(A + B) = \cos A\cos B - \sin A\sin B \tag{9}$$

$$\cos(A - B) = \cos A\cos B + \sin A\sin B \tag{10}$$

$$\frac{\sin\theta}{\cos\theta} = \tan\theta \tag{11}$$

$$\tan(A + B) = \frac{\tan A + \tan B}{1 - \tan A\tan B} \tag{12}$$

$$(\tan A - B) = \frac{\tan A\tan B}{1 + \tan A\tan B} \tag{13}$$

Replacing B by A in (7) and (9), we obtain the following:

In (7) $\sin 2A = \sin(A + A)$

$= \sin A\cos A + \cos A\sin A$

i.e.

$$\sin 2A = 2\sin A\cos A \tag{14}$$

In (9) $\cos 2A = \cos A\cos A - \sin A\sin A$

i.e.

$$\cos 2A = \cos^2 A - \sin^2 A \tag{15}$$

Similarly, in (12) we obtain

$$\tan 2A = \frac{2\tan A}{1 - \tan^2 A} \tag{16}$$

Substituting from (4), $\sin^2 A = 1 - \cos^2 A$ or $\cos^2 A = 1 - \sin^2 A$

in (15), we obtain

$$\cos 2A = 2\cos^2 A - 1 \quad (17)$$

and $$\cos 2A = 1 - 2\sin^2 A \quad (18)$$

From (17) $$2\cos^2 A = 1 + \cos 2A$$

$$\therefore \quad \cos^2 A = \tfrac{1}{2}(1 + \cos 2A) \quad (19)$$

and from (18) $$2\sin^2 A = 1 - \cos 2A$$

$$\therefore \quad \sin^2 A = \tfrac{1}{2}(1 - \cos 2A) \quad (20)$$

$$\sin A + \sin B = 2\sin\left(\frac{A+B}{2}\right)\cos\left(\frac{A-B}{2}\right) \quad (21)$$

$$\sin A - \sin B = 2\cos\left(\frac{A+B}{2}\right)\sin\left(\frac{A-B}{2}\right) \quad (22)$$

$$\cos A + \cos B = 2\cos\left(\frac{A+B}{2}\right)\cos\left(\frac{A-B}{2}\right) \quad (23)$$

$$\cos A - \cos B = 2\sin\left(\frac{A+B}{2}\right)\sin\left(\frac{B-A}{2}\right) \quad (24)$$

$$2\sin A\cos B = \sin(A+B) + \sin(A-B) \quad (25)$$

$$2\cos A\sin B = \sin(A+B) - \sin(A-B) \quad (26)$$

$$2\cos A\cos B = \cos(A+B) + \cos(A-B) \quad (27)$$

$$2\sin A\sin B = \cos(A-B) - \cos(A+B) \quad (28)$$

Triangle Formulae

The sine rule: $$a/\sin A = b/\sin B = c/\sin C$$

The cosine rule
$$a^2 = b^2 + c^2 - 2bc\cos A$$
$$b^2 = a^2 + c^2 - 2ac\cos B$$
$$c^2 = a^2 + b^2 - 2ab\cos C$$

Area of triangle $= \frac{1}{2}ab\sin C = \sqrt{s[(s-a)(s-b)(s-c)]}$

where a, b and c are the sides of the triangle and $s = \frac{1}{2}(a + b + c)$

Differentials

Some Common Derivatives

$y = ax^n$	dy/dx	$= anx^{n-1}$
$y = {}^1/x$	dy/dx	$= -{}^1/x^2$
$y = \sqrt{x}$	dy/dx	$= {}^1/2\sqrt{x}$
$y = \sin x$	dy/dx	$= \cos x$
$y = \cos x$	dy/dx	$= -\sin x$
$y = \tan x$	dy/dx	$= \sec^2 x$
$y = \operatorname{cosec} x$	dy/dx	$= -\operatorname{cosec} x \cot x$
$y = \sec x$	dy/dx	$= \sec x \tan x$
$y = a\, e^{bx}$	dy/dx	$= ab\, e^{bx}$
$y = a\, e^{-bx}$	dy/dx	$= -ab\, e^{-bx}$
$y = \log_e ax$	dy/dx	$= {}^1/x$
$y = \sinh x$	dy/dx	$= \cosh x$
$y = \cosh x$	dy/dx	$= \sinh x$
$y = \tanh x$	dy/dx	$= \operatorname{sech}^2 x$
$y = \operatorname{cosech} x$	dy/dx	$= -\operatorname{cosech} x \coth x$
$y = \operatorname{sech} x$	dy/dx	$= -\operatorname{sech} x \tanh x$
$y = \coth x$	dy/dx	$= -\operatorname{cosech}^2 x$
$y = \sin^{-1}(x/a)$	dy/dx	$= {}^1/\sqrt{(a^2 - x^2)}$
$y = \cos^{-1}(x/a)$	dy/dx	$= -{}^1/\sqrt{(a^2 - x^2)}$
$y = \tan^{-1}(x/a)$	dy/dx	$= {}^a/(a^2 + x^2)$
$y = a^x$	dy/dx	$= a^x \log_e a$
$y = \log_a x$	dy/dx	$= {}^1/(x \log_e a)$

Product Rule

If U and V are functions of x

$$^{d}/_{dx}(UV) = U(^{dV}/_{dx}) + V(^{dU}/_{dx})$$

Quotient Rule

If U and V are functions of x

$$^{d}/_{dx}(^{U}/_{V}) = \frac{V\,dU/dx - U\,dV/dx}{V^2}$$

Function of a function

Based on the identity $^{dy}/_{dx} = {}^{dy}/_{dz}\,{}^{dz}/_{dx}$

Integration

(In the following integrals the constant of integration is omitted)

$$\int x^n\,dx = \frac{x^{n+1}}{n+1} \qquad (n \neq -1)$$

$$\int \frac{dx}{x} = \log_e x$$

$$\int a^x\,dx = \frac{a^x}{\log_e a}$$

$$\int e^x\,dx = e^x$$

$$\int \sin x\;dx = -\cos x$$

$$\int \cos x\;dx = \sin x$$

$$\int \tan x\;dx = \log_e \sec x$$

$$\int \sec^2 x\;dx = \tan x$$

$$\int \sec x\;dx = \log_e(\sec x + \tan x) = \log_e \tan(^{\pi}/_4 + {}^{x}/_2)$$

$$\int \operatorname{cosec} x\,dx = -\log_e(\operatorname{cosec} x + \cot x) = \log_e \tan {}^{x}/_2$$

$$\int \cot x\,dx = \log_e \sin x$$

$$\int \sinh x\,dx = \cosh x$$

$$\int \cosh x\,dx = \sinh x$$

$$\int \tanh x\,dx = \log_e \cosh x$$

$$\int \coth x\,dx = \log_e \sinh x$$